Electricity, Fluid Power, and Mechanical Systems for Industrial Maintenance

Electricity, Fluid Power, and Mechanical Systems for Industrial Maintenance

◀ Thomas Kissell ▶

Terra Community College

Prentice Hall

Upper Saddle River, New Jersey ◆ *Columbus, Ohio*

Library of Congress Cataloging-in-Publication Data
Kissell, Thomas E.
    Electricity, fluid power, and mechanical systems for industrial maintenance / Thomas Kissell.
        p.   cm.
    Includes index.
    ISBN 0-13-896473-4
    1. Plant maintenance.   I. Title
TS192.K573   1999
658.2—dc21   98-25193
                                                    CIP

Cover photo: © Westlight
Editor: Ed Francis
Production Editor: Stephen C. Robb
Design Coordinator: Karrie M. Converse
Text Designer: Rebecca Bobb
Cover Designer: Rod Harris
Production Manager: Patricia A. Tonneman
Marketing Manager: Chris Bracken

This book was set in Century Book and Gill Sans by Carlisle Communications, Ltd. and was printed and bound by Courier/Kendallville, Inc. The cover was printed by Phoenix Color Corp.

© 1999 by Prentice-Hall, Inc.

Upper Saddle River, New Jersey 07458

Printed in the United States of America

10 9 8 7 6 5 4 3

ISBN: 0-13-896473-4

Prentice-Hall International (UK)Limited, *London*
Prentice-Hall of Australia Pty. Limited, *Sydney*
Prentice-Hall Canada Inc., *Toronto*
Prentice-Hall Hispanoamericana, S.A., *Mexico*
Prentice-Hall of India Private Limited, *New Delhi*
Prentice-Hall of Japan, Inc., *Tokyo*
Pearson Education Asia Pte. Ltd., *Singapore*
Editora Prentice-Hall do Brasil, Ltda., *Rio de Janerio*

*To Kathy
and our four granddaughters,
Amber, Cassidy, Shelby, and Christina*

◀ **Preface** ▶

Single-Trade Concept and Its Application

The latest trend in modern industry is for one person to troubleshoot and repair all of the systems on a machine. This requires a technician or repair person to fully understand electrical, hydraulic, pneumatic, and mechanical systems. *Electricity, Fluid Power, and Mechanical Systems for Industrial Maintenance* is written to help students readily learn the operational theory, installation, and troubleshooting of these three major systems found on nearly all industrial machines.

This book focuses specifically on the interactions and interdependence of these three systems. Traditionally, three or more individual textbooks are used to teach these subjects. Obviously, this artificially isolates the topics and ignores the critical interrelationship among the three systems. In contrast, this text is a comprehensive resource that enables students to learn not only the individual subjects but also how these technologies intertwine. Thus, one technician can learn to troubleshoot effectively and repair all three systems.

A troubleshooting technician who is responsible for all three of these trades needs to learn only the *most essential* theories and troubleshooting tips to become effective. Therefore, this book omits material that does not aid in understanding.

In addition, with the traditional instruction approach using multiple books, the individual authors of each might thoroughly know their subject but could lack understanding of the other technologies and the interactions. Individual authors might differ in their approaches or in the emphasis they place on each topic. Unfortunately, this often results in students receiving uneven instruction. This book is unique in that it is written by a single author who has over 20 years of experience in installation, troubleshooting, and repair of electrical controls, motors, electronics, hydraulics, and pneumatic and mechanical systems on a wide variety of machines. The author has devoted two decades to teaching the single trade concept to students of all ages in environments including the factory floor.

Features of the Book

Each chapter includes many photographs and diagrams to explain the theory of operation and show test points for troubleshooting individual components and complete systems. This text is designed to be simple enough for the beginner, yet comprehensive enough to be a complete reference for the technician who needs to review a technique or theory on the job.

The first chapter is devoted to all the aspects of safety that a student needs to understand to work safely in the laboratory and on the job. The remainder of the book is split into three parts to cover electrical, fluid power, and mechanical systems. Since electricity tends to be the most difficult and diverse of these three topics, it is presented in several subsections. Chapters 2 through 6 provide an introduction to the basic theory and fundamentals of electricity. Unlike traditional texts that delve deeply into classical calculations, these chapters show how the student must apply the basics of electricity to become a well-trained troubleshooting technician.

Chapters 7 through 12 explain in a comprehensive manner the operational theory for a wide variety of electrical control components such as relays and motors. Complete instructions are also included to help students learn how to troubleshoot these devices.

Chapter 13 features lock-out and tag-out procedures and introduces students to working safely around equipment when they are troubleshooting. Many texts provide this information in the initial chapters; unfortunately, at the beginning of instruction the students understand too little of the theory to fully appreciate the importance of these life-saving procedures. This text, instead, presents these topics *after* students understand the basic theory and operation of all the motors and controls, when the students are *fully aware* of the importance of strictly following these procedures.

Chapters 14 through 16 present detailed diagrams, theories of operation, and troubleshooting of motor control systems, programmable controllers, and electronics. These three chapters are perhaps the greatest strength of this text. They explain each of these complex electrical subsystems in a way that is easy to understand and utilize during troubleshooting. The author shows how modern machinery uses programmable controllers to integrate electronic controls with traditional motor controls. Coverage of these three subjects helps students to gain confidence as they see how *easy* it is to troubleshoot and repair these systems despite their complexity.

Chapter 17 comprises the "second section" of the text, which presents hydraulic and pneumatic components and systems. Even though it is a *single* chapter, it encompasses a comprehensive dialog with diagrams and photographs that fully explain the theory of operation and methods to troubleshoot these components and systems. In fact, this chapter is comprehensive enough to be used for *any* fluid power course.

The third section of the text includes two chapters that cover mechanical systems and power transmission. Over many years, the author has concluded that the main reason students have difficulty learning mechanical systems is their lack of knowledge about the theory of operation of the six simple machines. Chapter 18 provides in-depth coverage of how these simple machines become the basics of mechanical systems. Specifically, students will learn how each of the six is enhanced to become the complex components of a mechanical power-transmission system. The operational theories in this chapter make troubleshooting and repair of mechanical systems much easier. This chapter also introduces the scientific theories that are used to explain work, power, velocity, torque, momentum, and other concepts that students must understand to become troubleshooting technicians.

Chapter 19 covers comprehensively all the parts of power-transmission systems such as couplings, belts and pulleys, chain drives, gears, bearings, clutches, and brakes. These subjects are explained in regard to their theory of operation, installation techniques, and methods of troubleshooting. Even though these complex subjects are covered in one chapter, the chapter contains more than 60 figures to help students learn the necessary concepts.

The final chapter provides the vital essence missing from individual textbooks used to study separately the electrical, fluid power, and mechanical systems. Chapter 20 combines material from previous chapters to enable students to evaluate a complex machine that is malfunctioning and determine which of the systems is causing the problems. This chapter also reveals the important difference between "cause" and "effect." As most teachers and technicians know, the most important part of troubleshooting is to quickly analyze a system and distinguish the root cause of a problem from a mere symptom of the problem's effect on the remainder of the system. Instructors will be satisfied in how readily Chapter 20 teaches students to understand these important subtleties.

This text is designed to be used in a single semester course, or it can be used where basic electricity, electrical motors and controls, programmable controllers, electronics, fluid power, and mechanical power transmission are presented as separate courses. Each chapter begins with a list of objectives and concludes with a variety of multiple choice, true-and-false, and open-ended type questions. The questions can be used as homework assignments or as end-of-chapter tests.

◀ **Acknowledgments** ▶

I thank my wife, Kathleen Kissell, for all her help as graphic artist and project manager of this book in keeping the figures straight. I also thank her for all her support and for putting up with all the activities that are associated with writing a book. I thank Ed Francis for his encouragement and help at all stages of this book, as well as Steve Robb and Linda Thompson.

The following persons and their companies have helped locate photographs and diagrams for this book and graciously provided permission to use these materials:

Kathy McCoin, Acme Electric Corporation, Power Distribution Products Division

Mike Moore and Mr. Randy Randal, Rockwell Automation's Allen-Bradley Business

Craig Schuele, Boston Gear

Craig Meyer, Cooper Bussmann

Gary Forcey, Cutler-Hammer, Inc.

Steve Rosenthal, Danaher Controls

Herb Arum, Quality Transmission Corporation

Jerry Donovan, The Gates Rubber Company

Patricia Pruis, GE Industrial Systems, Fort Wayne, IN

Leslie Mantua, Honeywell Micro Switch Division

Mary Ellen Nunes, Ingersoll Rand

Thomas Risto, Leeson Electric Corporation

Mark McCullough, Lovejoy, Inc.

Chuck Ramsey, Magnetek

Diana Baxter, Motorola

Dennis Berry, National Fire Protection Association

Ray Hamilton, National Electrical Manufacturers Association

Terri Johnson, North American Capacitor Company

Russ Stroback, Parker Hannifin Corporation

Brenda Holt, Power Transmission Distributors Association

Ron Smith, Reliance Electric

Jeff Cipola, Shrader Bellows

Lori Eiler, Square D Company

Kip Larson, SymCom Inc.

John Mullen, Supco Sealed Unit Parts Co., Inc.

Dwight Keeney, The Torrington Company

Ted Suever, Triplett Corporation

Peter Parsons, Vickers, Inc.

Ray Freywald, Warner Electric, South Beloit, IL

Finally, I gratefully acknowledge the following reviewers for their insightful suggestions: James Shelton, Ivy Tech State College, and Lee Wilson, Rend Lake College.

◀ Contents in Brief ▶

◀ Contents ▶

Chapter 4
Series Circuits   27

Chapter 5
Parallel and
Series-Parallel Circuits   35

Chapter 6
Magnetic Theory   47

Chapter 7
Fundamentals of AC Electricity   55

Chapter 16
Electronics for Maintenance Personnel   191

Chapter 17
Basic Hydraulics and Pneumatics 209

Chapter 18
Basic Mechanics   231

Chapter 19
Power-Transmission Systems   243

Objectives   243

Chapter 20
Troubleshooting   269

Objectives   269

Index   277

Chapter 1

Shop Safety and Shop Practices

Objectives

After reading this chapter you will be able to:

1. List typical safety clothing and equipment you should use on the job.

2. Develop a plan for fire safety and implement it.

3. Inspect machines to ensure safety guards are in place and operating.

4. Identify problems with housekeeping to avoid accidents.

1.0
Overview of Shop Safety and Shop Practices

The most important part of working on the factory floor is personal safety. You must understand that you are responsible for your own safety and the safety of others working with you. In this chapter you will learn about protecting yourself and others and basic machine safety that will ensure a safe workplace. You will also learn about basic shop practices that are created to help maintain a safe environment in the workplace. You must also understand that your instructor will use many of the same safety rules and practices in your lab at school to help you learn them as well as to ensure your safety in that work environment.

1.1
Safety Glasses and Protective Clothing and Equipment

When you are working on the factory floor you will be required to wear a variety of safety gear, such as safety glasses and earplugs. Safety glasses with shatterproof glass lenses and side shields should be worn in a lab and the factory at all times to protect your eyes against foreign debris. Problems can also occur with electrical equipment if there is an electrical short, because the metal from electrical contacts can turn into molten metal and spray into your face and eyes. You can get safety glasses with prescription lenses to match your regular glasses so that you can feel comfortable wearing them all day long. Your safety glasses should be comfortable enough to wear over extended periods.

In most factory settings you will also need to wear ear protection, which may include earplugs, earmuff protectors, or both. The conditions in your factory can be tested to determine the decibel level of noise and the type of ear protection that will best protect your ears over a long period of time. The most important point to remember about ear protection is that ear damage occurs very slowly over time and it is almost impossible to detect without sophisticated testing equipment. Many noises in a factory will cause the ear to deteriorate very slowly over time. This means that on any given day, you will not realize the noise is causing any damage, but over a number of years you will find that you can no longer recognize conversation or tones in specific audio ranges. For example, if you are exposed to noise over a long period of time, you will begin to notice that you cannot understand certain conversations around you, or you may need to turn up the volume on the television or radio.

The important point to remember is that all excessive noise in a factory is dangerous and will cause permanent damage to your hearing over time. Certain frequencies of noise will cause the deterioration to occur more quickly. You should wear hearing protection as soon as you enter areas where you are exposed to noise and continue to wear it as long as you are in the exposed area.

You should also be aware of loose clothing and jewelry if you are working around equipment that has moving parts or if you work in electrical cabinets. Watches, finger rings, and other body rings need to be removed when you are working where these items can be snagged on equipment or if they come into contact with exposed electrical terminals. Many accidents occur each year when jewelry becomes snagged on equipment, causing lacerations to the skin or broken fingers. In some cases, jewelry becomes entangled in equipment such as gears or belts, and personnel are pulled deeper into the machine, where extreme injury or death may result.

Protective clothing may be required in applications where welding or cutting occurs. Other protective clothing and respirators may be required if spray painting or other similar activities occur. You will receive a briefing before beginning any job where protective clothing is required, and you should always follow the instructions provided.

1.1.1
Back Braces for Lifting and Carpal Tunnel Braces

Certain activities in industry require repeated activity that aggravate specific muscle groups. For example, if you must repeatedly lift loads larger than 10 lb., you should wear a back brace that is specifically designed to help prevent back-muscle strains. Other problems may occur if a job requires repetitive motions such as continually tightening nuts and bolts; the muscles in your hands and wrist will begin to weaken and they may begin to cramp. This condition is called *carpal tunnel syndrome* and it can be corrected or prevented by special braces that provide support for the muscles. In some instances, the muscles deteriorate so much that the person must be given a different job. Some employers recognize the activities that cause these types of injuries and provide protection, or they replace the job with

some type of automation. You can help prevent these types of injuries by wearing the appropriate braces. You should also request help in lifting larger weights and determine the maximum weight a person should lift without equipment.

1.1.2
Steel-Toed Shoes and Hard Hats

In some factories, you will be required to wear steel-toed shoes to prevent injuries to your toes or feet due to dropping heavy objects on your feet. Steel toes are available in a wide variety of shoes and boots. This allows you to wear shoes or boots that are both comfortable and safe. It is important to remember to wear steel-toed shoes at all times in areas where they are required and there is a danger of foot injuries. Some shoes are also specially fitted with steel soles to prevent punctures from walking on sharp objects. Hard hats or bump hats may also be required in some locations if material could drop from above or if you have a chance of bumping your head on an overhead object.

1.2
Safety Mats and Equipment to Prevent Fatigue

Some jobs in the factory require standing on concrete floors for long periods of time. In these conditions a cushion safety mat may be used to help absorb some of the shock and limit the fatigue from standing long periods of time. Other mats are provided to keep personnel from coming into contact with wet or damp surfaces. Some automation applications that have robots or presses with moving parts provide mats that have special sensors in them to ensure personnel are not in the area of the machine's moving parts when the machine starts. These mats may include safety light curtains that will automatically stop the machine if you move off the mat or break the light curtain.

1.3
Safety Light Curtains and Two-Hand-Start Buttons

Safety devices that are designed specifically to prevent machine operation where personnel may be injured are called *active safety devices*. Safety light curtains are used to ensure personnel are not in the work area of automated machines. Machine-start buttons may require two hands to ensure that the operator's hands are not near any moving parts of the machine.

Many machines have two pushbuttons that are located approximately 2 ft. apart. Both pushbuttons must be depressed at the same time to start the machine. Because the pushbuttons are mounted 2 ft. apart, the operator must have one hand on each switch in order to start the machine, ensuring that both of the operator's hands are clear of the machine so he or she will not be injured. Such starting controls are added to machines with moving parts that might injure operator's hands if they are not clear when the machine is in operation. The two-handed pushbuttons are designed so they cannot be held down or bypassed in some manner. When there are two switches, operators tend to try and hold one pushbutton down with a screwdriver or other tool to bypass the safety feature so they can start the machine with one hand. This is usually done to speed up loading and unloading the machine. The two-handed pushbutton circuit is specifically designed so that if one button is held down, the machine will not restart. Do not try to disable these types of safety circuits under any circumstances, because the operator can be seriously injured.

1.4
Safety Gates and Shields

Another safety device that is added to machines in a factory is a safety gate and fence around large-scale automation such as robot cells. In some factories one or more robots are used to stack boxes onto pallets or to do other jobs where personnel may be in close proximity. In these applications the equipment, robots, and machines are enclosed behind fencing. When personnel must work on the equipment inside the fence, they must enter through a safety gate that is wired so that all motion inside the fence is stopped when the gate has been opened. When personnel leave the fenced-in area, the gate is closed and the machines are reset.

1.4.1
Machine Guards

Machine guards are provided to keep personnel's hands and clothing out of rotating parts on machines such as shafts, gears, and belts. These guards are designed to be removed easily for maintenance and repairs. But it is important that these guards remain in place any time the machine is operating. If a guard is removed during

maintenance, it is essential that you do not try to restart the machine until the guard is back in place.

1.4.2
Glass Machine Guards

Some machines have see-through guards, such as glass guards on grinders and glass that is in doors on machines. Glass is designed so that you can see into the moving part of the machine without removing the guard completely. Glass guards on grinding machines are intended to be in place when the grinding operation is occurring, yet allow the operator to view through the guard.

Over time the glass will become dirty or may get scratched to the extent that you cannot see through it. If the safety glass is dirty or damaged, it should be replaced so that employees can utilize it as is intended.

1.5
Checking Guards and
Active Safety Devices

One of the most important activities that needs to be completed every day is a walk-around check to ensure safety guards are in place and that all active safety devices are operational. This type of check is called a daily PM (preventative maintenance) check. This check should be made each day to ensure the safety of all personnel as they work around equipment or in dangerous areas of the factory. A more complete test and cleaning of these safety devices should also be scheduled approximately twice a year. During this more in-depth test, the safety guards should be inspected for wear or deterioration of mounting mechanisms to ensure the guards will withstand normal use and wear over the next 6 months.

A formal procedure should be provided that allows employees to write up any piece of equipment that has a missing guard or a piece of safety equipment that is not operating correctly. The procedure should include a means for inspecting and repairing the problem as quickly as possible. A complete record should be kept on each piece of equipment to track these types of safety problems so that inspections and repairs can be made on similar equipment in the factory. This record is very important when a factory has more than one of any machine, such as multiple plastic presses. If a safety guard fails on one machine, it is a good bet that the same problem will occur on similar machines, especially if they are the same make and model.

1.6
Cleanliness and Shop Housekeeping

One of the major conditions that affect shop safety is the cleanliness of the shop. It is important to remember that many accidents are caused by a cluttered shop, where things are not put away or cleaned up. In these cases a routine should be developed to clean the shop on a regular basis. It is also important to have a proper storage place for everything that is used around machines on the shop floor. When you locate oil spills or water spills, you should make sure that they are cleaned up thoroughly and as quickly as possible. It is also important to remember that water may cause someone to slip and fall, and an electrical shock may occur if power cords from hand tools such as drills are allowed to lay in water.

Another problem with cleanliness is proper disposal of oily rags. It is important to have proper fire-rated storage cans for all oily rags. These types of containers have fire-rated lids so that fires cannot start in the containers. It is also important to empty these containers on a regular basis.

1.7
Leaks

Oil leaks and water leaks should be cleaned up and the source of the leak repaired as soon as possible to limit the amount of liquid that is leaked. The leaks can cause exposure to hazardous material as well as provide a potential for slips and falls. In most factories, periodic maintenance schedules identify specific parts of each machine that should be checked for leaks on a daily basis. When leaks are first identified, they should be fixed immediately before the leak becomes excessive.

1.8
Material Safety Data Sheets (MSDS)

At times in a factory or industry you will work with any number of chemicals and other types of materials that may be hazardous. In 1983 OSHA (Occupational Safety Health Administration) required manufacturing companies to maintain and distribute data sheets on any material used in the factory, and in 1987 OSHA expanded this to all employers. MSDS information may be stored in notebooks or on computer programs somewhere that is totally accessible to all personnel. Additional information is available by contacting the manufacturer of any material. When you are working where you may be exposed to any hazardous material, you should wear the proper safety equipment and clothing and locate the MSDS information and follow the instructions provided therein.

1.9
First Aid and CPR

It is important that as many people at every worksite have training in first aid and CPR (cardiopulmonary resuscitation). Local Red Cross agencies usually provide first aid and CPR courses for workers. If someone at the plant receives a severe electrical shock or collapses and stops breathing, he or she will need the aid of someone who is certified in CPR. Individuals who stop breathing from heart problems will also need such assistance.

First-aid safety includes the first response to problems such as severe cuts and other lacerations or broken bones. If people are injured by machines, it is important to turn off power to the machine as quickly as possible, determine if you can safely move the injured person, and call for help immediately. First-aid courses will show you how to apply pressure to deep wounds to stop bleeding. These courses will also show when it is unsafe to move someone with a head or neck injury.

1.10
Fire Safety

Fire safety is very important in industrial settings; it consists of prevention and fighting fires with fire extinguishers. The most important part of fire safety is to understand the basic principles of fire. Fire requires fuel, oxygen, heat, and a chain reaction (source of ignition) to start and be sustained. When you are trying to prevent a fire or fight a fire, you must separate these parts. Fuels may include paper, rubber, oils, gases, or any other material that can burn.

The National Electric Code (NEC) classifies the locations and types of materials that have the possibility of burning. They are classified as Class I, Class II, and Class III. Figure 1–1 shows an example of these classifications. Class I locations include petroleum-refining facilities, plants that have spray painting, and locations where solvents are used. Class II locations include locations that handle and store material that emits dust, such as grain elevators, and locations that handle coal. Class III locations include locations that work with material that emits fibers that can burn, such as rayon, cotton, and other textile mills, and facilities that use cutting and sanding operations on wood.

Figure 1–1 Typical hazardous locations by classifications.

Class I	Petroleum-refining facilities
	Petroleum distribution points
	Petrochemical plants
	Dip tanks containing flammable or combustible liquids
	Dry-cleaning plants
	Plants manufacturing organic coatings
	Spray-finishing areas (residue must be considered)
	Solvent-extract plants
	Locations where inhalation anesthetics are used
	Utility gas plants; operations involving storage and handling of liquefied petroleum and natural gas
	Aircraft hangers and fuel-servicing areas
Class II	Grain elevators and bulk-handling facilities
	Plants manufacturing and storing magnesium
	Plants manufacturing and storing starch
	Fireworks manufacturing and storage
	Flour and feed mills
	Areas for packaging and handling of pulverized sugar and cocoa
	Facilities for the manufacture of magnesium and aluminum powder
	Coal-preparation plants
	Spice-grinding plants
	Confectionary-manufacturing plants
Class III	Rayon, cotton, and other textile mills
	Combustible-fiber manufacturing and processing plants
	Cotton gins and cottonseed mills
	Flax-processing plants
	Clothing-manufacturing plants
	Sawmills and other woodworking locations

1.10.1
Sources of Ignition

One of the ways to prevent a fire is to remove the source of ignition from the fuel. Some sources of ignition that you should be aware of include open flames from torches and heaters, smoking and matches, electrical equipment and heating elements, friction, high-intensity lighting, combustion sparks from grinding and welding, hot surfaces, spontaneous ignition, and static electricity. You must also store flammable liquids in proper storage containers and keep them away from open flames and sources of ignition at all times.

Many companies have proactive programs when any of these sources are exposed. For example when you must use welding equipment or torches, safety covers and fire extinguishers are moved into place and additional personnel are used to watch for fire problems while the welding is completed. Other conditions are prevented by safety guards that keep fuels away from hot surfaces.

1.10.2
Fire Extinguishers

Fire extinguishers are rated by classification. **Class A** fire extinguishers can be used on ordinary combustible material, such as paper, wood, or clothing. This type of fire can easily be extinguished by water.

Class B fire extinguishers are designed for fires in flammable liquids, grease, and other material that can be extinguished by smothering or removing air (oxygen) from the chain reaction. *Class C* fire extinguishers are used on fires from live electrical equipment. The extinguishers for these types of fires must use material that is nonconducting, so that the person using the extinguisher is not exposed to additional hazard from electrical shock when fighting the fire. *Class D* fire extinguishers are used on fires in combustible metals such as sodium, magnesium, or lithium. Special extinguishers must be used to lower the temperature and remove oxygen from these fires.

Color codes for these fire extinguishers have been developed, and it is important that you recognize the type of fire and use the proper extinguisher. Class A fire extinguishers are colored green, Class B fire extinguishers are painted red, Class C fire extinguishers are painted blue, and Class D fire extinguishers are painted yellow. It is also important that you recognize the type of material in your shop area and have the proper fire extinguisher for each material.

1.10.3
Fire Extinguisher Safety

It is important to inspect fire extinguishers on an annual or semiannual basis. During the inspection each fire extinguisher should be checked to ensure that it is fully charged and ready for use. It is also important that all fire extinguishers are clearly marked for the type of fire for which they are specified and that they are clearly displayed where all employees can locate them when they are needed.

1.10.4
Personnel Safety During a Fire

When a fire occurs in your building it is important that several things occur. The most important point is that all personnel and the proper authorities are notified. All fire alarms in the building should be activated to warn the other personnel so that they can evacuate the area. Then, the plant fire brigade, if you have one, and the local fire company should be notified so that they can begin their response. If the fire is small in nature, the local plant fire brigade may begin to fight the fire with fire extinguishers. If the fire is larger, the most important point is to evacuate everyone safely from the building. Be sure you learn the proper evacuation routes for all areas of the factory in which you work. It is also important that procedure is in place to account for all personnel once they have evacuated.

1.11
Electrical Safety

When you are working in a factory it is important to completely understand how to work safely around electrical cabinets and circuits. This includes the inspection of electrical safety grounds to ensure they are connected correctly and are operational and working safely around circuits that are under power. The safest way to work around electrical circuits is to turn power off and *lock out and tag out* the electrical circuit. Lock-out and tag-out procedures are required by OSHA, and they are so important they are explained in detail in Chapter 13 of this text.

In some cases you may be required to test voltage and current in circuits that have electrical power applied. When you must work on circuits that are under power, be sure to wear heavy, rubber-soled shoes and try to work with only one hand exposed. If you work with both hands in an electrical panel, you may conduct electric current through one arm, through your body (near your heart), and out the other arm. This is the most dangerous type of shock, and it may cause severe damage or stop your heart. If you are working with only one hand in the electrical panel at a time, the chances of conducting electrical current directly through your body during an electrical shock is minimal.

1.11.1
Ground Fault Interruption
Circuit Breakers (GFIs)

If you are using electrical hand tools or if you must work in areas that are damp or have standing water, it is important that all circuits have ground fault interruption circuit breakers, GFIs. The GFI-type breaker measures all the electrical current that is supplied to a circuit or power tool and compares it to the amount of current returning from the tool. If there is no electrical short circuit to ground and the power tool is working correctly, the amount of current returning from the circuit will be the same. If less current is returning, the GFI assumes a short circuit is occurring and a dangerous condition will result, so it will trip to the open position. For example, if you are using a power tool that has a faulty power cord, as soon as a short circuit occurs and you start to get an electrical shock, the GFI will interrupt all current in the circuit and prevent you from becoming shocked. GFI protection is also available as a built-in feature for some extension cords or power cords that you will use to provide power for your portable electrical power tools. Figure 1–2 shows a typical ground fault circuit breaker.

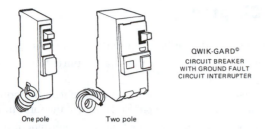

QWIK-GARD©
CIRCUIT BREAKER
WITH GROUND FAULT
CIRCUIT INTERRUPTER

One pole   Two pole

Figure 1–2 Examples of ground fault interruption (GFI) circuit breakers.

(Courtesy Square D Company)

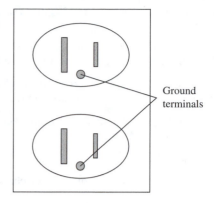

Ground terminals

Figure 1–3 Ground terminals on a duplex receptacle.

1.11.2
Ground Wires

All electrical circuits and metal power tools require an electrical ground wire. The ground wire is a bare wire or a wire that is identified by its green color, and it is used to provide a direct path to ground, which makes electrical fault current flow through fuses or circuit breakers, causing them to trip.

Figure 1–3 shows the location of the ground terminal on the outlet receptacle. If a power tool requires a ground circuit, it will have a third prong on its plug end. It is also important that you use these tools only in an electrical outlet receptacle that also has a ground terminal. If you have a power tool with a three-prong plug, you should never cut the ground prong off the plug so that it will fit a two-prong (nongrounded) receptacle. You should also never use a two-prong to three-prong adapter with a power tool that requires a three-prong plug. Some power tools have a plastic case, and these tools have a two-prong plug that does not need to be grounded when it is used. You will learn more about short circuits, grounding circuits, fuses, and circuit breakers in a later chapter. It is important to inspect the power cord of any electrical power tool that has a metal case to ensure that it has a ground wire provided.

1.11.3
Ground Indicators on Electrical Panels

Some electrical panels have a set of two white lights on the front that are used to indicate that the cabinet or equipment is properly grounded. If the equipment ground wire is faulty, the lights will not be illuminated. If you work on an electrical cabinet where the ground-indicating lights are not illuminated, it may indicate that the ground wire is faulty; you should not touch the cabinet until you can determine if the ground wire is effective.

1.12
Working Safely with Hand Tools and Power Tools

At times you will be required to work with hand tools and power tools, which increases the chances of having an accident. It is important that you learn the safe operation of these tools and practice safe working habits when you must use them. Always read and follow the directions for safe operation prior to using tools. It is also important to have proper safety equipment, such as safety glasses, when you are operating hand tools. Do not use tools if you have not received the proper instructions for using them safely.

1.13
Safety Briefings and Safety Meetings

Many companies use safety meetings to provide basic safety information on a weekly basis. These meetings provide a forum to identify and discuss safety problems throughout the plant. Other plants use a safety committee that identifies safety areas of concern and also provides safety consciousness by posting safety warnings and posters throughout the plant.

Some companies have designated personnel who inspect the various shop areas for safety compliance, whereas other companies empower all employees to keep an eye on safety conditions throughout the plant. The important point to remember is that you are always responsible for your own safety at all times. Make working safely a habit.

Questions for This Chapter

1. Explain why you should wear safety glasses with side shields at all times while you are working in a factory.

2. Explain why you should always wear earplugs or other types of ear protection while you are working in a factory.

3. List two safety items you can wear to protect you against falling debris in a factory.

4. Explain what MSDS is.

5. Identify a type of fire for each of the three classifications of locations and types of material.

True or False

1. Class I fires include petroleum and solvent-type fires.

2. The reason a two-hand pushbutton station is used on a stamping press is so that you can operate it with either switch.

3. You can use a Class A fire extinguisher on an electrical fire.

4. A Class B fire extinguisher is painted red.

5. A GFI (ground fault interruption) circuit breaker should be used with an extension cord if it is used in a damp or wet location.

Multiple Choice

1. The Class A type fire extinguisher is painted_____and it is used for paper or wood fires.
   a. red
   b. green
   c. yellow

2. A GFI (ground fault interruption) receptacle is _____.
   a. a special receptacle that has a third terminal for a ground wire.
   b. a special receptacle for power tools that allows more than one power tool to be plugged into it at the same time.
   c. a special circuit breaker that can determine if a short circuit occurs and automatically opens the circuit breaker to protect the circuit.

3. A two-handed pushbutton station is _____.
   a. a safety circuit that requires an operator of a stamping press to have both hands on buttons for the press to begin its cycle.
   b. a convenience circuit that allows an operator to operate the press from two different locations.
   c. a single pushbutton that has an oversize spring that requires the operator to use both hands to depress it.

4. Ear protection should be used _____.
   a. at all times even in locations with moderate or small amounts of noise because this noise is damaging over a long period of time.
   b. Only in high-noise areas.
   c. Only if a person has been tested and found to have some hearing loss.

5. If an electrical power tool has a power cord with a three-prong plug and the only electrical outlet has two prongs, you should _____.
   a. cut the third prong (ground lug) off the power cord so that you can use the outlet.
   b. find an adapter that allows you to plug a three-prong plug into a two-prong outlet.
   c. not use this outlet because the power tool needs to be grounded for safe operation, and the two-prong outlet is not grounded.

Problems

1. Make a fire safety plan for your lab area and classroom that includes an evacuation plan.

2. Make a safety plan that includes a method of accounting for all personnel if evacuation must occur.

3. Make a safety plan that lists all the potential fire safety problems in your lab.

4. Make an electrical safety plan that includes a weekly inspection of all the power cords used for power tools to ensure they are properly grounded.

5. Make a safety plan that includes the proper periodic inspection of all safety guards for all equipment in your lab.

◄ Chapter 2 ►

Fundamentals of Electricity

Objectives

After reading this chapter you will be able to:

1. Explain the term *electricity*.
2. Describe the term *voltage*.
3. Explain the term *current*.
4. Describe the term *resistance*.
5. Understand the sources of electricity.
6. Understand Ohm's law.

2.0
Overview of Electricity

As a maintenance technician, you must thoroughly understand electricity so that you are able to troubleshoot the electrical components of systems, such as motors and controls. This chapter provides you with the basics of electricity, which will be the building blocks for more complex circuits that you will encounter in the field. You do not need any previous knowledge of electricity to understand this material. The simplest form of electricity to understand is *direct current* (DC) electricity, so many of the examples in this chapter use DC electricity. It is also important to understand that the majority of electrical components and circuits in field equipment use *alternating current* (AC) electricity, so some of the examples use AC electricity. This chapter begins by defining basic terminology such as voltage, current, and resistance and showing simple relationships between them.

2.1
Example of a Simple Electrical Circuit

It is easier to understand the basic terms of electricity if they are connected with a working circuit with which you can identify. For example, we will use the motor that moves a conveyor belt as an example. The motor must have a source of voltage and current. *Current* is a flow of electrons and *voltage* is the force that makes the electrons flow. This is just an introduction; much more detail about the terms is provided later in this chapter.

This circuit is shown in Figure 2–1. It shows the source of voltage is 110 V, which is identified by the letters L1 and N. The switch is a manually actuated switch for this circuit, so the switch provides a way to turn the conveyor motor on and off. When the switch is moved to the on position, or closed, voltage and current move through the wires to the motor. When voltage and current reach the motor, its shaft begins to turn. The wires in the circuit are called *conductors*. The conductors provide a path for voltage and current, and they provide a small amount of opposition to the current flow. This opposition is called *resistance*.

After the motor has run for a while, the switch is opened, and the voltage and current are stopped at the switch and the motor is turned off. When the switch is closed again, the voltage and current again reach the motor. Now that you have an idea of the basic terms voltage, current, and resistance, you are ready to learn more about them.

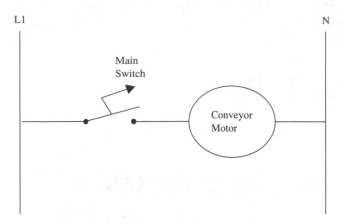

Figure 2–1   Electrical diagram of a manually operated switch controlling a conveyor motor.

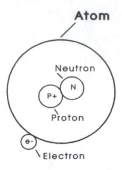

Figure 2–2   An atom with its electron, proton, and neutron identified.

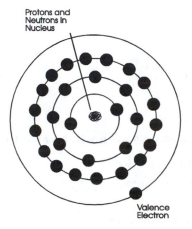

Figure 2–3   A copper atom has 29 electrons. This atom has one electron in its valence shell that can break free to become current flow.

The source of energy that makes the shaft in the motor turn is called electricity. *Electricity* is defined as a flow of electrons. You should remember from the previous discussion that the definition of *current* is also a flow of electrons. In many cases the terms *electricity* and *current* are used interchangeably. *Electrons* are the negative parts of an atom, and *atoms* are the basic building blocks of all matter. All matter can be broken into one of 103 basic materials called *elements*. Two of the elements that are commonly used in electrical components and circuits are copper, which is used for wire, and iron, which is used to make many of the parts, such as motors.

To understand electricity at this point, we need to understand only that the atom has three parts: electrons, which have a negative charge; protons, which have a positive charge; and neutrons, which have a no charge (also called a neutral charge). Figure 2–2 shows a diagram of the simplest atom, which has one electron, one proton, and one neutron. From this diagram you can see that the proton and neutron are located in the center of the atom. The center of the atom is called the *nucleus*. The electron moves around the nucleus in a path called an *orbit*, in much the same way the earth revolves around the sun.

Because the electron is negatively charged and the proton is positively charged, an attraction between these two charges holds the electron in orbit around the nucleus until sufficient energy, such as heat, light, or magnetism, breaks it loose so that it can become free to flow. Any time an electron breaks out of its orbit around the proton and becomes free to flow, it is called electrical current. Electrical current can occur only when the electron is free from the atom so it can flow. The energy to create electricity can come from burning coal, nuclear power reactions, or hydroelectric dams.

The atom shown in Figure 2–2 is a hydrogen atom; it has only one electron. In the circuit presented in Figure 2–1, the copper wire is used to provide a path for the electricity to reach the conveyor motor. Because copper is the element most often used for the conductors in circuits, it is important to study the copper atom. The copper atom is different from the hydrogen atom in that the copper atom has 29 electrons, whereas the hydrogen atom has only 1. The copper wire that provides a path for the electricity for the fan motor contains millions of copper atoms.

Figure 2–3 shows a diagram of the copper atom with its 29 electrons, which move about the nucleus in a series of orbits. Because the atom has so many electrons, they will not all fit in the same orbit. Instead, the copper atom has four orbits. These orbits are referred to as *shells*. There are also 29 protons in the nucleus of the copper atom, so the positive charges of the protons will attract the negative charges of the electrons and keep the electrons orbiting in their proper shells. The attraction for the electrons in the three shells closest to

Figure 2–4 Voltage is the force that knocks one electron out of its orbit. In this diagram the valence electron in the atom on the far right breaks free from its orbit and leaves a hole. The electrons in the atoms to the left of the first atom, move from hole to hole to create current flow.

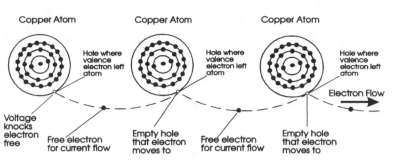

the nucleus is so strong that these electrons cannot break free to become current flow. The only electrons that can break free from the atom are the electrons in the outer shell. The outer shell is called the *valence shell*, and the electrons in the valence shell are called *valence electrons*. Notice that the copper atom only has one electron in the valence shell. All good conductors have a low number of valence electrons. For example, gold—which is one of the best conductors—has only one valence electron, and silver—which is a good conductor—has three valence electrons.

In the motor circuit we studied earlier, a large amount of electrical energy, called *voltage*, is applied to the copper atoms in the copper wire. The voltage provides a force that causes a valence electron in each copper atom to break loose and begin to move freely. The free electrons actually move from one atom to the next. This movement is called *electron flow;* more precisely, it is called *electrical current*. Because the copper wire has millions of copper atoms, it will be easier to understand if we study the electron flow between three of them. Figure 2–4 shows a diagram of the electrons breaking free from three separate copper atoms.

When the electron breaks loose from one atom, it leaves a space, called a *hole*, where the electron was located. This hole will attract a free electron from the next nearest atom. As voltage causes the electrons to move, each electron moves only as far as the hole that was created in the adjacent atom. This means that electrons move through the conductor by moving from hole to hole in each atom. The movement will begin to look like cars that are bumper to bumper on a freeway.

2.2
Example of Voltage in a Circuit

Voltage is the force that moves electrons through a circuit. Figure 2–5 shows the effects of voltage in a circuit where two identical lightbulbs are connected to separate voltage supplies. Notice that the bulb connected to the 12-V battery is much brighter than the bulb connected to the 6-V battery. The bulb connected to the 12-

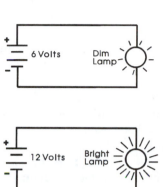

Figure 2–5 Two identical lightbulbs have different voltages applied to them. The lightbulb with 12 V applied to it is brighter than the one with 6 V applied to it.

V battery is brighter because 12 V can exert more force on the electrons in the circuit than 6 V, and higher voltage will also cause more electrons to flow.

In the original diagram of the motor, the amount of voltage is 110 V. The energy that produces this force comes from a generator. The shaft of the generator is turned by a source of energy such as steam. Steam is created from heating water to its boiling point by burning coal or from a nuclear reaction. The energy to turn the generator shaft can also come from water moving past a dam at a hydroelectric facility. In the lightbulb example, the voltage comes from a battery. The voltage in a battery was previously stored in a chemical form when the battery was originally charged. The larger the amount of voltage in a circuit, the larger the force that can be exerted on the electrons, which makes more of them flow. The scientific name for voltage is *electromotive force* (EMF).

2.3
Example of Current in a Circuit

The definition of *current* is a flow of electrons. The actual number of electrons that flow at any time in a circuit can be counted, and their number is very large. The unit of current flow is the *ampere* (A). The word

Figure 2–6 A number of electrons passing a point where they are counted. When the number reaches 6.24×10^{18}, 1 A is measured.

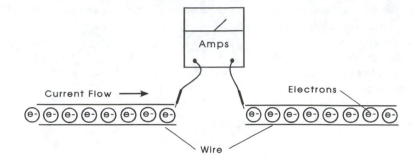

ampere is usually shortened to amp. One amp is equal to 6.24×10^{18} electrons flowing past a point in 1 s. Figure 2–6 shows an example of the number of electrons flowing that equal 1 A. In this diagram you can see that each electron that is flowing is counted; when 6,240,000,000,000,000,000 electrons pass a point in 1 s, it is called an *ampere* (A). The ampere is used in all electrical equipment, such as motors, pumps, and switches, as the unit for electrical current.

2.4
Example of Resistance in a Circuit

Resistance is defined as the opposition to current flow. Resistance is present in the wire that is used for conductors and in the insulation that covers the wires. The unit of resistance is the ohm, abbreviated Ω (the capital Greek letter omega, which represents the letter *O*). When the resistance of a material is very high, it is considered to be an insulator, and if its resistance is very low, it is considered to be a conductor.

It is important for conductors to have very low resistance so that electrons do not require a lot of force to move through them. It is also important for insulation that is used to cover wire to have very high resistance so that the electrons cannot move out of the wire into another wire that is nearby. The insulation also ensures that a wire on a hand tool such as a drill or saw can be touched without electrons traveling through it to shock the person using the tool. Examples of materials that have very high resistance and can be used as insulators are rubber, plastic, and air.

The amount of resistance in an electrical circuit can also be adjusted to provide a useful function. For example, the heating element of a furnace is manufactured to have a specific amount of resistance so that the electrons will heat it up when they try to flow through it. The same concept is used in determining the amount of resistance in the filament of a lightbulb. The lightbulb filament of a 100-W lightbulb has approximately 100 Ω of resistance; 100 Ω is sufficient resistance to cause the

electrons to work harder as they move through the filament, which causes the filament to heat up to a point where it glows. When the amount of resistance in a circuit is designed to convert energy, such as a heating element or motor, it is referred to as a *load*.

2.5
Identifying the Basic Parts of a Circuit

It is important to understand that each electrical circuit must have a *supply of voltage*, *conductors* to provide a path for the electrons to flow, and at least one *load*. The circuit may also have one or more controls. Figure 2–7 shows an example of a typical electrical circuit with the voltage supply, conductors, control, and load identified. In this circuit, the voltage source is supplied through terminals L1 and N, the load is a motor, and the two wires connecting the motor to the voltage source are the conductors. The manual switch is a control, be-

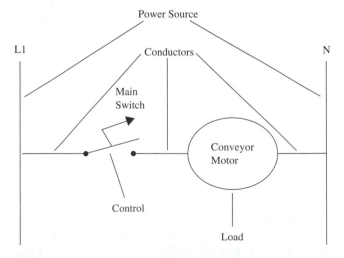

Figure 2–7 The basic parts of a circuit are identified in this example. They are the power source, conductors, control, and load.

cause it controls the current flow to the motor. In a large system in a factory, it is possible to have more than one load. For example, the system may have multiple conveyors or multiple hydraulic pumps. You will learn more about these components later.

2.6
Equating Electricity to a Water System

It may be easier to understand electricity if you think of it as a system such as a water system, with which you may be more familiar. If you examine water flowing through a hose, for example, the flow of water would be equivalent to the flow of current (amps), and the water pressure would be equal to voltage. If you stepped on the hose or bent the end over, it would create a resistance that would slow the flow, just as resistance in an electrical circuit slows the flow of current. Figure 2–8 provides a diagram that shows the similarities between the water system and an electrical circuit.

By definition, 1 V is the amount of force required to push the number of electrons equal to 1 A through a circuit that has 1 Ω of resistance. Also, 1 A is the number of electrons that flow through a circuit that has 1 Ω of resistance when a force of 1 V is applied, and 1 Ω is the amount of resistance that causes 1 A of current to flow when a

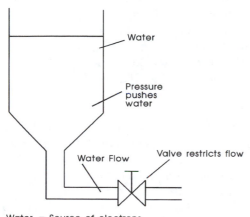

Water = Source of electrons
Water pressure = Voltage that pushes electrons
Water flow = Current flow
Valve restrics flow = Electrical resistance

Figure 2–8   A diagram of a water system that shows the similarities to an electrical system. The flow of water is similar to the flow of current (amperes), the pressure on the water is similar to voltage, and the resistance from the valve causes the flow of water to slow, just as resistance in an electrical system causes current flow to slow.

force of 1 V is applied. You can see that voltage, current, and resistance can all be defined in terms of each other.

2.7
Using Ohm's Law to Calculate Volts, Amperes, and Ohms

There are times when you are working in a circuit that you need to calculate or estimate the volts, amperes, or ohms the circuit should have. In most cases when you are troubleshooting an electrical problem, you will also need to make measurements of the volts, amperes, and ohms to determine if the circuit is operating correctly. The problem with taking a measurement is that you will not be sure if the values are too high or too low or if they are just about right. A calculation can tell you if the measurements are within the estimated values the circuit should have.

A relationship exists between the volts, amperes, and ohms in every DC circuit that can be identified by Ohm's law. Ohm's law states simply that the amount of voltage in a DC circuit is always equal to the amperes multiplied by the resistance in that circuit. If a circuit has 2 A of current and 4 Ω of resistance, the total voltage is 8 V.

The formula can be changed to determine the amount of amperes or ohms. Because we know that $2 \times 4 = 8$, it stands to reason that 8 divided by 2 equals 4, and 8 divided by 4 equals 2. Thus, if you measure the volts in a circuit and find that you have 8 V and if you measured the amperes in a circuit and find that you have 2 A, you can determine that you have 4 Ω by dividing 8 by 2. You can also determine the number of amperes you have in the circuit by measuring the volts and ohms and calculating; 8 V divided by 4 Ω is 2 A.

2.8
Ohm's Law Formulas

When scientists, engineers, and technicians do calculations, they use letters of the alphabet to represent units such as volts, amps, and ohms. The letters are accepted by everyone, so when a calculation is passed from one person to another, everyone uses the same standard abbreviations.

In Ohm's law formulas, the letter E is used to represent voltage. The E is derived from first letter of the word *electromotive force*. The letter R is used to represent *resistance*. The letter I is used to represent current (amperes); it has been derived from the first letter of the word *intensity*.

Ohm's law is represented by the formula $E = IR$. When you use Ohm's law the basic rule of thumb is that you should use E, I, and R to represent the unknown values in a formula when you begin calculations and you should use V for volts, A for amps, and Ω for ohms as units of measure when you have determined an answer or anytime the values are known or measured.

2.9
Using Ohm's Law to Calculate Voltage

When you are solving for voltage, you need to know the amount of current (amperes) and the amount of resistance (ohms). If you do not know these two values, you cannot calculate voltage. For example, if you are asked to calculate the amount of voltage when the current is 20 A and the resistance is 40 Ω, you can multiply 20 by 40 to get an answer of 800 V. To solve a problem, start with the formula and substitute the values of I and R. Then you can continue the calculation. Notice that the units A (for amperes) and Ω (for ohms) are added when the values are known.

$$E = IR$$
$$E = 20\,\text{A} \times 40\,\Omega$$
$$E = 800\,\text{V}$$

2.10
Using Ohm's Law to Calculate Current

The formula for calculating an unknown amount of current in a circuit is $I = \frac{E}{R}$. To calculate the current, you divide the voltage by the resistance. For example if you have measured the circuit voltage and found that it is 50 V and if you have measured the resistance and found that it is 10 Ω, the current is found by dividing 50 V by 10 Ω, which equals 5 A.

$$I = \frac{E}{R}$$
$$I = \frac{50\,\text{V}}{10\,\Omega}$$
$$I = 5\,\text{A}$$

2.11
Using Ohm's Law to Calculate Resistance

The formula for calculating an unknown amount of resistance in a circuit is $R = \frac{E}{I}$. To calculate the resistance of a circuit, you divide the voltage by the current.

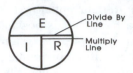

Figure 2–9   Ohm's law pie that shows a memory tool for remembering all the Ohm's law formulas.

For example, if you have measured voltage of 60 V and current of 12 A, the resistance is found by dividing the voltage by the current. Again, start with the formula and substitute values to get an answer:

$$R = \frac{E}{I}$$
$$R = \frac{60\,\text{V}}{12\,\text{A}}$$
$$R = 5\,\Omega$$

2.12
Using the Ohm's Law Wheel to Remember the Ohm's Law Formulas

A learning aid has been developed to help you remember all the Ohm's law formulas. Figure 2–9 shows this aid, and you can see that it is in the shape of a wheel, or pie. The top of the pie has the letter E, representing voltage, the bottom left has the letter I, representing current (amperes), and the bottom right has the letter R, representing resistance (ohms).

If you want to remember the formula for voltage, put your finger over the E and the letters I and R show up side by side. The vertical line separating the I and R is called the *multiply line* because that is the math function you use with I and R. This represents the formula $E = IR$.

If you want to know the formula for current (I), put your finger over the I; the remaining letters are E over R. The horizontal line that separates the E and R is called the divide-by line, because to determine I you divide E by R. This represents the formula $I = \frac{E}{R}$.

If you want to know the formula for resistance (R), you put your finger over the R; the remaining letters are E over I. This represents the formula $R = \frac{E}{I}$.

2.13
Calculating Electrical Power

Electrical power (P) is work that electricity can do. The units for electrical power are *watts* (W). Watts are determined by multiplying amps times volts ($P = IE$).

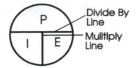

Figure 2–10 The formula pie for Watt's law formulas, which are used to calculate electrical power.

Most electrical loads that are resistive in nature, such as heating elements, are rated in watts. For example if a heating element is rated for 5 A and 110 V, it uses 550 W of power. You can calculate the amount of power by using Watt's law, $P = IE$: 5 A $\times$ 110 V = 550 W.

This basic formula can be rearranged in the same way as the Ohm's law formula. You can calculate current by dividing the power (550 W) by the voltage (110 V), and you can calculate voltage by dividing the power (550 W) by the current (5 A). Figure 2–10 shows a pie that is a memory aid for remembering all of the formulas for Watts Law. From this pie you can see that the original formula, $P = IE$, can be found by placing your finger over the P. The formula for solving for current $\left(I = \frac{P}{E}\right)$ can be found by placing your finger over the I, and the formula to solve for voltage $\left(E = \frac{P}{I}\right)$ can be found by placing your finger over the E.

2.14
Presenting All of the Formulas

When you are troubleshooting, you may be able to get a voltage reading easily. If you can find a data plate that provides the wattage for a heating element or other load, you will be able to use one of the formulas that was presented to calculate (estimate) the current or the resistance. This estimate will give you an idea of what the values should be so that you can determine if the system is working correctly or not. When you are in the field making these measurements, it may be difficult to

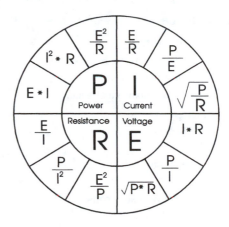

Figure 2–11 Formulas for calculating power, current, voltage and resistance.

remember all of the formulas, so a chart has been developed that provides all the formulas (Figure 2–11). From the figure you can see that the chart is in the shape of a wheel with spokes emanating from an inner ring. A cross is shown in the middle of the inner wheel, and it separates the inner wheel into four sections. These sections are represented by the letters P (power, in watts), I (amps), E (volts) and R (ohms).

The formulas in each section of the outer ring correspond to either P, I, E, or R. The formulas for wattage are shown emanating from the inner section marked with P. All the formulas for current are shown emanating from the inner section marked with I. All the formulas for voltage are shown emanating from the inner section marked with E, and all the formulas for resistance are shown emanating from the inner section marked with R.

If you are trying to calculate the power in a circuit or the power used by any individual component, you could use any of the three formulas shown on the wheel emanating from the section identified by the letter P. Thus, you could use $P = EI$, $P = I^2R$, or $P = E^2/R$. Your choice of formula would depend on the two values that are given.

Questions for This Chapter

1. Define the term electricity.

2. What is voltage?

3. What is current?

4. Explain what a valence electron is and why it is important to current flow.

5. What is resistance?

6. Define the term *power*.

True or False

1. The insulation coating on wire has high resistance.

2. Voltage is the flow of electrons.

3. Resistance is the opposition to current flow.

4. Watts are the units for current.

5. A good conductor has low resistance.

6. The negative part of an atom is an electron.

Multiple Choice

1. A valence electron is _____.
   a. the electron closest to the nucleus.
   b. the electron in the outer shell.
   c. an electron with a positive charge.
2. Wattage is _____.
   a. the unit for power.
   b. the unit for current.
   c. the unit for voltage.
3. Resistance is _____.
   a. the force that moves electrons.
   b. wattage.
   c. the opposition to current flow.

4. Current will _____ when voltage increases and resistance stays the same.
   a. increase
   b. decrease
   c. stay the same
5. Wattage will _____ when voltage increases and current stays the same.
   a. increase
   b. decrease
   c. stay the same

Problems

1. Use Ohm's law to calculate the amount of voltage if current in a DC circuit is 5 A and resistance is 8 Ω.

2. Use Ohm's law to calculate the amount of current in a DC circuit if voltage is 25 V and resistance is 50 Ω.

3. Use Ohm's law to calculate the amount of resistance in a DC circuit if voltage is 40 V and current is 10 A.

4. Use the power formula to determine the number of watts for Problem 1.

5. Use the power formula to determine the number of watts for Problem 2.

Chapter 3

Voltmeters, Ammeters, and Ohmmeters

Objectives

After reading this chapter you will be able to:

1. Use a VOM meter to measure voltage.
2. Use a VOM meter to measure resistance.
3. Use a VOM meter to measure milliamperes.
4. Understand safety warnings for using ohmmeters and ammeters.
5. Explain the operation of a clamp-on ammeter.

3.0
Measuring Volts, Amps, and Ohms

It is important to be able to measure the amount of voltage (volts), current (amperes), and resistance (ohms) in an industrial system to be able to determine when it is working correctly or if something is faulty. If you are checking a tire on your automobile, you can see if it is flat by simply looking at it. Because the flow of electrons through a wire is invisible, it is not possible to simply look at the wire and determine if it has voltage applied to it or if it has current flowing through it, so a meter must be used to make voltage, current, and resistance measurements.

3.1
Measuring Voltage

The meter that is used to measure voltage is called a voltmeter. Figures 3–1 and 3–2 show pictures of two types of voltmeters. The meter in Fig. 3–1 is called an *analog-type voltmeter* because it uses a needle and a scale to indicate the amount of voltage being measured. The meter in Fig. 3–2 is called a *digital-type voltmeter* because the display uses numbers to indicate the amount of voltage that is being measured.

The voltmeter has very high internal resistance, which is approximately 20,000 Ω per volt, so its probes can be safely placed directly across the terminals of the power source without damaging the meter. The amount of voltage that is measured by the voltmeter is shown on the display. The range-selector switch adjusts the internal

Figure 3–1   An analog voltmeter.

(Courtesy of Triplett Corporation)

Figure 3–2   A digital voltmeter.
(Courtesy of Triplett Corporation)

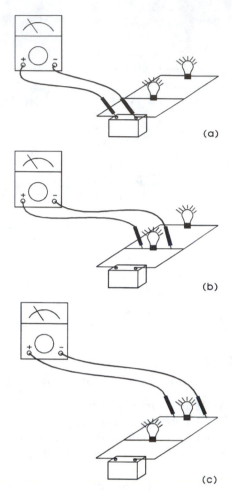

Figure 3–3   A diagram showing (a) the proper location for placing the probes of a voltmeter to measure the amount of voltage available at the battery; (b) the proper location for placing the voltmeter probes to measure the voltage available at lamp 1; (c) the proper location for placing the voltmeter probes to measure the voltage available at lamp 2.

resistance of the meter to set the maximum amount of voltage the meter can measure. The voltmeter can be used to make many different voltage measurements. For example, it is used to measure the amount of supply voltage in a system as well as measuring the amount of voltage that is provided to each load. Figure 3–3 shows three examples of where the probes of a voltmeter should be placed to make voltage measurements in a circuit safely. In the diagram in Fig. 3–3a, the voltmeter probes are shown across the terminals of a battery to measure the amount of battery voltage supplied. Figure 3–3b shows the proper location for placing the voltmeter probes to measure the voltage available to lamp 1, and Fig. 3–3c shows the proper location for placing the voltmeter probes to measure the voltage available to lamp 2. In each case the voltmeter probes are placed across the terminals where the voltage is being measured.

SAFETY NOTICE
It is important to remember that the voltage to a circuit must be turned on to make a voltage measurement. This presents an electrical shock hazard. You must take extreme caution when you are working around a circuit that has voltage applied so that you do not come into contact with exposed terminals or wires when you are making voltage measurements.

Typical meter ranges for digital voltmeters are 0–3 V, 0–30 V, 0–300 V, and 0–600 V. Special high-voltage probes must be used to measure voltages above 600 V. Typical voltages that are found in industrial systems are 24 V, 120 V, 208 V, 240 V, and 480 V. The actual amount of voltage in your area may very slightly different; for example, the 120 V may actually be measured as 110 V, and the 480 volts may be closer to 440 V.

The ranges for analog-type voltmeters are typically 0–3 V, 0–12 V, 0–60 V, 0–300 V, and 0–600 V. If you try to measure 120 V when the selector switch is set at 60 V, the needle will try to go past the maximum reading and the meter will be damaged. If you are using a digital voltmeter and you try to read a voltage that exceeds the setting of the range-selector switch, the display will indicate you are over the range but the meter will not be damaged.

Figure 3–4 (a) A digital volt-ohm-milliampere meter, or VOM meter.
(Courtesy of Triplett Corporation)
(b) A clamp-on ammeter that is used to read high currents.
(Courtesy of Triplett Corporation)

(a)

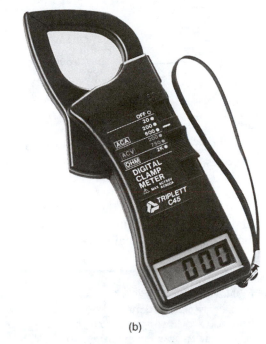

(b)

3.2
Measuring Electrical Current

An ammeter is used to measure electrical current. Figure 3–4 shows an example of two types of digital ammeters. (It should be noted that ammeters can also be digital- or an analog-type meters.) The ammeter in Fig. 3–4a is part of a volt-ohm ammeter, and it is used to measure current that is less than 10 A. In many cases the ammeter is actually built into the same meter with the voltmeter, and the selector switch allows you to set the meter to be a voltmeter, ohmmeter, or ammeter. When a meter can measure volts, ohms, or amperes it is called a *VOM meter* (volt-ohm-milliampere meter). The ammeter is actually designed to read very small amounts of current, such as 1/1000 A. The quantity 1/1000 A is called a *milliamp* (mA), which can also be written 0.001 A. The word *milli* is a prefix that means *one thousandth*. The VOM-type meter has a single meter movement that causes a needle to deflect when it senses current flow. This means that when a voltage is applied to the VOM meter, it must be routed through resistors that limit the amount of current that actually flows through the meter movement. When the VOM meter is set to measure current or resistance, the same meter movement is used. The only part of the meter that changes as it is switched between voltmeter, ammeter, and ohmmeter is the arrangement of resistors to limit the amount of current flow through the meter movement.

The meter shown in Fig. 3–4b is called a clamp-on ammeter, and it is used to measure current in larger AC circuits. From the figure you can see this type of meter has two *claws* at the top that form a circle when they are

closed. A button on the side of the meter is depressed to cause the claws to open so that they can fit around a wire without the wire having to be disconnected at one end. The claws of the clamp-on ammeter are actually a transformer that measures the amount of current flowing through a wire by *induction*. When current flows through a wire, it creates a magnetic field that has flux lines. Because AC voltage reverses polarity once every 1/60 s, the flux lines collapse and cause a small current to flow through the transformer in the claw of the ammeter. This small current is measured by the meter movement, and the display indicates the actual current flowing in the conductor that the meter is clamped around. Induction is explained in more detail in Chapters 6 and 7.

3.3
Measuring Milliamps

At times you will need to measure the milliamperes in a circuit. For example, you will need to measure the amount of current that a solid-state control is using. Figure 3–5 shows that an open should be made in the circuit when a milliampere measurement is made, and each terminal of the ammeter should be placed on the wires where the hole has been made in the circuit. This places the meter in series with the circuit, so all the electrons flowing in the circuit must also go through the meter because the ammeter has become part of the circuit.

Because the ammeter must become part of the circuit to measure the current, it must have very low internal resistance so that it does not change the total

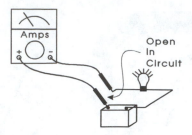

Figure 3–5   Diagram showing the proper way to place the meter probes to measure current.

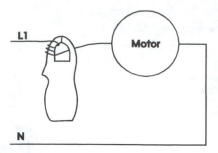

Figure 3–6   A clamp-on ammeter with wire twisted around its claws to create a multiplier.

resistance of the circuit. When the VOM meter is set to read milliamps, it is very vulnerable to damage because it has very small internal resistance.

SAFETY NOTICE
If you mistakenly place the meter leads across a power source (as you would to take a voltage reading) when the meter is set to read current, the meter will be severely damaged. You could also be severely burned—the meter could explode because it is a very low-resistance component when it is set to read current. Placing the meter leads across a power source would be similar to taking a piece of bare wire and bending it so it could be inserted into the two holes of an electrical outlet.

As a technician you will be expected to use both the milliampere meter and the clamp-on-type ammeter. Milliampere meters are used to measure small amounts of current, and clamp-on meters are used to measure larger currents, such as the amount of current a motor is using. If a motor is using (pulling) too much current, you will be able to determine that it is overloaded and that it will soon fail. Later chapters in this book explain how to use the clamp-on ammeter to measure motor current.

Digital-type VOM meters and some analog VOM meters have circuits that measure both DC and AC current. These types of meters have several ranges for measuring very small current in the range of milliamps up to 10 A.

3.4
Creating a Multiplier with a Clamp-on-Type Ammeter

Sometimes the amount of current that is measured with the clamp-on ammeter is very small, and the needle will not deflect sufficiently to create an accurate measurement. Because the claws of the clamp-on ammeter are actually a transformer, the amount of current in the reading can be multiplied by wrapping the wire in which you are measuring the current around the claw several

times to cause the reading to be multiplied. For example, if you wrap the wire around the claw two times and the wire has 1.25 A flowing through it, the meter reading will indicate 2.5 A, which is double the actual amount. You will need to divide the meter reading by 2 to get the exact current. If you wrap the wire around the claw four times, the meter reading will be 5 A, and you will need to divide the meter reading by 4 to get the true current flow in the wire. Creating a multiplier will make very small current measurements easier to read more accurately. Figure 3–6 shows an example of a wire twisted around the claws of a clamp-on ammeter to create a multiplier.

3.5
Measuring Electrical Resistance

The VOM-type meter is also able to measure resistance when the selector switch is moved to one of the ohms positions, such as R×1, R×10, or R×1000. When the VOM meter is set in the ohms configuration, it uses an *internal battery* to provide voltage to make the meter movement deflect the needle. When this voltage is applied to a resistance, a small amount of current will flow, which the meter detects.

SAFETY NOTICE
Because the meter provides the power source for the current that the meter movement will read, it is important the components being tested are not connected to any other source of voltage. It is also a good practice to remove the component from the circuit when resistance measurements are made. When a component is disconnected from a circuit and all sources of power, it is said to be *isolated*.

The ohmmeter can be used to measure the resistance of a wire or fuse to ensure that the resistance is near zero ohms (0 Ω). A wire is said to have *continuity*

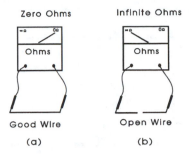

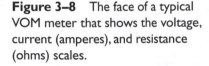

Figure 3–7 (a) An ohmmeter indicates 0 Ω when a good wire is tested. (b) An ohmmeter indicates infinite ohms when a wire with an open is tested.

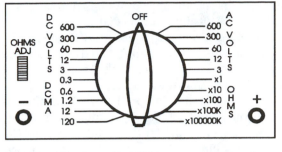

Figure 3–9 The face of a VOM meter, showing the ranges and the knob attached to the switch that connects the meter movement to different circuits of resistors inside the meter.

Figure 3–8 The face of a typical VOM meter that shows the voltage, current (amperes), and resistance (ohms) scales.

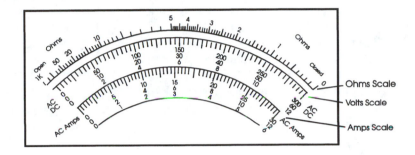

and it is considered to be a good wire when it has near 0 Ω. If the wire has extremely high resistance (several million ohms), it is considered to be bad, because it must have an *open* that is causing the high resistance. The same terms apply to fuses. If a fuse is good, it will be described as having continuity, and if it is bad, it will be described as an *open fuse*. When a wire or fuse is open, it is also said to have *infinite resistance*. The word *infinite* indicates the amount of resistance is very high due to the fact the wire or fuse has an open. Because a typical VOM meter cannot measure resistance that is more than 2 million ohms, its scale is marked with the sign for infinity (∞) at the highest point on its scale. Digital-type VOM meters flash an over-range display when an infinite resistance reading is made.

Figure 3–7a shows an example of the ohmmeter indicating 0 Ω when a good wire is tested, and Fig. 3–7b shows the ohmmeter indicating infinite ohms when the meter is testing a wire with an open in it.

The electrical loads in industrial systems such as motors or heating elements may have some amount of resistance that falls between 0 Ω and infinite ohms. For example a motor may have 12 Ω of resistance in its run winding and 30 Ω of resistance in its start winding. The ohmmeter can be used to measure each winding accurately and determine whether it has 12 Ω or 30 Ω. The ohmmeter can be used to test motor leads and determine if they are open; if they are not open, the run and start windings can be distinguished by the amount of resistance each has. You will see in later chapters that the start winding of an

AC induction motor will always have more resistance than its run winding, and the ohmmeter allows you easily to tell the run winding from the start winding.

3.6
Reading the Scales of a VOM Meter

The scale of a typical VOM meter is shown in Fig. 3–8. From this figure you can see that the face has four separate scales (lines that are shown in the shape of an arc). The bottom scale is used to measure decibels (dB), and it will not be used at this time. The second scale from the bottom is used to measure AC amps, and the third scale from the bottom, identified by the words AC DC, is used to measure AC and DC voltages and DC milliamps. The top scale is identified by the word *ohms* and is used to measure resistance. The needle that is attached to the meter movement should rest at the far left corner of the meter face when voltage, current, or resistance is not being measured.

The VOM meter can measure volts, ohms, and milliamperes with the same meter movement because a number of resistors are switched into and out of the circuit that connects the two probes to the meter movement. The switch is moved by a large knob on the front of the meter. Figure 3–9 shows the knob and the range selections on the face of the meter. If we use the hands of a clock to explain the range selections, you can see that the AC voltage circuit is located between 1 and

3 o'clock. The AC voltage circuit has five ranges: 0–3 V, 0–12 V, 0–60 V, 0–300 V, and 0–600 V.

The ohms circuits that are used to measure resistance are located between the 3 and 6 o'clock positions. Notice that × precedes each of the ranges, because you will need to multiply the reading on the scale of the meter by the multiplier factor that is identified by the switch. You should get into the habit of saying the word *times* when you encounter the ×. For example, the first range of the ohms scale is identified as ×1. You will call this range the *times 1 range*. The other ohms ranges are ×10 (times 10), ×100 (times 100), ×1000 or ×1K (times 1,000), and ×100000 or ×100K (times 100,000).

When the range is ×1000 or ×100000, note that K is substituted for three zeros. The letter K stands for *kilo*, which represents 1,000. The K is often used when there is not enough space to write in all the zeros.

The DC milliamp circuits are located between the 6 and 9 o'clock positions. The ranges for milliamps are 0 to 0.06, 0 to 1.2, 0 to 12, and 0 to 120. The DC voltage ranges are located between the 9 and 12 o'clock positions. The ranges for DC voltage are 0–3 V, 0 to 12 V, 0 to 60 V, 0 to 300 V, and 0 to 600 V.

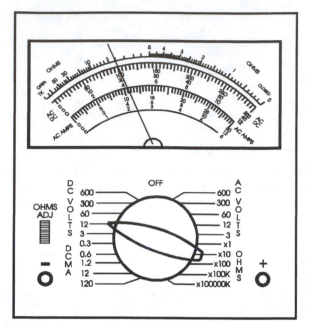

Figure 3–10 Voltmeter with its selector switch set on the 0–12-V range and its needle indicating 4 V is being measured.

3.7
Measuring DC Volts and Reading the VOM Scales

When you are making a DC voltage measurement, you place the meter leads (probes) in the sockets identified as + and −, and you should set the selector switch on the face of the meter to the highest range. For the meter in Fig. 3–9, the selector switch would be set to 600 V. Setting the selector switch to 600 V protects the meter when you are not sure how large the voltage is that you are measuring. When you touch the tips of the probes to a voltage source, the needle that is attached to the meter movement begins to move from the position at the far left, where it rests when no voltage is being sensed, toward the middle. If you have the positive meter probe touching the negative voltage source, the meter movement will try to deflect farther to the left. If this occurs, simply switch the two probes so that the positive meter probe is touching the positive voltage source. This is called *changing the polarity* of the meter.

If the voltage that is being measured is small, the needle will not move very far, and you can change the switch on the meter face to the next lower scale. If the needle still does not move sufficiently, you can continue to move the switch to a lower scale. Figure 3–10 shows an example of a meter being used to measure the amount of DC voltage in a circuit that has approximately 4 V. Because

the needle would not move sufficiently when the switch was set on the 0–600-V scale, the switch was changed down through the lower ranges until it was switched to the 0–12-V range and the needle moved to the position shown in the example.

When you are ready to read the position of the needle on the face of the meter, you start by looking at the position of the switch. Because the switch in this example is at the 12-V DC setting, you must use the scale marked AC DC for this reading. Figure 3–11 shows an enlarged diagram of only the face of the meter. The AC-DC scales are located directly below the ohms scale, which is the top scale. The voltage scales are highlighted in this figure so they are easier to identify. Notice that the scale marked AC DC has three numbers shown at the far right of the scale: 300, 60, and 12. Because the selector switch is set on 12 V, we will use the scale that has the numbers 0–12 marked on it. These numbers are also highlighted in this example.

When you look closely at the meter face, you can see that the needle points to the 4 on this scale. This means that the meter is measuring 4 V DC. You may be confused at first because the numbers 100 and 20 also occur at this same position. You can be sure that the reading is 4 V by starting at the far right side of the scale and locating the number 12. Because the number 12 is the bottom number, you should use the bottom set of numbers for this scale. (*A good habit to get into is to start with the number on the right side of the scale and*

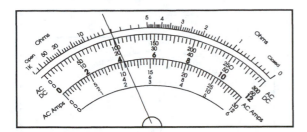

Figure 3–11 Example of the face of a VOM meter with the 0–12-V DC scale highlighted.

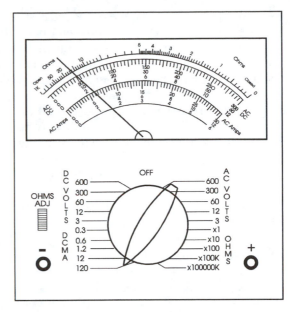

Figure 3–12 Example of a meter reading 20 DC mA.

count backward as you locate the exact position to which the needle points. For this example you would start at 12 and count backward 8, 6, 4 as you move to the left, where the needle is pointing.)

3.8
Measuring and Reading the Milliamp Scales on the VOM

When you are making a current reading with a milliampere meter, you must open the circuit and connect the meter probes so that the meter becomes part of the circuit. You should set the selector switch to the highest milliampere setting (the 120 DC mA range). If the needle does not deflect enough so that you can read it easily, you can switch the meter to the next-lower range. When the selector switch is set on this setting, you will need to read the amount of current from the same AC-DC scale that you used for the DC voltage reading.

> **NOTICE!**
> The scale marked AC AMPS is used only when measuring AC current; it cannot be used for DC current readings.

In the example shown in Fig. 3–12 the selector switch is set on the 120 DC MA setting. Because there is not a scale that ends with 120, you will have to use the 0 to 12 scale and move the decimal point or multiply the reading by 10.

For example, in the figure you can see that the needle is pointing to the number 2. You can determine this by finding the 0 to 12 scale and begin counting backward (12, 10, 8, 6, 4, 2) until you reach the place where the needle is resting. Because the selector switch is set to the 120-mA setting, you must treat the 12 as 120. You

can do this by multiplying the original reading by 10 or by counting backward from 120 to 20.

If the needle is pointing to the same position and the selector switch is pointing at 1.2 mA, the value the meter is measuring is 0.2 mA. The major point to remember when reading the milliampere scales is that all three ranges use the numbers 1 and 2; the difference between each range is in the placement of the decimal point.

3.9
Measuring and Reading Resistance Scales on the VOM

When you are making a resistance measurement with the VOM meter, you must ensure that the device or component that you are measuring does not have power connected to it and that it is isolated from other components. You should remember that this is a necessary step, because the VOM meter uses an internal battery as the power source when it is used to measure resistance. When the meter is set to read resistance, the voltage from the battery is rather small, 3 V, 9 V, or 30 V, depending on the brand name of meter. The meter will be seriously damaged if you touch its probes to a power source, such as 110 V, by mistake when the meter is set for resistance.

Also, when using the VOM meter to read resistance, you must zero the meter. Because the meter uses a battery for the power source, the voltage in the battery must be able to move the needle to the 0-Ω mark on the scale when the two probes are touched together. This process is called *zeroing the meter*, and it should be done prior to all resistance readings. If the battery is weak, the needle will not be able to move to the zero point, and all resistance readings will be faulty. This step is necessary because the actual voltage in the battery changes constantly as the battery is used and it starts to drain down.

The first step in the process of zeroing the meter is to place the selector knob on the ohms range you want to use. Next you need to touch the two meter probes together, and the meter needle should swing toward the right side of the scale. Because the VOM meter is being used to measure resistance, the very top scale—identified with the word OHMS—is used. If the meter is zeroed correctly, the needle will rest directly on the 0-Ω setting. If the needle does not move to this position, you can adjust the needle position by changing the OHMS ADJ knob, located to the left of the selector switch, just above the terminal for the negative probe. The OHMS ADJ knob is actually a potentiometer, and when you adjust it, you are actually changing the amount of resistance in the meter movement circuit, which changes the amount of current the battery is sending the meter movement. If you change the OHMS ADJ and the meter needle does not reach all the way to the 0-Ω position, the meter has a weak battery, and it should be replaced.

NOTICE!
You must zero the VOM meter any time you switch resistance scales or any time you change from one resistance scale to another, because a different amount of internal resistance is used for each circuit. You do not need to zero the VOM meter when it is used to make voltage or current measurements, because these types of readings do not use the meter's internal battery.

Figure 3–13 shows an example of the VOM meter measuring resistance. In this example, the meter selector switch is set to the ×1 (times 1) setting. Because the meter is being used to measure resistance, the top scale is used. The needle is resting on the value 10 on the top scale. The selector knob is on the ×1 (times 1) scale, so you will need to multiply the reading by 1. This means that the VOM meter is measuring 10 Ω.

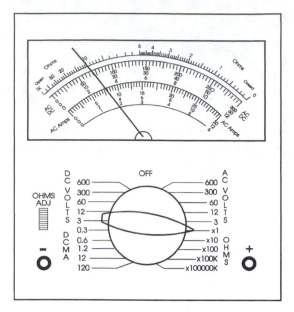

Figure 3–13 VOM meter used to measure resistance. The scale indicates the needle is resting on the 10-Ω value and the selector is set to the ×1 (times 1) range, so the meter is measuring 10 Ω.

If the selector switch is on the ×10 (times 10) range and the needle is resting in the same position, the meter is measuring 100 Ω, because you need to multiply the scale reading by 10.

3.10
Making Measurements with Digital VOM Meters

When you make a measurement with a digital VOM meter, you should follow the same rules as when you use an analog VOM meter. The major difference is that the measurement is presented as a digital number on the meter's display. All you need to do is observe the range-selector switch to determine the units that the meter is measuring and use the value that is displayed as the measurement. Be sure to observe polarity (+/−) if the voltage or current is DC. Some digital meters also provide an audible signal for resistance measurements, which is useful when you are locating wires in multiconductor cables.

Questions for This Chapter

1. Explain the term *infinite resistance*.

2. Use the diagrams in Fig. 3–3 to explain where you would place the probes of a voltmeter when making voltage measurement.

3. Use the diagram in Fig. 3–5 to explain where you would place the probes of an ammeter when making a current measurement.

4. List two things that you should be aware of when making resistance measurements.

5. Explain why you should zero an ohmmeter every time you change scales or make a measurement.

6. Explain why you should set the voltmeter and ammeter to the highest setting when you first make a measurement of an unknown voltage or current.

True or False

1. All voltage should be turned off when making a resistance measurement.

2. Infinity is the highest resistance reading for the ohms meter.

3. The VOM meter must be zeroed prior to making voltage, current, or resistance measurements.

4. The voltage should be turned off for safety reasons when making a voltage measurement.

5. The ohmmeter uses an internal battery to supply the power for all resistance readings.

Multiple Choice

1. Milliamps are _____ .

   a. one thousandth of an ampere.
   b. one millionth of an ampere.
   c. one million amperes.

2. When a current measurement is made with a VOM-type meter you should _____ .

   a. place the meter probes across the load.
   b. turn off all power to the circuit during the current reading so you don't get shocked.
   c. create an open in the circuit and place the meter probes so that the meter is in series with the load.

3. When a voltage measurement is made you should _____ .

   a. place the meter probes across the load.
   b. turn off all power to the circuit during the voltage reading so you don't get shocked.
   c. create an open in the circuit and place the meter probes so that the meter is in series with the load.

4. When a resistance measurement is made you should _____ .

   a. keep all voltage applied to the component you are measuring so you can determine the true resistance.
   b. turn off all power to the circuit and isolate the component you want to measure.
   c. create an open in the circuit and place the meter probes so that the meter is in series with the load.

5. When you are measuring AC current with a clamp-on-type ammeter you should _____ .

   a. open the claws of the meter and place them around the wire where you want to measure the current.
   b. create an open in the circuit and place the meter probes so that the meter is in series with the load.
   c. turn off all power to the circuit so that you don't get shocked.

Problems

1. Use the diagram in Fig. 3–11 to determine the voltage reading for the meter if the selector switch is set on 60 V.

2. Use the diagram in Fig. 3–11 and determine the current reading for the meter if the selector switch is set on 12 mA.

3. Use the diagram in Fig. 3–13 and determine the resistance reading for the meter if the selector switch is set for ×1K.

4. Determine the amount of current flowing in a wire if a clamp-on ammeter has four turns of the wire twisted around its claw and the reading on the scale is 2 A.

5. List the kind of problems that could be present in the circuit if a milliamp meter reading indicates 0 mA are present in the circuit.

◄   Chapter 4   ►

Series Circuits

Objectives

After reading this chapter you will be able to:

1. Explain the operation of switches that are connected in series.
2. Explain the operation of electrical loads that are connected in series.
3. Utilize Ohm's law for series circuit calculations.
4. Calculate the voltage, current, resistance, and power in a series circuit.

4.0
Introduction

Simple circuits consisting of resistors will be used to explain the relationship between voltage, current, and resistance in series circuits. Resistors are not common loads in equipment found on the factory floor, but they are commonly used as heating elements in injection molding machines, and you will also find them in modern electronic circuit boards used to control a wide variety of equipment found in factories. The motors and other loads commonly found in factory equipment have coils of wire that react in some part as resistance, so it is important for you to understand the relationship between multiple resistors in series circuits. Understanding resistance in series circuits and in parallel circuits will also provide the initial theory that you will require to understand solid-state components that are commonly used in solid-state controls.

As a technician you must understand the effects of switches in series circuits and electrical loads in series circuits. The first part of this chapter explains the relationships between switches that are connected in series. The second part of this chapter explains the effects of electrical loads that are connected in series.

The relationship between voltage, current, and resistance in series circuits is introduced in the last part of this chapter. It will be easier to understand these relationships when a known value of resistance is used. A component called a *resistor* will be used to provide the resistance for these simple series circuits. Each resistor has a color code consisting of colored bands that are painted on the resistor to identify the amount of resistance the resistor has. Ohm's law is also introduced to help you understand how voltage and current should react in all types of series circuits. The basic rules provided in this chapter will be used to help you troubleshoot larger circuits that are commonly found in heating, air conditioning, and refrigeration systems.

4.1
Examples of Series Circuits

All electrical circuits have at least a power source, one load, and conductors to provide a path for current. As circuits become more complex, switches are added for control. These switches can provide a variety of functions, depending on the way they are connected in the circuit. In Fig. 1–1 we saw a diagram of a simple circuit that included a power source, a switch for control, and a motor as the load. This circuit is called a *series* circuit because there is only one path for the current to travel to the load and back to the power source.

Figure 4–1 Example of a series circuit that has three switches connected in series. If any of the three switches are open, the current flow to the motor is stopped and the motor is deenergized.

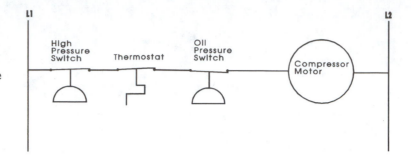

When additional switches are added to the circuit to control the load, they can be connected so that all current continues to have only one path travel around the circuit. Figure 4–1 shows an example of a typical circuit that controls power to a compressor for a small air conditioner used to provide cooling to an electrical cabinet that has a controller inside it. This circuit has a thermostat that controls the temperature where the compressor turns on and off, a high-pressure switch that protects the compressor against excessive pressure if the condenser coils should get dirty, and an oil-pressure switch that protects the compressor against loss of oil pressure.

This circuit is called a series circuit because the current has only one path to follow from L1 through the switches to the load and back to L2. If any of the three switches are opened, current is interrupted and the motor is turned off. When all the switches are closed, the compressor motor will run. When the electrical cabinet cools to 80°, the thermostat (temperature switch) will open and turn the compressor motor off. When the cabinet becomes warm again, the thermostat will close and energize the compressor again. This means the thermostat is an *operational control*, because its job is to operate the system at the setpoint temperature. If the condenser becomes dirty and causes the high-side pressure (head pressure) to become excessive, the high-pressure switch will open and turn the compressor motor off. The same is true of the oil-pressure switch. If the oil pressure becomes too low, the oil-pressure switch will open and turn the motor off. The high-pressure switch and oil-pressure switches are in the circuit for safety purposes.

Because all three of these switches are connected in series, each is capable of opening and causing current to stop flowing to the compressor motor. If additional operational or safety switches need to be added to this circuit, they would be connected in series with the original switches. It is important to remember that when switches are connected in series, any one of them can open and stop current flow in the entire circuit.

Another example of switches connected in series is the door safety switches that are used in many of the machines found on the factory floor. If any of the doors are opened while the machine is running, the door switch will open and cause current flow to the machine

to stop, and the machine will turn off. The important point to remember is that any switch in a series circuit can stop current flow through the entire circuit.

4.2
Adding Loads in Series

In industrial systems, loads such as motors are not connected in series with other loads; instead they are connected in parallel with each other so that they all receive the same voltage. Even though loads such as motors are not connected in series, you should understand the problems that exist if motor loads were connected in series. This section explains the problems associated with connecting motor type loads in series, and Section 4.3 explains how resistors are connected in series to create useful voltage-drop circuits for electronic circuit boards and components used in factory electronic controls. Parallel circuits are explained in Chapter 5.

The reason loads are not generally connected in series is that they will split the amount of applied voltage. For example, if you connected two lightbulbs that have the same amount of resistance in their filaments in series with each other and supplied the circuit with 110 V, each lightbulb would receive 55 V. Each bulb would be glowing half as brightly as if only one light were connected so that it would receive the entire 110 V.

There are some exceptions to connecting loads in series, such as electric heaters used on plastic molding machines, where two heating elements are connected in series on purpose. Figure 4–2 shows an example of two heating elements connected in series. If the amount of resistance for each heater is the same, the supply voltage will be split equally between them. In this example the supply voltage to the heaters is 240 V, and each heater will receive 120 V. This means that each heater's rating is 120 V. This type of circuit allows smaller heaters to be used on larger-voltage sources. In some applications two heater elements, each rated for 240 V, can be connected in series and will use a supply voltage of 480 V. In theory you could connect four 120-V heaters together in a series circuit and supply 480 V, and each heater would only see 120 V.

Figure 4–2 Electrical diagram of two 120-V heaters connected in series so that a 240-V power supply is divided equally between them.

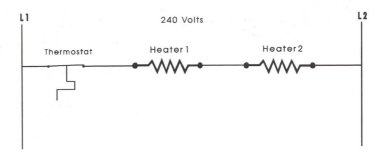

Another problem with connecting loads in series is that if one of the loads has a defect and develops an open circuit, the current to all the loads in the circuit will be interrupted. An example of this type of circuit is the strings of lights that are used to decorate a Christmas tree. If all the lights are connected in series, they all go out if one of the lights burns out. The advantage of connecting 50 lights in series is that the light bulbs will each receive approximately 2 V when the circuit is plugged into a 110-V power source.

4.3
Using Resistors as Loads

Some circuits that you encounter have resistors that are used to change the amount of voltage and current in a circuit. These types of circuits are commonly found in the solid-state control boards that are now used in many industrial systems. These solid-state boards are used to control motor speeds, for time-delay functions, and to convert small-signal voltages from sensors to larger voltages that can energize motor controls.

Figure 4–3 shows several resistors. A resistor is a component that is specifically manufactured with a precise amount of resistance. The amount of resistance will be identified on each resistor by color stripes or bands. Since most resistors are very small, a color code

system has been designed to show the amount of resistance each resistor has.

4.4
Resistor Color Codes

The resistor color bands are shown in their respective positions in Fig. 4–4. From this figure you can see that the resistor has four basic color bands that are grouped at one end of the resistor. (*Note:* A few resistors have fifth and sixth bands that represent other manufacturing information.) The color bands are used to represent numbers that indicate the amount of resistance each resistor has. The first color band represents the first digit of the number, and the second color band represents the second digit of the number. Each digit can be a value from 0 to 9. The third band is a multiplier band. This is basically the number of zeros that you place with two-digit numbers to show numbers greater than 99. The fourth band is called the *tolerance band.* The actual amount of resistance in the resistor is close to the amount identified by the color bands, but it may be slightly more or less depending on the amount of tolerance that is used. Typical tolerances are 5% and 10%. This means that the actual amount of resistance can be plus or minus the percentage identified by the tolerance band.

The colors that are assigned to the color code are shown in Fig. 4–5, and you can see each number from 0 to 9 is represented by a color. The tolerances are also represented by the colors silver and gold. A silver band means the resistor has a tolerance of 10% and a gold band means the resistor has a tolerance of 5%.

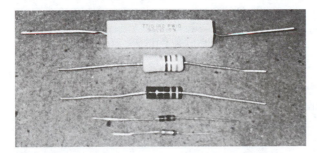

Figure 4–3 Examples of resistors that are commonly found in electronic circuit boards.

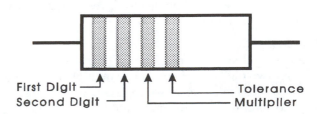

Figure 4–4 The location of color bands on a resistor.

Color	Digit Value	Multiplier	Tolerance
Black[1]	0	1	
Brown	1	10	
Red	2	100	
Orange	3	1,000	
Yellow	4	10,000	
Green	5	100,000	
Blue	6	1,000,000	
Violet[2]	7		
Gray[2]	8		
White[2]	9		
Silver		0.01	10%
Gold		0.1	5%

[1]Cannot be used in the first band (first digit).

[2]Maximum resistance is 22 M. Therefore, these colors cannot be used as multipliers.

Figure 4–5   The numerical value assigned to each of the colors for the resistor color code.

4.5
Decoding a Resistor

A resistor with a brown first band, blue second band, and red third band is decoded as follows: The first two colors, brown and blue, mean that the first two numbers are 1 and 6. The third band, which is the multiplier band, is red, so you need to add two zeros to the first two digits. Thus, the total resistance is 1,600 Ω. To use the multiplier for red, simply multiply the first two digits by 100 ($16 \times 100 = 1,600 \ \Omega$).

A second resistor has a violet first band, a white second band, and a green third band. This resistor is decoded as follows: The first two colors, violet and white, mean the first two numbers are 7 and 9. The third band, which is the multiplier band, is green, so you need to add five zeros to the first two digits. This makes the total resistance 7,900,000 Ω. Again, to use the multiplier for green, multiply 79 by 100,000 ($79 \times 100,000 = 7,900,000 \ \Omega$).

▶ **Example 4–1**
Determine the total resistance of a resistor that has a color code of yellow, brown, orange.

Solution:
The first and second colors are yellow and brown, so the first two digits are 4 and 1. The multiplier is orange, so three zeros must be added to 41, giving 41,000 Ω. To use the multiplier for orange, multiply $41 \times 1,000 = 41,000 \ \Omega$. ◀

4.6
Calculating the Resistor Tolerance

The first resistor that was decoded had color bands of brown, blue, and red and a total resistance of 1,600 Ω. If this resistor has a fourth band of silver, the resistor tolerance is calculated as $\pm 10\%$. The first step in calculating this tolerance is to multiply the total resistance by 10%.

$$1,600 \ \Omega \times 0.10 = 160 \ \Omega$$

The second step in calculating the tolerance is to add 160 Ω to and subtract 160 Ω from 1600 Ω.

1,600 Ω	1,600 Ω
+ 160 Ω	− 160 Ω
1,760 Ω	1,440 Ω

These calculations show that the total resistance of the resistor with color bands of brown, blue, red, and silver can be from 1,760 Ω to 1,440 Ω. If you were measuring this resistor with an ohmmeter and its total resistance fell between 1,440 Ω and 1,760 Ω, it would be considered a good resistor. If the total resistance were less than 1,440 Ω or greater than 1,760 Ω, the resistor would be considered out of tolerance and it would not be usable. If the fourth band was gold instead of silver, the multiplier would be 5% instead of 10%. The calculation would be completed the same way as with a tolerance of 10%, except you would add 5% and subtract 5% from the total to get the upper and lower limits.

▶ **Example 4–2**
Determine the total resistance, including tolerance, of a resistor whose color code is yellow, red, orange, and gold.

Solution:
Because the first two colors are yellow and red, the first two digits are 4 and 2. The multiplier is orange, so you need to add three zeros to make the total resistance 42,000 Ω.

The tolerance band is gold, so the tolerance is $\pm 5\%$. 5% of 42,000 Ω is found by multiplying 42,000 $\Omega \times 0.05$:

$$42,000 \times 0.05 = 2,100$$

The next step requires that you add 2,100 Ω to 42,000 Ω to get the upper value and subtract 2,100 Ω from 42,000 Ω to get the lower value:

42,000 Ω	42,000 Ω
+2,100 Ω	−2,100 Ω
44,100 Ω	39,900 Ω

The upper value of this resistor is 44,100 Ω and the lower value is 39,900 Ω.   ◀

4.7
Using Ohm's Law to Calculate Ohms, Volts, and Amps for Resistors in Series

In some electronic circuits all the loads are resistors. These resistors are sized so that the supply voltage will be dropped into smaller increments. These types of circuits are widely used in electronic circuits for industrial equipment, and they are also useful for learning the concepts of how voltage and current are affected by changes in resistance. Several circuits that have resistors connected in series will be used to explain these concepts. Figure 4–6 shows an example of three resistors connected in series with a voltage source (battery). Because the resistors are connected in series, there is only one path for the current. The resistors are numbered R_1, R_2, and R_3, and the supply voltage is identified as E_T. The formula for calculating total resistance in a series circuit is $R_T = R_1 + R_2 + R_3$. If more than three resistors are used in the circuit, the additional resistors are added to the first three to get a total. The arrows show conventional current flow. (Electron flow would show the flow of electrons (current) moving from the negative battery terminal to the positive terminal. This text uses conventional current flow in its explanations and diagrams.)

Figure 4–7 shows a series circuit that has the size of each resistor identified. Resistor R_1 is 30 Ω, R_2 is 20 Ω, and R_3 is 50 Ω. The total resistance for this circuit can be calculated as follows:

$$R_T = R_1 + R_2 + R_3$$
$$R_T = 30\ \Omega + 20\ \Omega + 50\ \Omega$$
$$R_T = 100\ \Omega$$

4.8
Solving for Current in a Series Circuit

The total current can be calculated by the Ohm's law formula

$$I = \frac{E}{R}$$
$$I = \frac{330\ V}{100\Omega}$$
$$I = 3\ A.$$

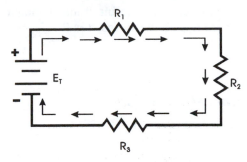

Figure 4–6   A series circuit that has three resistors connected so current can flow in only one path.

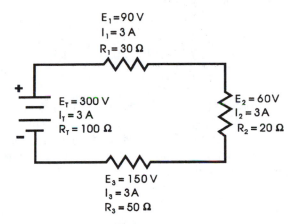

Figure 4–7   A series circuit with the total resistance calculated and the voltage and current for each resistance are calculated.

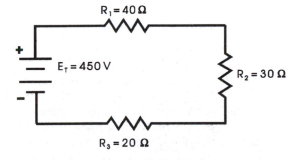

Figure 4–8   Series circuit for Problem 4–3.

The total current in a series circuit is the same everywhere in the circuit, because there is only one path for the current. This means that once you find the total resistance of a series circuit, you also have determined the current that flows through each resistor. The current (3 A) has been listed with each resistor on the diagram in Fig. 4–8. Because the current is the same everywhere in the series circuit, it may be identified as I_T or I_1 where it is shown at resistor R_1.

4.9
Calculating the Voltage Drop across Each Resistor

The voltage drop across each resistor can be calculated by the Ohm's law formula ($E = IR$). It is important to remember that the current in the series circuit is the same in all places, so I_T is equal to 3 A and I_1, I_2, and I_3 also equal 3 A. In this case the voltage drop across resistor R_1 is calculated by

$$E_1 = I_1 R_1$$
$$E_1 = 3\,A \times 30\,\Omega$$
$$E_1 = 90\,V$$

If you placed the probes from a voltmeter on each side of resistor R_1, you would measure 90 V. This is the actual voltage that the resistor is causing to *drop* when current is flowing through it. The voltage drop across resistor R_2 can be calculated by

$$E_2 = I_2 \times R_2$$
$$E_2 = 3\,A \times 20\,\Omega$$
$$E_2 = 60\,V$$

The voltage drop across resistor R_3 can be calculated by

$$E_3 = I_3 \times R_3$$
$$E_3 = 3\,A \times 50\,\Omega$$
$$E_3 = 150\,V$$

From these calculations you can see that the voltage dropped across R_1 is 90 V, that across R_2 is 60 V, and that across R_3 is 150 V. If you add all these voltage drops, you will find that the sum equals the supply voltage. This means that $E_1 + E_2 + E_3 = E_T$ (90 V + 60 V + 150 V = 300 V).

▶ *Example 4–3*

Figure 4–8 shows three resistors connected in series. Resistor R_1 is 40 Ω, R_2 is 30 Ω, and R_3 is 20 Ω. Calculate the total resistance for this circuit. After you have calculated the total resistance, calculate the total current. When you have determined the total current for this circuit, calculate the voltage drop that would be measured across each resistor.

Solution:
R_T can be calculated as follows:

$$R_T = R_1 + R_2 = R_3$$
$$R_T = 40\,\Omega + 30\,\Omega + 20\,\Omega$$
$$R_T = 90\,\Omega$$

I_T is calculated by

$$I_T = \frac{E_T}{R_T}$$
$$I_T = \frac{450\,V}{90\,\Omega}$$
$$I_T = 5\,A$$

Because current is the same in all parts of the circuit;

$$I_T = 5\,A \quad I_1 = 5\,A \quad I_2 = 5\,A \quad I_3 = 5\,A$$

The voltage that is dropped across each resistor from the current flowing through it is calculated as follows:

$E_1 = R_1 \times I_1$	$E_2 = R_2 \times I_2$	$E_3 = R_3 \times I_3$
$E_1 = 40\,\Omega \times 5\,A$	$E_2 = 30\,\Omega \times 5\,A$	$E_3 = 20\,\Omega \times 5\,A$
$E_1 = 200\,V$	$E_2 = 150\,V$	$E_3 = 100\,V$

You can check your answers by adding the drops to see that they equal the total supply voltage.

$$E_T = E_1 + E_2 + E_3$$
$$E_T = 200\,V + 150\,V + 100\,V$$
$$E_T = 450\,V \qquad ◀$$

4.10
Calculating the Power Consumption of Each Resistor

The power consumed by each resistor or the power consumed by the total circuit can be calculated by the formulas for power (wattage): $P = EI$, $P = I^2 R$, and $P = E^2/R$. In the circuit for Example 4–3, we calculated the voltage, current, and resistance for each resistor. This means that we can use any of the three formulas to calculate the power for any resistor. For this example we will use all three formulas to show that they all work. We will use the values for R_1: $E_{R1} = 200$ V, $I_{R1} = 5$A, and $R_1 = 40\,\Omega$

$$P = EI \qquad P = I^2R \qquad P = \frac{E^2}{R}$$

$$P = 200\ \text{V} \times 5\ \text{A} \quad P = (5\ \text{A})^2 \times 40\ \Omega \quad P = \frac{(200\ \text{V})^2}{40\ \Omega}$$

$$P = 1{,}000\ \text{W} \qquad P = 1{,}000\ \text{W} \qquad P = 1{,}000\ \text{W}$$

Because all three formulas give the same results, you can use any of the three that you choose. The major factor in deciding which formula to use will be the values that you have given or that you can determine by measuring.

4.11
Calculating the Power Consumption of an Electrical Heating Element

The power formula can also be used to calculate the power consumption of any other type of resistance used in a circuit. The electric heating elements used in an electric furnace are resistances, so if you know the amount of resistance in the element and the amount of voltage applied to the circuit, you can calculate the current the element uses and the amount of power it consumes. For example, if the heating element has 10 Ω of resistance and it is connected to 220 V, it would draw 22 A. This element would consume 4,840 W of power.

▶ *Example 4–4*

A heating element in an electric furnace has resistance of 2.5 Ω. Calculate the current and the power consumption of this heating element if the furnace is connected to 240 V AC.

Solution:
The amount of current is 240 V/2.5 Ω = 96 A. The power can be calculated by $P = I^2R$ or $P = IE$: Using $P = I^2R$, $(96)^2 \times 2.5\ \Omega = 23{,}040$ W. Using $P = IE$, $96 \times 240 =$

23,040 W. (*Note:* This answer can be expressed as 23.040 kW (kilowatts).)   ◀

4.12
Review of Series Circuits

In a series circuit all the loads are connected in such a way that there is only one path for current to follow. If one or more switches are used to control current flow to the loads in the circuit, they must also be connected so that only one path exists. The total resistance in a circuit can be calculated by adding all the resistor values together ($R_T = R_1 + R_2 + R_3 + \ldots$). (Notice that the three dots mean that any number of resistors can be used in this formula because they are all added.)

The total current of a series circuit can be calculated using the formula $I = E/R$, and it is important to remember that the current will be the same in all parts of the circuit. This means that once you measure or calculate current at any point or through any component, it will be the same at all other components in the circuit.

The voltage that is measured across each component in the series circuit can be different. If the resistance values of any two resistors in a circuit are the same, the amount of voltage measure across them will be the same. The amount of voltage that is measured across each resistor is caused by the value of the resistor and the amount of current flowing through it. The formula for calculating voltage is $E = IR$. The amount of voltage measured across each resistor in a series circuit is also called the *voltage drop* for that resistor. The total of all the voltage drops in a series circuit will always be equal to the supply voltage for that circuit.

The power consumed by each component in a series circuit can be calculated by the formula $P = IE$. The total power consumed by the components in a series circuit can be calculated by the same formula or by adding the amount of power consumed by each component. For example, $P_T = P_1 + P_2 + P_3$.

Questions for This Chapter

1. Discuss the advantage of connecting switches in series with a load.

2. Explain why loads are typically not connected in series with each other.

3. In some cases loads may be connected in series with each other. Explain when this would be advantageous and what type of loads would be used.

4. Explain what a resistor is and where you would find it.

5. Explain what each of the four color bands on a resistor is used to indicate.

True or False

1. When a series circuit has an open, current flow stops in all parts of the circuit.

2. The main reason the switches in Fig. 4–1 are connected in series is so that if any one of the switches open, current to the load will be interrupted.

3. It is a good practice to connect two electrical loads such as motors in series so that when one quits the other will stop.

4. Two heating elements should be connected in series if you want them to split the amount of voltage applied to them.

5. All resistors have the same amount of resistance if they are the same physical size.

Multiple Choice

1. In a series circuit the current in each part of the circuit is _____ .
   a. always zero.
   b. the same.
   c. equal to the voltage.

2. The amount of resistance a resistor has _____ .
   a. is the same if the resistors are the same size.
   b. can be determined by color codes.
   c. changes as the amount of current that flows through it changes.

3. When three resistors are connected in series, the amount of voltage that is measured across each one is _____ .
   a. determined by the amount of current flowing through it and the amount of resistance it has.
   b. determined by the wattage rating of each resistor.
   c. impossible to determine by calculations.

4. If a thermostat, high-pressure switch, and oil-pressure switch are connected in series with a compressor motor and the oil-pressure switch is opened because of low oil pressure, the compressor motor will _____ .
   a. still run because two of the other switches are still closed.
   b. not be affected because no other loads are connected to it in series.
   c. stop running because current flow will be zero.

5. If the resistance of a heating element increases from 2 Ω to 5 Ω, the current it draws will _____ .
   a. increase if the voltage stays the same.
   b. decrease if the voltage stays the same.
   c. not change if the voltage does not change.

Problems

1. Decode the following resistors and indicate their tolerances; the first color band is listed on the left.

	First Band	Second Band	Third Band	Fourth Band	
a.	Red	Blue	Brown	Gold	_____
b.	Green	Gray	Red	Silver	_____
c.	Orange	Black	Blue	Gold	_____
d.	Brown	Yellow	Violet	Gold	_____

2. Calculate the amount of voltage for a circuit that has 6 A and 20 Ω.

3. Calculate the amount of current for a circuit that has 100 V and 25 Ω.

4. Calculate the wattage for a circuit that has 120 V and 8 Ω.

5. Calculate the wattage for a circuit that has 15 Ω total and draws 25 A.

◄ Chapter 5 ►

Parallel and Series-Parallel Circuits

Objectives

After reading this chapter you will be able to:

1. Explain the operation of components connected in parallel.
2. Calculate the voltage, current, resistance, and power in a parallel circuit.
3. Explain Ohm's law as it applies to parallel circuits.
4. Calculate the voltage, current, resistance, and power in a series-parallel circuit.

5.0

Introduction

Parallel circuits are used frequently in factory electrical control. Figure 5–1 shows a sample circuit, which shows the difference between the series circuit and the parallel circuit: all the loads in a parallel circuit have the same voltage supplied, where as in a series circuit, each load has a different amount of voltage. The load components in systems such as conveyor motors and fan motors must all be provided with the same voltage, so they must be connected in parallel when they are in the same circuit. Each point where a resistor is connected in parallel in this circuit is called a *branch circuit*. The parallel circuit can have any number of branch circuits. The current in a parallel circuit allows current along more than one path to return to the power supply. These paths are identified by the arrows in the diagram.

The current in a parallel circuit is additive; the formula for calculating total resistance is $I_T = I_1 + I_2 + I_3 + \ldots$ (remember the dots mean that additional cur-

rents are added to the total). Another point to understand is that the current in a parallel circuit gets larger as more loads are added to the circuit. The parallel circuit also provides a means by which any load can be disconnected from the power source while still allowing voltage to remain supplied to the remaining loads. This is accomplished by placing a switch in each branch circuit just ahead of each resistor.

Series-parallel circuits are used in electrical systems where some parts of the circuit are series in nature and other parts of the circuit are parallel. For example, the fuse and disconnect for the circuit must be able to interrupt all power that is supplied to the loads in a circuit, so the fuse and disconnect must be connected in series with the power supply. If the system has two conveyor motors that are both rated for 220 V, they must have the same amount of voltage supplied to each of them, so they must be connected in parallel with each other and the power supply. The important point to remember when working with series-parallel circuits is that the part of the circuit that is series uses the formulas for series circuits and the part of the circuit that is parallel uses the formulas for parallel circuits. It is also

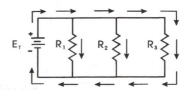

Figure 5–1 Example of a parallel circuit. Notice in this circuit that all the resistors have the same amount of voltage supplied to each.

possible to combine parts of the circuit to make it a simplified series or simplified parallel circuit.

5.1
Calculating Voltage, Current, and Resistance in a Parallel Circuit

Voltage, current, and resistance can be calculated in a parallel circuit just as in a series circuit by using Ohm's law. Figure 5–2 shows a sample circuit with the voltage, current, and resistance calculated at each point in the circuit. From this diagram you can see that the supply voltage is 300 V. Because the supply voltage is 300 V, you can determine that the voltage of each branch circuit (across each resistor) is also 300 V, because the voltage at each branch circuit in a parallel circuit is the same as the supply voltage.

The current that is flowing through each resistor can be calculated by using the Ohm's law formula for current, $I = E/R$. The current in each branch circuit is calculated from this formula, and it is placed in the diagram beside each resistor.

$$I_1 = \frac{E}{R_1} \qquad I_2 = \frac{E}{R_2} \qquad I_3 = \frac{E}{R_3}$$

$$I_1 = \frac{300\ V}{30\ \Omega} \qquad I_2 = \frac{300\ V}{20\ \Omega} \qquad I_3 = \frac{300\ V}{60\ \Omega}$$

$$I_1 = 10\ A \qquad I_2 = 15\ A \qquad I_3 = 5\ A$$

$$I_T = \frac{E_T}{R_T} \qquad\qquad I_T = I_1 + I_2 + I_3$$

$$I_T = \frac{300\ V}{10\ \Omega} \qquad\qquad I_T = 10\ A + 15\ A + 5\ A$$

$$I_T = 30\ A \qquad\qquad I_T = 30\ A$$

If the voltage and current were given and the resistance needed to be calculated, the Ohm's law formula for resistance, $R = E/I$, can be used:

$$R_1 = \frac{E}{I_1} \qquad R_2 = \frac{E}{I_2} \qquad R_3 = \frac{E}{I_3} \qquad R_T = \frac{E}{I_T}$$

$$R_1 = \frac{300\ V}{10\ A} \qquad R_2 = \frac{300\ V}{15\ A} \qquad R_3 = \frac{300\ V}{5\ A} \qquad R_T = \frac{300\ V}{30\ A}$$

$$R_1 = 30\ \Omega \qquad R_2 = 20\ \Omega \qquad R_3 = 30\ \Omega \qquad R_T = 10\ \Omega$$

If the total resistance and the total current in a circuit are known, you can calculate the voltage for the circuit. If you know the current and resistance at any branch circuit, you can also calculate the voltage using the Ohm's law formula $E = IR$. The nice part about calculating voltage is that once you determine the voltage at any branch circuit, the same amount of voltage occurs at every other branch circuit and at the supply. The same is true if you calculate the voltage at the supply; you do not need to calculate the voltage at any other branch, because it is the same.

$$E_T = I_T R_T \qquad\qquad E_1 = I_1 R_1$$

$$E_T = 30\ A \times 10\ \Omega \qquad E_1 = 10\ A \times 30\ \Omega$$

$$E_T = 300\ V \qquad\qquad E_1 = 300\ V$$

$$E_2 = I_2 R_2 \qquad\qquad E_3 = I_3 R_3$$

$$E_2 = 15\ A \times 20\ \Omega \qquad E_3 = 5\ A \times 60\ \Omega$$

$$E_2 = 300\ V \qquad\qquad E_3 = 300\ V$$

▶ *Example 5–1*

Use the circuit in Fig. 5–3 to calculate the individual branch current, total current, and total resistance.

Figure 5–2   Example of a parallel circuit with the voltage, resistance, and current shown at each load.

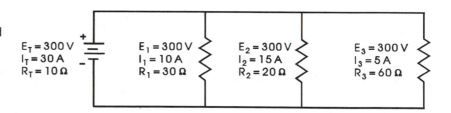

Figure 5–3   Parallel circuit for Example 5–1.

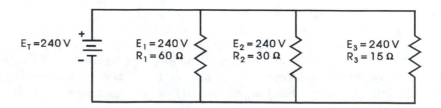

Solution:

Use the following formulas to calculate the current for each branch.

$$I_1 = \frac{E}{R_1} \qquad I_2 = \frac{E}{R_2} \qquad I_3 = \frac{E}{R_3}$$

$$I_1 = \frac{240\text{ V}}{60\ \Omega} \qquad I_2 = \frac{240\text{ V}}{30\ \Omega} \qquad I_3 = \frac{240\text{ V}}{15\ \Omega}$$

$$I_1 = 4\text{ A} \qquad I_2 = 8\text{ A} \qquad I_3 = 16\text{ A}$$

Use the following formula to calculate the total current, I_T.

$$I_T = I_1 + I_2 + I_3$$
$$I_T = 4\text{ A} + 8\text{ A} + 16\text{ A}$$
$$I_T = 28\text{ A}$$

Use the following formula to calculate total resistance R_T.

$$R_T = \frac{E_T}{I_T}$$

$$R_T = \frac{240\text{ V}}{28\text{ A}}$$

$$R_T = 8.57\ \Omega \qquad \blacktriangleleft$$

5.2
Calculating Resistance in a Parallel Circuit

At times you will need to calculate the total resistance of a parallel circuit when only the branch resistance and supply voltage are provided. You can calculate the individual currents at each branch circuit and then calculate the total with the formula $I_T = I_1 + I_2 + I_3$. After you have determined the total current, you can use the formula $R_T = E_T/I_T$.

Another method, called the *product over the sum method*, requires you to use a formula for calculating total resistance in a parallel circuit. This method is called the product over the sum method because you multiply the two resistors to get the product, and then you add

the two resistors to get the sum. The third step in the calculation includes dividing the product by the sum (product over sum). The formula is

$$R_T = \frac{R_1 \times R_2}{R_1 + R_2}$$

With this formula you can only calculate the total resistance of two resistors at a time. Because this circuit has three resistors, you need to find the total of the first two resistors in the branch and then use the formula again with that total and the third resistance to find the *grand total* resistance. We will use this method to calculate the total resistance for the resistors shown in the circuit in Fig. 5–4.

$$R_{T_{R1R2}} = \frac{R_1 \times R_2}{R_1 + R_2}$$

$$R_{T_{R1R2}} = \frac{30\ \Omega \times 20\ \Omega}{30\ \Omega + 20\ \Omega}$$

$$R_{T_{R1R2}} = \frac{600\ \Omega}{50\ \Omega}$$

$$R_{T_{R1R2}} = 12\ \Omega$$

$$R_{T_{R1R2}} = \frac{R_{T_{R1R2}} \times R_3}{R_{T_{R1R2}} + R_3}$$

$$R_{T_{R1R2}} = \frac{12\ \Omega \times 60\ \Omega}{12\ \Omega + 60\ \Omega}$$

$$R_{T_{R1R2}} = \frac{720\ \Omega}{72\ \Omega}$$

$$R_T = 10\ \Omega$$

From these calculations you can see that the total parallel resistance is 10 Ω. *It is important to understand that in all parallel circuits, the total resistance will always be smaller than the smallest branch circuit resistance.* You can see that in this circuit, the smallest resistance in the branch circuits is 20 Ω, and the total resistance is 10 Ω.

The next calculations show the total resistance, determined from the formula

$$R_T = \frac{1}{\dfrac{1}{R_1} + \dfrac{1}{R_2} + \dfrac{1}{R_3}}$$

Figure 5–4 Parallel circuit for calculating total resistance.

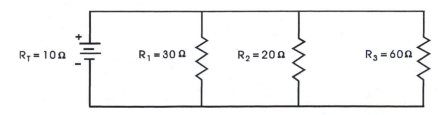

This formula is designed to be used with a calculator. If you do not have a calculator, it is recommended that you use the previous method. If you use a calculator, you should use the specified keystrokes to get an answer from this formula.

$$R_T = \cfrac{1}{\cfrac{1}{R_1} + \cfrac{1}{R_2} + \cfrac{1}{R_3}}$$

$$R_T = \cfrac{1}{\cfrac{1}{20\ \Omega} + \cfrac{1}{10\ \Omega} + \cfrac{1}{30\ \Omega}}$$

Keystrokes

(The boxes indicate keys on the calculator to use and the numbers indicate the number keys to use.)

30 $\boxed{\frac{1}{X}}$ $\boxed{+}$

20 $\boxed{\frac{1}{X}}$ $\boxed{+}$

60 $\boxed{\frac{1}{X}}$ $\boxed{=}$

$\boxed{\frac{1}{X}}$ $\boxed{=}$

$$R_T = 10\ \Omega$$

▶ **Example 5–2**

Use the parallel circuit in Fig. 5–5 to calculate the current through each resistor, the total current, and the total resistance for this circuit. The supply voltage is 100 V.

Solution:

$$I_1 = \frac{E}{R_1} \qquad I_2 = \frac{E}{R_2} \qquad I_3 = \frac{E}{R_3}$$

$$I_1 = \frac{100\ \text{V}}{20\ \Omega} \qquad I_2 = \frac{100\ \text{V}}{10\ \Omega} \qquad I_3 = \frac{100\ \text{V}}{5\ \Omega}$$

$$I_1 = 5\ \text{A} \qquad I_2 = 10\ \text{A} \qquad I_3 = 20\ \text{A}$$

$$I_T = I_1 + I_2 + I_3$$

$$I_T = 5\ \text{A} + 10\ \text{A} + 20\ \text{A}$$

$$I_T = 35\ \text{A}$$

$$R_T = \frac{E_T}{I_T}$$

$$R_T = \frac{100\ \text{V}}{35\ \text{A}}$$

$$R_T = 2.86\ \Omega$$

$$R_T = \cfrac{1}{\cfrac{1}{R_1} + \cfrac{1}{R_2} + \cfrac{1}{R_3}}$$

$$R_T = \cfrac{1}{\cfrac{1}{20\ \Omega} + \cfrac{1}{10\ \Omega} + \cfrac{1}{5\ \Omega}}$$

$$R_T = 2.86\ \Omega \qquad \blacktriangleleft$$

5.3
Calculating Power in a Parallel Circuit

The formula for calculating power in a parallel circuit is the same as the formula for a series circuit: $P_T = P_1 + P_2 + P_3$. The formula for power at each individual resistor is found from the original Watt's law formula, $P = IE$. This means that you must calculate the power consumed by each branch resistor and then add them all together to get the total power consumed. For example, in the parallel circuit that was shown in Fig. 5–2, the voltage at R_1 is 300 V and the current is 10 A, the voltage at R_2 is 300 V and the current is 15 A, and the voltage at R_3 is 300 V and the current is 5 A. The following calculations determine the power consumed by each individual resistor and the total power used by the whole circuit.

Figure 5–5 Parallel circuit for Example 5–2.

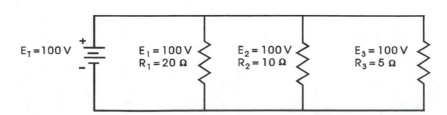

$$P_1 = I_1E_1 \qquad P_2 = I_2E_2 \qquad P_3 = I_3E_3$$
$$P_1 = 10\text{ A} \times 300\text{ V} \quad P_2 = 15\text{ A} \times 300\text{ V} \quad P_3 = 5\text{ A} \times 300\text{ V}$$
$$P_1 = 3{,}000\text{ W} \qquad P_2 = 4{,}500\text{ W} \qquad P_3 = 1{,}500\text{ W}$$
$$P_T = P_1 + P_2 + P_3$$
$$P_T = 3{,}000\text{ W} + 4{,}500\text{ W} + 1{,}500\text{ W}$$
$$P_T = 9000\text{ W}$$

You can also calculate the total power consumed by this circuit as follows:

$$P_T = I_T E_T.$$
$$P_T = 30\text{ A} \times 300\text{ V}$$
$$P_T = 9000\text{ W}$$

5.4
Reviewing the Principles of Parallel Circuits

Parallel circuits are used in electrical systems because these types of circuits provide the same amount of voltage to each motor in the circuit. For example, if a system has a conveyor motor that requires 240 V and a hydraulic pump motor that requires 240 V, a parallel circuit will provide both motors this voltage if they are connected in parallel. A parallel circuit provides more than one path for current to return to the power source. Each path is called a branch circuit. The amount of current in each branch of the parallel circuit can be determined by dividing the amount of voltage supplied to that branch by the amount of resistance for the load in that branch. The amount of total resistance in a parallel circuit will always be smaller than the smallest branch resistance. This also means that as additional loads (resistances) are added to a parallel circuit, the total amount of resistance will become smaller. Since the amount of resistance is reduced as loads are added, the total amount of current will increase.

5.5
Using Prefixes with Units of Measure

At times in previous examples you have seen the amount of resistance exceed 1,000 Ω or the amount of power exceed 1000 W. In such cases it is useful to substitute a letter for the large number of zeros. For example, we learned that the word for 1000 is kilo, so the letter k can be substituted for the value 1,000. This means that if the answer to a problem is 1,000 Ω, we can use the term 1 kΩ to indicate the same amount. The letter k in this case replaces three zeros.

When resistor values are calculated, the amount of resistance can be as large as 300,000 Ω. This amount can be represented by the term 300 kΩ, because the letter k represents 1,000. If the value of a resistor is determined to be 5,000,000 (5 million), the six zeros can be replaced by the letter M. The letter M is the abbreviation for the word mega, which is the word for million. Thus 5,000,000 Ω can be represented by the value 5 MΩ. When you see the value 5 MΩ, you say that you have measured 5 megohms.

If the amount of current is one thousandth ampere, you can show this value as 0.001 A or 1/1,000 A. The prefix *milli* is used to represent one thousandth. This prefix is represented by the lowercase letter m. This means that the value 0.001 A can be written as 1 mA. (Notice that a lowercase m is used to specify milli and a capital A is used to represent amperes.)

If the amount of current is measured as one millionth of ampere, it can be shown as 0.000001 A or 1/1,000,000 A. The prefix *micro* and the Greek letter μ are used to represent this amount. Thus the value 0.000001 A can be represented as 1 μA.

Figure 5–6 lists the common prefixes you will use most often in electricity. The exponential value of each prefix is also shown. The exponent is another way of expressing the number without showing a large number of zeros. For example, the prefix mega represents the value 1 million. To show the value 6 kilohms, you can use 6 kΩ. Because k represents 10^3, 6 kΩ = $6 \times 10^3\ \Omega$ = $6 \times 10 \times 10 \times 10\ \Omega$ = 6,000 Ω. Thus, to multiply 6 kΩ times 20 A, you can simply multiply 6,000 $\times$ 20 to get 120,000 V.

The exponential value for a million is 10^6, which means that 1,000,000 is 10 times itself six times ($10 \times 10 \times 10 \times 10 \times 10 \times 10$). The representation of the value, 1,000,000 as 1×10^6 makes it easier to see that the value has six zeros. The other advantage of exponents is that when you multiply numbers that have exponents in them, you can simply add the exponents to get the cor-

Prefix	Symbol	Number	Exponent
Mega-	M	1,000,000	10^6
Kilo-	k	1,000	10^3
Milli-	m	0.001	10^{-3}
Micro-	μ	0.000001	10^{-6}

Figure 5–6 Table of common prefixes and their numerical and exponential values.

rect answer. For example, to multiply 5,000,000 times 4,000, you can do it longhand as shown;

$$
\begin{array}{r}
5{,}000{,}000 \\
\times\ 4{,}000 \\
\hline
20{,}000{,}000{,}000
\end{array}
$$

You can also do it with exponents: simply multiply 5×4 to get 20, and then add the exponents in 10^6 and 10^3, giving 10^9. Thus, the answer is 20 followed by 9 zeros (20,000,000,000)

$$
\begin{array}{r}
5 \times 10^6 \\
\times\ 4 \times 10^3 \\
\hline
20 \times 10^9
\end{array}
$$

Most calculators have exponent buttons. This button is identified as

$$\boxed{\text{EE}}.$$

The EE on this button stands for *enter exponent*. If you used a calculator to solve the previous problem, you would use the following key strokes:

5 $\boxed{\text{EE}}$ 6 $\boxed{\times}$

4 $\boxed{\text{EE}}$ 3 $\boxed{=}$ and the answer would show 20^9.

Note: For some calculators the exponent key is identified as

$$\boxed{\text{EXP}}.$$

(It is important to remember that the exponent will be shown in the upper right corner of the calculator's display as a number only; there is not room to display the $\times$ 10.)

If the value 2 mA is shown as a number of amperes, using exponents, the exponent will be displayed as a negative number (2×10^{-3}). The negative number (-3) that is used to indicate the exponent means that the decimal point should be moved three places to the left. Because the original number is 2, you can write 2.0 to show where the decimal point is. Then, the exponent -3 means the decimal point should be moved three places to the left, as shown; 0.002. (It is also important to understand that the first zero that is shown to the left of the decimal point does not add any value to the number. It is used so that you do not confuse the decimal point with a period.) The value 7 μA (microamps) can be shown in terms of amps using an exponent of -6 (7×10^{-6}A). To multiply 15 mA times 4 MΩ, you can use a calculator and

use exponents instead of trying to remember how many zeros to use and where to place the decimal point. The key strokes are as follows:

15 $\boxed{\text{EE}}$ -3 $\boxed{\times}$

4 $\boxed{\text{EE}}$ 6 $\boxed{\times}$ and the answer would show 60^3.

The answer of 60^3 indicates 60,000 V. Because the exponent is 3, you can also give the answer as 60 kV.

When you are working with electrical circuits you will find the prefixes commonly used to save space. On most meters, the kilohms scale is shown using kΩ and the megohm scale is shown with MΩ. The milliamp scale is shown with mA, and the microamp scale is shown with μA. Some companies that manufacture meters use the capital letter K to represent kilo instead of lowercase k because it is easier to read. The correct symbol for the prefix kilo is lowercase k.

It is also important to understand that when you perform calculations with your calculator using units involving micro, milli, kilo, and mega, your answer may contain either positive or negative exponential values from 0 to 6. The problem with an answer that has an exponential value of 2 or 5 is that the VOM meter has scales that recognize only the exponents of 10^{-6}, 10^{-3}, 10^{+3} and 10^{+6}. Thus you must write all the answers to your calculations in terms of one of these exponents if you plan to make the measurement with the VOM.

5.6
Series-Parallel Circuits

As a technician or maintenance person you will work on circuits for a wide variety of factory systems. These systems will have a number of switches and loads that are connected in series and parallel. A typical circuit is shown in Fig. 5–7. In this diagram you can see that a conveyor motor and a hydraulic pump motor are connected in parallel. These two motors are connected in parallel so that they will each receive the same amount of voltage.

These loads are also connected in series with three control switches, a high-pressure switch, a thermostat, and an oil-pressure switch. These switches are designed to control problems with the hydraulic part of the system. Any time the hydraulic part of the system is deenergized, it is also important to stop the conveyor. If any of these three switches open, the current to both loads will be stopped. The switches are connected in series for this reason. Because this circuit has some switches connected in series and two loads connected in parallel, it is called a *series-parallel circuit*.

Figure 5–7 An electrical diagram of a hydraulic pump motor and conveyor motor connected in parallel and a high-pressure switch, thermostat, and oil-pressure switch in series.

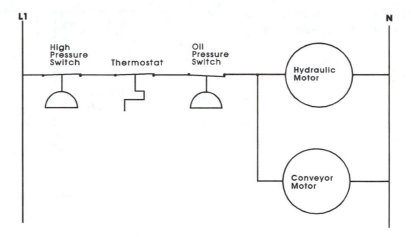

Figure 5–8 A series-parallel circuit consisting of six resistors.

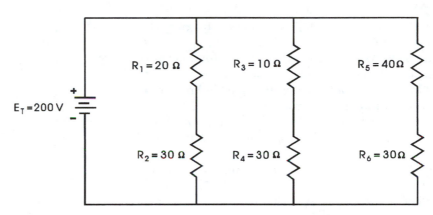

The parts of the circuit that are connected in series will follow all the rules for series circuits, and the parts of the circuit that are connected in parallel will follow all the rules for parallel circuits. This series-parallel circuit will provide the best of series circuits and the best of parallel circuits to supply voltage to the loads and control their operation. This type of series-parallel circuit is very common in field equipment that you will encounter.

It will be easier to see the operation of series-parallel circuits by following the voltage, current, and resistance of a circuit that has resistors connected in series-parallel. Figure 5–8 shows an example of a typical series-parallel circuit that uses six resistors. You should notice that the overall shape of this circuit is similar to a large parallel circuit. You can better envision the outline of the parallel circuit if you combined R_1 and R_2 into one equivalent resistance, combine R_3 and R_4 into one equivalent resistance, and combine R_5 and R_6 into one equivalent resistance.

The first branch of this circuit consists of a 20-Ω resistor connected in series with a 30-Ω resistor. The second branch consists of a 10-Ω resistor with a 30-Ω resistor, and the third branch consists of a 40-Ω resistor with a 30-Ω resistor. The supply voltage for this circuit is 200 V. This

means that each branch has 200 V applied, but each individual resistor will have a different voltage applied, because each is connected in series with one other resistor.

5.6.1
Calculating the Amount of Resistance for Each Branch

The amount of voltage that is measured across each resistor and the amount of current flowing through each resistor can be calculated by using the formulas that you have previously learned for series and parallel circuits. Figure 5–9 shows the next step in calculating the voltage and current for each individual resistor. This step includes calculating the total resistance for each branch circuit and the total resistance for the circuit. In this diagram you can see that the equivalent resistance of each branch is calculated and shown at each branch. Equivalent resistance is very similar to making change for a dollar bill. For example, if you have two 50¢ pieces, you can replace them with a dollar bill if you are solving a problem. The same is true if you have four quarters; you can show an equivalent amount of money by using two 50¢ pieces or you could use a dollar bill. When you

Figure 5–9   The series-parallel circuit that shows the series resistors in each branch combined. The value of the equivalent resistance is identified as Req_1, Req_2, and Req_3.

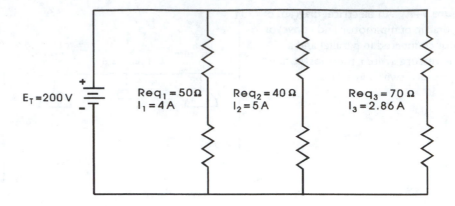

$E_T = 200 \text{ V}$

$Req_1 = 50 \, \Omega$
$I_1 = 4 \text{ A}$

$Req_2 = 40 \, \Omega$
$I_2 = 5 \text{ A}$

$Req_3 = 70 \, \Omega$
$I_3 = 2.86 \text{ A}$

are working with change and dollar bills, you basically use the form that is the simplest to handle for each different operation.

The same is true of combining resistor values to make a circuit easier to understand. For example, in Fig. 5–9 when you combine the 20-Ω and 30-Ω resistors, you can substitute the equivalent value of 50 Ω; the two resistors (R_1 and R_2) can be replaced with one resistance called Req_1 (*resistance equivalent* 1) that has the same value as the two resistors in series (50 Ω). Resistors R_3 (10 Ω) and R_4 (30 Ω) are combined to form the equivalent resistance Req_2 equal to 40 Ω. Resistors R_5 (40 Ω) and R_6 (30 Ω) are combined to form the equivalent resistance Req_3 equal to 70 Ω.

5.6.2
Calculating the Current for Each Branch

The current in each branch circuit of the circuit in Fig. 5–9 can be calculated using Ohm's law ($I = E/R$). The following calculations show how the branch circuit current was determined. You should notice that because the overall design of the circuit is a parallel circuit, the supply voltage of 200 V is also supplied to each of the branch circuits.

$$I_1 = \frac{E_1}{Req_1} \quad I_2 = \frac{E_2}{Req_2} \quad I_3 = \frac{E_3}{Req_3} \quad I_T = I_1 + I_2 + I_3$$

$$I_1 = \frac{200}{50 \, \Omega} \quad I_2 = \frac{200}{40 \, \Omega} \quad I_3 = \frac{200}{70 \, \Omega} \quad I_T = 4\text{A} + 5\text{A} + 2.86\text{A}$$

$$I_1 = 4 \text{ A} \quad I_2 = 5 \text{ A} \quad I_3 = 2.86 \text{ A} \quad I_T = 11.86 \text{ A}$$

5.6.3
Calculating the Total Resistance for the Circuit

The total resistance for this circuit can be calculated using the formula

$$R_T = \frac{E_T}{I_T}$$

$$R_T = \frac{200}{11.86 \text{ A}}$$

$$R_T = 16.86 \, \Omega$$

or the formula

$$R_T = \frac{1}{\dfrac{1}{Req_1} + \dfrac{1}{Req_2} + \dfrac{1}{Req_3}}$$

$$R_T = \frac{1}{\dfrac{1}{50 \, \Omega} + \dfrac{1}{40 \, \Omega} + \dfrac{1}{70 \, \Omega}}$$

$$R_T = 16.87 \, \Omega$$

You should notice a slight difference in the total resistance when you use different formulas, but this is due to rounding the numbers and it really does not affect the circuit. Regardless of which method you use, the total resistance will determine the total current the circuit requires. The total resistance is determined by the connections of each load (resistor). Because the circuit looks overall like a parallel circuit, the more branch circuits that are added, the more current is required for the loads. It should also be obvious that as more loads are added to this circuit in parallel, the total resistance will continue to drop.

5.6.4
Calculating the Voltage Drop across Each Resistor

The voltage drop across each resistor can be calculated by determining the current through each branch circuit and multiplying it by the resistance in each branch circuit. Because the two resistors in each branch circuit

Figure 5–10 Series-parallel circuit with voltage, current, and resistance calculated for each resistor and for the total circuit.

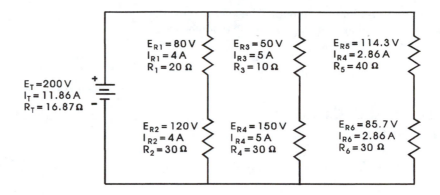

are connected in series, the Ohm's law formulas are valid. Figure 5–10 shows the current and resistance for each resistor in each of the three branch circuits. The following formula will be used to calculate the voltage that is measured across each resistor. As you know, the voltage across each resistor is also called the *voltage drop* for that resistor. The total current and total resistance is also shown in this circuit.

In the first branch circuit the current (I_1) is 4 A. When this current flows through the 20 Ω of resistor R_1, the voltage drop across R_1 is 80 V.

$$E_{R1} = I_1 \times R_1$$
$$E_{R1} = 4\,A \times 20\,\Omega$$
$$E_{R1} = 80\,V$$

Because the second resistor (R_2) is connected in series with R_1, it will also see 4 A of current. The voltage drop across R_2 is 120 V. It is found by multiplying 4 A times 30 Ω.

$$E_{R2} = I_1 \times R_2$$
$$E_{R2} = 4\,A \times 30\,\Omega$$
$$E_{R2} = 120\,V$$

In the second branch circuit the current is 5 A. When the 5-A current flows through the 10-Ω resistor R_3, the voltage drop across the resistor is 50 V.

$$E_{R3} = 12 \times R_3$$
$$E_{R3} = 5\,A \times 10\,\Omega$$
$$E_{R3} = 50\,V$$

Because the fourth resistor (R_4, 30 Ω) is connected in series with R_3, the current flowing through R_4 will also be 5 A. The voltage drop across R_4 is 150 V, and it is found by multiplying 5 A × 30 Ω.

$$E_{R4} = I_2 \times R_4$$
$$E_{R4} = 5\,A \times 30\,\Omega$$
$$E_{R4} = 150\,V$$

In the third branch circuit the current is 2.86 A. When this current flows through the 40-Ω fifth resistor (R_5), the voltage drop across the resistor is 114.4 V.

$$E_{R5} = I_3 \times R_5$$
$$E_{R5} = 2.86\,A \times 40\,\Omega$$
$$E_{R5} = 114.4\,V$$

Since the sixth resistor (R_6, 30 Ω) is connected in series with R_5, the current flowing through R_6 will also be 2.86 A. The voltage drop across R_6 is 85.8 V.

$$E_{R6} = I_3 \times R_6$$
$$E_{R6} = 2.86\,A \times 30\,\Omega$$
$$E_{R6} = 85.8\,V$$

After all the voltage current and resistance are calculated for each part of the circuit, you can begin to see patterns that will help you when you troubleshoot a series-parallel circuit that has motors and other types of loads instead of simple resistors. For example, in the previous circuit, a 30-Ω resistor was placed in each of the branches to show that in a series-parallel circuit the voltage drop across each of these resistors will not necessarily be the same. The only time the voltage drop across two resistors that have the same resistance value will be the same is when they have an equal amount of current flowing through them. This could occur in a series-parallel circuit only if the total resistance of two branch circuits were the same. The main point to remember is that it is generally possible to measure the voltage, current, and resistance of all components in a circuit. If you know some of these values, it is possible to use the Ohm's law formulas to calculate other values in the circuit to predict its behavior. These concepts are built upon in later chapters when specific troubleshooting examples are provided.

5.7
Series-Parallel Circuits in Factory Electrical Systems

Series-parallel circuits exist in a variety of factory electrical equipment. These types of circuits are also used in heating equipment. You can use the basic information that you learned about series-parallel circuits to make some simple observations about the size of wires and switches in regard to the amount of voltage and current for which they will need to be rated. Figure 5–11 shows an example of a series-parallel circuit for a plastic injection molding machine with two conveyors that move parts from the machine. From this diagram you can see that the circuit has some switches in series with motors, and the motors are connected in parallel with each other. For example, the disconnect switch at the top of the circuit in both L_1 and N is connected in series with the entire circuit. This means that if you open the disconnect switch, all power the entire circuit is disconnected.

The main control switch is connected in series with the hydraulic pump motor and the parallel circuit that has the conveyor 1 and conveyor 2 motors connected in parallel in this circuit. The conveyor 1 motor draws 5 A, the hydraulic pump motor draws 20 A, and the second conveyor motor draws 7 A; therefore, the main control switch must be rated for at least 32 A, because all this current will flow through it. The high-pressure switch and oil-pressure switch are connected in series with the hydraulic pump motor and the conveyor 2 motor, so they must be rated to carry the 27 A used by these two motors.

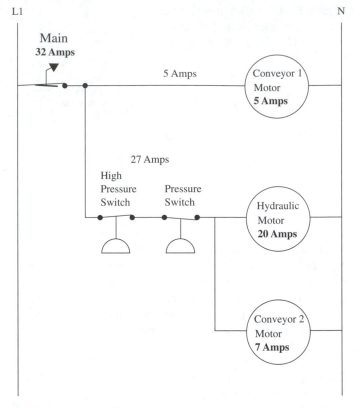

Figure 5–11   Electric diagram of a typical series-parallel circuit for a system.

This also means the current to these motors will be interrupted any time either of these switches is opened. All three motors are connected in parallel with each other, so they will all receive the same amount of voltage, but the amount of current each will draw will be different unless their load characteristics are identical. The amount of total current that is needed to supply the three motors can be calculated by the formula $I_T = I_1 + I_2 + I_3$.

Questions for This Chapter

1. Explain why the voltage at each branch of a parallel circuit is the same as the supply voltage.

2. What is the advantage of using exponents when you do Ohm's law calculations?

3. Explain what happens to the voltage at conveyor motor 2 if the hydraulic pump motor has an open in the diagram in Fig. 5–11.

4. Determine the total current that flows through the main switch in the diagram in Fig. 5–11.

5. Explain the prefixes M, k, m, and μ and provide an example of how each is used.

True or False

1. Current in each part of a parallel circuit is always the same.

2. Voltage in the branches of parallel circuits is always the same.

3. Total resistance in a parallel circuit becomes smaller as more resistors are added in parallel.

4. Total current in a parallel circuit becomes smaller as more resistors are added in parallel.

5. The prefix milli (m) means one millionth.

Multiple Choice

1. Current in a parallel circuit _____ .
   a. increases as additional resistors are added in parallel.
   b. decreases as additional resistors are added in parallel.
   c. may increase or decrease when resistors are added in parallel depending on their size.

2. Voltage in a parallel circuit _____ .
   a. increases as additional resistors are added in parallel.
   b. decreases as additional resistors are added in parallel.
   c. stays the same across parallel branches as additional resistors are added in parallel.

3. Resistance in a parallel circuit _____ .
   a. increases as additional resistors are added in parallel.
   b. decreases as additional resistors are added in parallel.
   c. may increase or decrease when resistors are added in parallel, depending on their size.

4. When one branch circuit of a multiple branch parallel circuit develops an open, voltage in other branch circuits _____ .
   a. decreases to zero.
   b. increases because fewer resistors are using up the voltage.
   c. stays the same as the supply voltage.

5. Electrical meters and circuits use exponents 10^6, 10^3, 10^{-3} and 10^{-6} because _____ .
   a. these numbers are easier to use than other exponents.
   b. these exponents are the values for the prefix M, k, m, and μ.
   c. exponents must be multiples of 3 or 6.

Problems

Note: Please show the formula and all work for each problem.

1. Draw a circuit that has three resistors (20 Ω, 40 Ω, and 80 Ω) connected in parallel and calculate the total resistance for this circuit.

2. Calculate the current for each branch of the circuit that you made for Problem 1.

3. Calculate the total current for the circuit that you made for Problem 1.

4. Calculate the power consumed by each resistor in the circuit you made for Problem 1. Voltage for the circuit is 160 V.

5. Determine the current capacity for each switch in the circuit for Fig. 5–11 if conveyor motor 1 uses 8 A, the hydraulic pump motor uses 15 A, and conveyor motor 2 uses 5 A.

◀ **Chapter 6** ▶

Magnetic Theory

Objectives

After reading this chapter you will be able to:

1. Describe a magnet and explain how it works.
2. Explain the difference between a permanent magnet and an electromagnet.
3. Identify ways to increase the strength of an electromagnet.
4. Explain what flux lines are and where you would find them.
5. Explain electromagnetic induction.

6.0
Introduction to Magnetic Theory

The operation of all types of transformers, motors, and relay coils can be explained with several simple magnetic theories. As a technician or maintenance person that works on equipment on the factory floor, you will need to fully comprehend all magnetic theories so that you will understand how these components operate. You must understand how a component is supposed to operate before you can troubleshoot it and perform tests to determine if it has failed. Understanding magnetic theory will make this job easier. It is also important to understand that some of the magnetic theories rely on AC voltage. These theories are introduced in this chapter, and more detail about AC voltage and magnetic theories that use AC voltage follows in the next chapter. In this chapter you learn about permanent magnets first and then you learn about electromagnets.

Magnet is the name given to material that has an attraction to iron or steel. This material was first found naturally about 4000 years ago as a rock in a city called Magnesia. The rock was called magnetite and was not usable at the time it was discovered. Later it was found that pieces of this material could be suspended from a wire and it would always orient itself so that the same ends always pointed the same direction, which is toward the earth's North Pole. Scientists soon learned from this phenomenon that the earth itself is magnetic. At first the only use for magnetic material was in compasses. It was many years before the forces caused by two magnets attracting or repelling could be utilized as part of a control device or motor.

As scientists gained more knowledge and equipment became available to study magnets more closely, a set of principles and laws evolved. The first of these showed that every magnet has two poles, called the north pole and south pole. When two magnets are placed end to end so that similar poles are near each other, the magnets repel each other. It does not matter if the poles are both north or both south, the result is the same. When the two magnets are placed end to end so that the south pole of one magnet is near the north pole of the other magnet, the two magnets attract each other. These concepts are called the *first and second laws of magnets*.

When sophisticated laboratory equipment became available, it was found that this phenomenon is due to the basic atomic structure and electron alignments. When the atomic structure of a magnet was studied, it was found that its atoms were grouped in regions called *domains*, or *dipoles*. In material that is not magnetic or

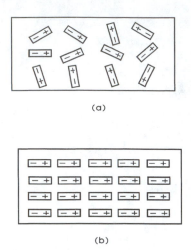

(a)

(b)

Figure 6–1   (a) Example of metal in which dipoles are randomly placed, which makes it a very weak magnet or not magnetic at all; (b) example of metal in which the dipoles are aligned to make a very strong magnet.

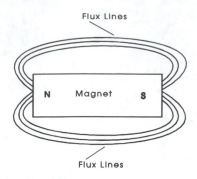

Figure 6–2   Example of a bar magnet. Notice that the poles are identified as north (N) and south (S). Flux lines are shown emanating from the north pole to the south pole.

that cannot be magnetized, the alignment of the electrons in the dipoles is random and usually follows the crystalline structure of the material. In material that is magnetic, the alignments in each dipole are along the lines of the magnetic field. Because each dipole is aligned exactly like the ones next to it, the magnetic forces are additive and are much stronger. In material where the magnetic forces are weak, it was found that the alignment of the dipole was random and not along the magnetic field lines. The more closely this alignment is to the magnetic field lines, the stronger the magnet is. Today we refer to a piece of soft iron in which all dipoles are aligned as a *permanent magnet*. The name permanent magnet is used because the dipoles remain aligned for very long periods of time, which means the magnet will retain its magnetic properties for long periods of time. Figure 6–1a shows a diagram of nonmagnetic metal, in which the dipoles are randomly placed, and Figure 6–1b shows a piece of metal that is magnetic, in which the dipoles align to make a strong magnet.

6.1
A Typical Bar Magnet and Flux Lines

Figure 6–2 shows a bar magnet that is made of soft iron that has been magnetized. The magnet is in the shape of a bar, and its north and south poles are identified. Because the bar remains magnetized for a long period of time, it is a permanent magnet. The magnet produces a strong magnetic field because all its dipoles are aligned. The magnetic field produces invisible *flux lines* that move from the north pole to the south pole along the

outside of the bar magnet. The diagram in Fig. 6–2 shows these flux lines are lines of force that form a slight arc as they move from pole to pole.

Because the flux lines are invisible, you will need to perform a simple experiment to allow you to see that the flux lines do exist and what they look like as they surround the bar magnet. For this experiment you will need a piece of clear plastic film, such as the plastic sheets used for overhead transparencies, and some iron filings. Place the plastic sheet over a bar magnet, making the plastic as flat as possible, and sprinkle iron filings over it. The filings will be attracted by the invisible flux lines as they extend in an arc from the north pole to the south pole along the outside of the magnet. Because the flux lines begin at one pole and stretch to the other, the highest concentration of flux lines will be near the poles. The iron filings will also concentrate around the poles, but a definite pattern of flux lines can be seen along each side of the bar magnet. If an overhead projector is available, the image of the flux lines can be projected onto a projector screen or blackboard so that they can be seen more easily. The pattern of these filings will look similar to the diagram in Fig. 6–2. The number of flux lines around a magnet is directly related to the strength of the magnet. A stronger magnet will have more flux lines than a weaker magnet. The strength of a magnet's field can be measured by the number of flux lines per unit area. Because the strength of a magnet's field is based on the alignment of the magnetic dipoles, the number of flux lines will increase as the alignment of the magnetic dipoles increases.

Some materials, such as alnico and Permalloy, make better permanent magnets than iron, because the alignment of their magnetic domains (dipoles) remains consistent even after repeated use. You may find these materials used in some expensive controls and motors, but normally the permanent magnet will be made of soft iron. The reason permanent magnets are useful in many

Figure 6–3 (a) Example of flux lines around a straight wire that is carrying current; (b) flux lines around a coil of wire that is carrying current. Notice that the number of flux lines increases when the wire is coiled.

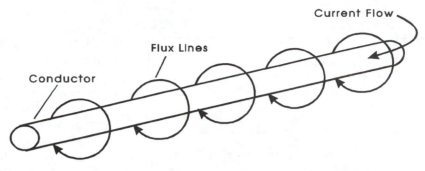

(a) Few flux lines around conductor that is not coiled

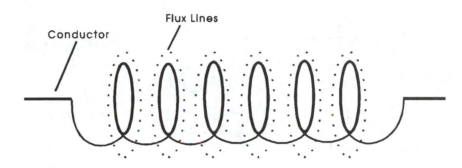

(b) Flux lines become more concentrated when wire is coiled

types of controls, especially in motors and generators, is because the soft iron produces residual magnetism for long periods of time over many years. Permanent magnets have several drawbacks, however. One of these problems is that the magnetic force of a permanent magnet is constant and cannot be turned off, if it is not needed. This means that if something is attracted to a magnet, it will remain attracted until it is physically removed from the force of the flux lines. Another problem with a permanent magnet's flux field being constant is that it cannot easily be made stronger or weaker if circumstances so require.

6.2
Electromagnets

An electromagnet is made by connecting a coil of wire to an electric cell (battery). The electromagnet has properties that are similar to a permanent magnet. When a wire conductor is connected to the terminals of the battery, current will begin to flow, and magnetic flux lines will form around the wire like concentric circles. If the wire is placed near a pile of iron filings while current is flowing through it, the filings will be attracted to the wire just as if the coil were a permanent magnet. Figure 6–3 shows several diagrams indicating the location of magnetic flux lines around conductors. Figure

6–3a shows flux lines will occur around any wire when current is flowing through it. You can set up several simple experiments to demonstrate these principles. In one experiment you can insert a current-carrying conductor through a piece of cardboard and place iron filings around the conductor on the cardboard. When current is flowing in the wire, the filings will settle around the conductor in concentric circles, showing where the flux lines are located. As the amount of current is increased, the number of flux lines will also increase. The flux lines will also concentrate closer and closer to the wire until the current reaches *saturation*. When the flux lines reach the saturation point, additional increase of current in the wire will not produce any more flux lines.

When a straight wire is coiled up, the flux lines will concentrate and become stronger. Figure 6–3b shows an example of flux lines around a coil of wire that has current flowing through it. Because the flux lines are much stronger in a coil of wire, most of the electromagnets that you will encounter will be in the form of coils. For example, coils are used in transformers, relays, solenoids, and motors.

One advantage an electromagnet has over a permanent magnet is that the magnetic field can be energized and deenergized by interrupting the current flow through the wire. The strength of the magnetic field can also be varied by varying the strength of the current flow through the conductor that is used for the

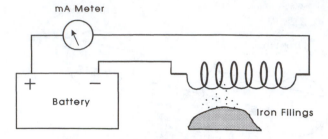

a) Few filings are attracted when small current passes through the conductor.

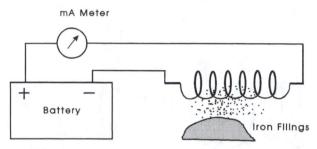

(b)   Large number of filings are attracted as current is increased.

Figure 6–4   (a) A small amount of current flowing in a coil of wire creates a small number of flux lines that attract iron filings. (b) A large amount of current flowing in a coil of wire creates a large number of flux lines that attract iron filings.

electromagnet. This theory is perhaps the most important theory of magnetism, because it is used to change the strength of magnetic fields in motors, which causes the motor shaft to produce more torque so it can turn larger loads. Because the flux lines are invisible, you may need to perform an experiment to prove that the magnetic field becomes stronger as current flow through the coil is increased. Figure 6–4 shows how to set up this experiment. Wrap a wire into a coil, connect the ends to a dry cell battery, and place it near a pile of iron filings. Add a variable resistor to the circuit to increase or decrease the current. When the current is set to a minimum, as in Fig. 6–4a, the magnetic field around the coil of wire will attract only a few filings. As the current is increased, as in Fig. 6–4b, the number of filings the magnetic field will attract also increases, until the current causes the magnetic field in the wire to reach *saturation*. When the saturation point is reached, any additional current flowing in the wire will not produce additional flux lines. When a switch is added to this circuit, the magnetic field can be turned on and off by turning the switch on and off to interrupt the flow of current in the coil. When the switch is opened, current is interrupted and no flux lines are produced, so the magnetic field will not exist.

Components such as electromechanical relays and solenoids use the principle of switching the magnetic field on and off. When current flows in the coil, the magnetic field causes the contacts in the relay or the valve in the solenoid to close. When the current to the coil in these devices is interrupted and the magnetic field is turned off, springs cause the relay contacts or valves to open. More information about relays and solenoids is provided in later chapters.

This principle is also used to turn motors on and off. When current flows through the coils of a motor, its shaft will turn. When current is interrupted, the magnetic field will diminish and the motor will stop rotating. This point is also important to remember if the wire that is used in the coil develops an open. When current stops flowing in the coil due to the open circuit, the shaft of the motor stops turning.

6.3
Adding Coils of Wire to Increase the Strength of an Electromagnet

Another advantage of the electromagnet is that its magnetic strength can be increased by adding coils of wire to the original single coil of wire. The increase of the magnetic field occurs because the additional coils of wire require a longer length of wire, which provides additional flux lines. The magnetic field will be stronger when the coil is more tightly wound because the flux lines are more concentrated. Thus very fine wire is used in some electromagnets to maximize the number of coils. As smaller wire is used, however, the amount of current flowing through it must be reduced so that the wire is not burned open.

You will learn that some motors use this principle to increase their horsepower and torque ratings. These motors have more than one coil that can be connected in various ways to affect the torque and speed of the motor's shaft. *Torque* is defined as the amount of rotating force available at the shaft of a motor. You will also learn that coils can be connected in series or in parallel to affect the torque and speed of a motor.

6.4
Using a Core to Increase the Strength of the Magnetic Field of a Coil

The strength of a magnetic field can also be increased by placing material inside the helix of the coil to act as a core. The farther the core is inserted into the coil, the stronger the flux field becomes. When the core is re-

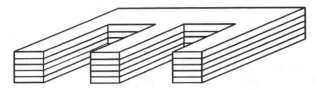

Figure 6–5   Example of pieces of steel pressed together to form a laminated steel core for use in electromagnets.

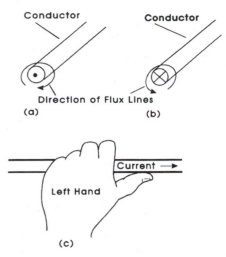

Figure 6–6   (a) The direction of flux lines around a wire when current is flowing in the wire toward you. The dot indicates the direction of current flow is toward the observer. (b) The direction of flux lines around a wire when current is flowing in the wire away from you. The X indicates current is flowing in the wire away from the observer. (c) An example of the left-hand rule. The thumb is pointing in the direction of the current flow in the wire, and the fingers are pointing in the direction of the flux lines as they move around the wire when current is flowing.

moved completely from the coil, it is considered to be an *air coil magnet,* and the magnetic field is at its weakest point. If a soft iron is used as the core, it will strengthen the magnetic field, but it also creates a problem because it has excessive residual magnetism, which is unwanted. Residual magnetism means the core will retain magnetic properties when current is interrupted in the coil, which will make it like a permanent magnet. This problem can be corrected by using laminated steel for the core. The laminated steel core is made by pressing sheets of steel together to form solid core. Figure 6–5 shows an example of layers of laminated steel pressed together to form a core. When current flows through the conductors in the coil, the laminated steel core enhances the magnetic field in much the same way as the soft iron, and when current flow is interrupted, the magnetic field collapses rapidly because each piece of the laminated steel does not retain sufficient magnetic field.

6.5
Reversing the Polarity of a Magnetic Field in an Electromagnet

When current flows through a coil of wire, the direction of the current flow through the coil will determine polarity of the magnetic field around the wire. The polarity of the magnetic field around the coil of wire is important because it determines the direction a motor shaft turns in an AC or DC motor. If the direction of current flow is reversed, the polarity of the magnetic field is reversed, and the direction a motor shaft is turning is reversed. In some motors used in factory systems, such as fan motors and pump motors, the direction of rotation is very important. In these applications you will be requested to change the connections for the windings in the motor or the supply voltage for a three-phase motor to make the motor rotate in the opposite direction. The changes you are making take advantage of changing the direction of current flow through a coil or changing the polarity of the supply voltage with respect to the other phases so that the motor will reverse its rotation.

As you are learning advanced theories about motors and other electromagnetic components, diagrams will be presented to explain more complex concepts. When these diagrams are used to explain the direction of current flow through a wire, they will use a method of identifying the direction of current flow that has been developed and universally accepted. Figure 6–6a shows a diagram that indicates the location of the flux lines in a coil of wire and shows the direction the flux lines travel around a straight, current-carrying conductor. A dot or a cross (X) is used to mark the conductor to indicate the direction current is flowing. In Fig. 6–6a you should notice that a dot is used to indicate that current is flowing toward the observer. The flux lines in this diagram show their flow is in a clockwise motion. In Fig. 6–6b, a cross is used to indicate that current is flowing away from the observer and the flux lines are shown moving in the counterclockwise direction. If you know the direction of current flow, the advanced theories will allow you to determine the polarity of the magnetic field and predict the direction a motor's shaft will rotate.

Figure 6–6c shows another way to determine the directions of the flux lines, the direction of current flow, and which pole is the north pole. This method is called the *left-hand rule.* From this diagram you can see that you need to know either the direction of the current flow or the direction of the flux lines to determine the polarity of the coil. Normally the direction of the current flow is easy to determine with a voltmeter by determining which end of the wire is positive and which is negative. Current (electron flow) is from negative to positive.

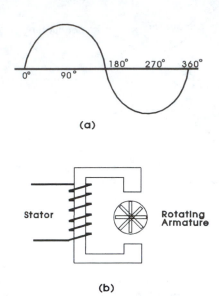

(a)

(b)

Figure 6–7   (a) A sine wave for AC voltage; (b) a simple alternator that has a moving coil and a stationary coil. The alternator produces the AC sine wave.

6.6
An Overview of AC Voltage

Before you can understand more about magnetic theories you will need to know more about AC voltage and current. This section provides an overview of AC voltage, and the next chapter provides an in-depth explanation of AC voltage. *AC stands for alternating current.* This name is derived from the fact that AC voltage alternates, being positive for half of a cycle and then negative for the other half of the cycle. Figure 6–7a shows the characteristic AC waveform. The AC waveform shown in this figure is called a *sine wave.*

The AC waveform is created by an alternator that looks similar to a motor. Figure 6–7b shows a diagram of an alternator. The alternator has a rotating coil of wire that is mounted on a shaft and a stationary coil that has current flowing through it so that it creates a magnetic field. The shaft of the alternator is rotated by an energy source such as a steam turbine. When the rotating coil of wire passes through the magnetic field, an electron flow is created in the coil of wire. Because the coil of wire rotates, it will pass the positive magnetic field and then the negative magnetic field during each complete rotation of 360°. This action causes the voltage sine wave to be produced, and a complete sine wave occurs in 360°. In the United States, the speed of rotation of the alternator is maintained at a constant rate so that the sine wave will have a frequency of 60 cycles per second, which is also called 60 Hz (hertz).

6.7
Electromagnetic Induction

Electromagnetic induction is another magnetic principle that is used in transformer and motor theory. This theory states that if two individual coils of wire are placed in close proximity to each other and a magnetic field is created in one of the coils so that flux lines are created around it, the flux lines will collapse across the second coil when the current is interrupted in the first coil. When these flux lines collapse across the coils of the second coil, they will cause a current to begin to flow in the second coil. This current flow will be 180° out of phase with the current flow that created the magnetic field in the first coil.

Because AC voltage periodically builds voltage to a positive peak and interrupts it when the sine wave reaches 180°, it is ideal for creating the magnetic field that builds and collapses. When the current flows, it causes the flux lines to build in the coil, and when the current flow is interrupted it causes the flux lines to collapse. Because AC voltage in America is generated at 60 Hz, the magnetic field in the first coil will build and collapse 60 times a second. The flux lines will also build and collapse across the windings of the second coil 60 times a second.

One component that uses two coils is called a *transformer.* The first coil in the transformer is called the *primary winding,* and the second coil in the transformer is called the *secondary winding.* The ratio of the number of turns of wire in the primary winding and to the number in the secondary winding will determine the ratio of the voltage in the primary winding and secondary winding. Another important point to remember is that because the two coils are electrically isolated, the voltage produced in the second coil by induction is totally isolated from the voltage in the first coil. The isolation allows the number of turns of wire in the two coils to be different so that a different amount of voltage can be created in the second coil. If the number of turns of wire in the first coil and second coil are identical, the amount of voltage induced in the second coil will be approximately the same as the voltage in the first coil.

In motors, electromagnetic induction is used to create a second magnetic field in a part of the motor that rotates. Because the shaft of the motor must become magnetic and rotate freely from the stationary part of the motor, it would not be practical to make a permanent connection with a wire to the rotating member so that it could receive current to create its magnetic field. Electromagnetic induction creates the second magnetic field without any connections. In future chapters you will learn more about transformer turns ratios and magnetic theories for motors.

6.8
Review of Magnetic Principles

When you are learning about components that utilize electromagnetic principles, you can return to this section to review the basic concepts about them. The important electromagnetic principles are as follows:

1. The dipoles of a magnet are aligned.
2. Every magnet has a north pole and a south pole.
3. Like magnetic poles repel, and unlike magnetic poles attract.
4. A permanent magnet is useful because its magnetic field is residual and can remain strong over a number of years.
5. When current passes through a conductor, magnetic flux lines form around the conductor in concentric circles.
6. The strength of the magnetic field around a conductor is proportional to the amount of current flowing through it until the amount of current reaches the saturation point of the electromagnet.
7. The magnetic field of an electromagnet can be turned on and off by interrupting the current flow through the coil of wire.
8. The polarity of an electromagnetic coil is determined by the direction of current flow through the conductor.
9. If the conductor is wrapped in coils, the strength of the magnetic field is increased.
10. When more coils of wire are added to an electromagnet, the magnetic field strength is increased.
11. A core will cause the field of an electromagnet to be strengthened.
12. The core must be made from laminated steel when AC voltage is used to power the coil so that the magnetic field can quickly build and collapse in the core as the AC voltage changes polarity every 1/60 s.

Questions for This Chapter

1. Explain the difference between a permanent magnet and an electromagnet.

2. Define the term *saturation* as it applies to an electromagnet.

3. Explain the principle of electromagnetic induction.

4. Identify two ways to increase the strength of an electromagnet.

5. Explain why it is important that the magnetic field of an electromagnet can be turned on and off.

True or False

1. When dipoles are aligned in a material, it is a good magnet.

2. Any time current flows through a conductor, magnetic flux lines are created around the conductor.

3. The polarity of the magnetic field of an electromagnet is determined by the direction of current flow through the conductor.

4. Like poles of magnets attract.

5. The strength of the magnetic field for a permanent magnet can be changed easily.

Multiple Choice

1. _____ can have the strength of its field changed easily.
   a. A permanent magnet
   b. An electromagnet
   c. A dipole

2. The polarity of an electromagnet can be changed by _____ .
   a. changing the direction of current flow through the coil of wire.
   b. changing the amount of current flow through the coil of wire.
   c. changing the frequency of the current flow through a coil of wire.

3. A laminated steel core is used in electromagnets that have AC voltage applied to them because _____ .
   a. laminated steel is more economical to use than soft iron.
   b. laminated steel is easily formed to any shape which makes it more usable in complex components.
   c. laminated steel allows the magnetic field to build and collapse quickly.

4. A magnetic field is created in a coil of wire when _____ .
   a. current flows through the wire.
   b. current stops flowing through a wire.
   c. the coil has a core.

5. The left-hand rule is used to _____ .
   a. determine the amount of magnetic flux in a coil.
   b. determine the number of coils in an electromagnet.
   c. determine the polarity of the magnetic field in an electromagnet.

Problems

1. Draw a sketch of a permanent magnet and show where flux lines occur.

2. Draw a sketch of an electromagnetic coil and show where flux lines occur.

3. Draw a sketch of an electromagnetic coil with an open somewhere in the coil of wire. Explain what affect the open will have on the magnetic field.

4. Draw two permanent magnets with their poles placed so the two magnets will attract each other.

5. Perform the experiment presented in Section 6.1 and explain why the flux lines are concentrated around the ends of the permanent magnet.

◀ Chapter 7 ▶

Fundamentals of AC Electricity

Objectives

After reading this chapter you will be able to:

1. Explain the term *alternating current*.
2. Explain the terms *peak-to-peak* and *RMS voltage*.
3. Calculate frequency and period of an AC sine wave.
4. Explain the effects of a capacitor and inductor in an AC circuit.

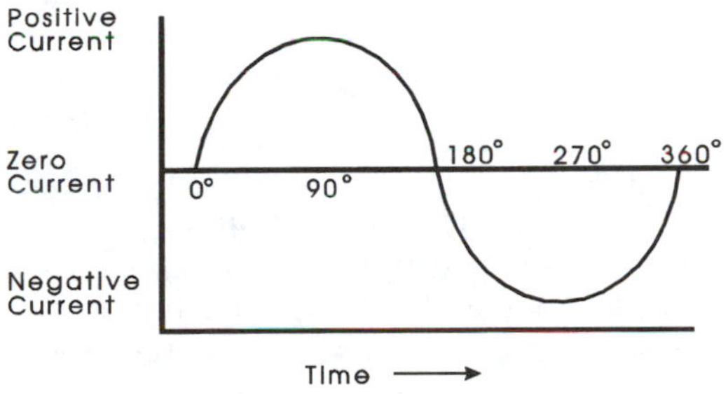

7.0
What Is Alternating Current?

This section reviews some of the principles of AC voltage that you learned in the previous chapter. *Alternating current* (AC) is identified by its characteristic wave form. The AC wave form is shown in Fig. 7–1. The sine wave has a positive half-cycle and a negative half-cycle. The device that creates AC electricity is called an *alternator*. The alternator has coils of wire mounted on its rotating part, which is called the *rotor*. The alternator also has coils of wire that are used to create strong magnetic fields. These coils of wire are mounted in the *stationary part* of the alternator, which is called the *stator*. The magnetic field in the stator, like all other magnets, has flux lines that move between its north and south poles. When the shaft of the alternator is rotated, the coils of wire in the rotor move past the flux lines from the strong magnetic field in the stator. Because the coils pass the north pole of the magnetic field and then the south pole, they create a sine wave in the rotor, which becomes the AC voltage. This voltage is used to power a wide variety of electrical components in the factory.

Figure 7–1   AC sine wave shown moving through 360°.

AC voltage is the primary voltage used in factory equipment. AC electricity is generated by utility companies and transmitted to commercial and residential users.

7.1
Where Does AC Voltage Come From?

AC voltage is generated at a number of power stations around the country. The voltage that is used in your city is generated within a range of 600 mi. In some areas the energy source that turns the rotor of the alternator to produce this voltage comes from burning coal, burning fuel oil, or burning other fossil fuels to make steam. In some areas of the country the steam is generated by nuclear power stations. The steam is used to turn a turbine wheel that turns an alternator shaft. In other parts of the country the turbine wheel is turned by water that is

Frequency = Number of cycles in 1 second

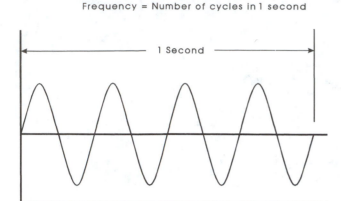

Figure 7–2   The frequency of AC voltage is calculated from the number of cycles that occur in 1 s.

stored behind hydroelectric dams. When the alternator shaft turns, AC voltage is produced.

The voltage that is produced at large utility power stations is generated from three equal fields in the alternator. Thus the generated voltage is *three-phase voltage*. The voltage is sent to transformers at the generating station, where it is stepped up to several hundred thousand volts so it can be transmitted over long distances. When the voltage arrives at a city it is transformed down (stepped down) to a value of approximately 40,000 V. When the voltage reaches an industrial or commercial area, the voltage is stepped down again to 480 or 230 V. The voltage that is used in residential areas is stepped down to 230 V.

7.2
Frequency of AC Voltage

The most important feature of AC voltage that is different from DC voltage is that AC voltage has a frequency. The typical frequency for AC voltage in the United States and much of North America is 60 Hz. The frequency of 60 Hz is determined by the rotating speed of the alternator when the voltage is generated. The frequency of AC voltage that is used to supply voltage for heating and air conditioning equipment is constant. *Frequency* is defined as the number of cycles (sine waves) that occur in 1 s. Figure 7–2 shows a number of sine waves occurring in 1 s.

The *period* of a sine wave is the time it takes one sine wave to start from the zero point and pass through 360°, as seen in Fig. 7–3. A period represents one complete cycle, which is one complete revolution of the alternator. Because the shaft of the alternator rotates one full circle to produce the sine wave, we equate one complete cycle to 360°. Thus the sine wave can be described in terms of 360°. In Fig. 7–3 you can see the sine wave

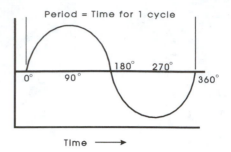

Period = Time for 1 cycle

Figure 7–3   The period of AC voltage is calculated as the time it takes for one cycle to occur.

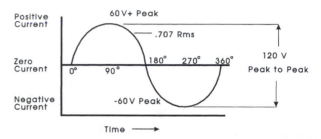

Figure 7–4   Sine wave with positive and negative peaks identified. The peak-to-peak (PP) voltage is also identified.

starts at 0°, it reaches a positive peak at 90°, returns to zero at 180°, reaches a negative peak at 270°, and finally returns to 0 at 360°. The number of degrees will be used to identify points of the sine wave in the future discussions. The period of a sine wave can be described as $P = \frac{1}{\text{frequency}}$, and frequency can be defined as $F = \frac{1}{\text{period}}$.

▶ **Example 7–1**

Determine the period of a 60-Hz sine wave.

Solution:
    Use the formula $P = \frac{1}{F}$ (where F = frequency):

$$P = \frac{1}{F} = \frac{1}{60} = 0.016 \text{ s}$$ ◀

7.3
Peak Voltage and RMS Voltage

Figure 7–4 shows a sine wave, including the point where peak voltage occurs. *Peak voltage* is the highest point of the sine wave. You can see that the peak voltage is 60 V for this example. The negative peak voltage is also identified as −60 V in this example. The total voltage peak to peak (PP) is 120 V. The peak voltage can be measured only with a peak-reading voltmeter or an oscilloscope.

The voltage that you read with a VOM type meter is called *RMS voltage*. The term RMS stands for *root-*

mean-square. Root, mean, and square are the mathematical functions used to calculate the voltage that a typical RMS meter reads. The major difference between root-mean-square voltage and peak voltage is that the meter compensates for the RMS voltage being less than peak at various times in any given cycle. The formula for calculating RMS voltage from peak voltage (V_p) is

$$RMS = 0.707 \times \text{peak voltage}$$

As you can see, the RMS voltage is approximately 70% of the peak voltage. This means that if the peak voltage is 100 V, the RMS voltage is 70 V. In the example in Fig. 7–4, the peak voltage is 120, so the RMS voltage is 84 V.

▶ **Example 7–2**

Calculate the RMS voltage if the peak voltage is 156 V.

Solution:
$$RMS = 0.707 \times \text{peak}$$
$$RMS = 0.707 \times 156$$
$$RMS = 110 \text{ V} \quad ◀$$

▶ **Example 7–3**

Find the peak voltage if RMS voltage is 120 V.

Solution:
Because RMS = 0.707 × peak voltage, peak = 1/0.707 × RMS = 1.414 × RMS.

$$\text{Peak} = 1.414 \times RMS$$
$$\text{Peak} = 1.414 \times 120$$
$$\text{Peak} = 169.68 \text{ V} \quad ◀$$

7.4
The Source of AC Voltage for Factory Electrical Systems

When you are working on a job in a factory or in a commercial building, you will need to locate the source of incoming voltage. The source of incoming voltage at these locations is typically a *circuit breaker panel.* The circuit breaker panel is also called a *load center,* and it is shown in Figure 7–5. The load center in a larger commercial building is typically three-phase, and the load center in a smaller commercial building is typically single-phase.

Voltage from the load center is sent to a *disconnect box,* which is mounted near the equipment on which you are working. Figure 7–6 shows a picture of a disconnect box. If the disconnect box has fuses, it is called a *fused disconnect.* The disconnect is used to turn power off to the unit when you are working on the equipment.

Figure 7–5 A typical load center.
(Courtesy of Cutler-Hammer, Inc.)

Figure 7–6 A fused disconnect that is mounted near the equipment.
(Courtesy of Cutler-Hammer, Inc.)

SAFETY NOTICE!
You should always use a lockout, tag-out procedure when you are working on a machine. This procedure includes placing your personal padlock on the disconnect after it is turned to the off position. The padlock will also have a card with it that has your name and picture on it so that others will know you are working on this system. The other part of the lockout, tag-out procedure includes dropping all hydraulic and pneumatic pressure to zero and using the proper blocks on vertical and horizontal dies to ensure that you are not in danger of being pinched in a machine.

The fuses in the disconnect are generally cartridge-type fuses that provide protection against too much current for the equipment that connected to them.

7.5
Measuring AC Voltage in a Disconnect

When you are installing or troubleshooting equipment in the field, you will need to measure the amount of AC voltage available at the load center or at the disconnect. You must be extremely careful when you take these readings because full voltage will be applied to the system at the time you are taking the measurement. Figure 7–7 shows the points in the disconnect where you should place the meter probes to measure high voltage and low voltage. Figure 7–7a shows the points to place the meter leads when you are reading voltage line to line (L_1 to L_2). From this figure you can see that the meter probes are placed on the line 1 (L_1) and line 2 (L_2) incoming terminals, and the applied voltage will be 230 V if you are in a residence. If the fused disconnect is located in a commercial loca-

tion, you may measure 208 V when you place the meter on L_1 and L_2 or you may measure 480 V. These terminals are called the *line-side terminals*.

Figure 7–7b shows the meter probes on line 1 (L_1) and neutral. The meter should measure 115 V if the disconnect is in a residence. This is the lower voltage, which is used on smaller equipment. You may measure a higher voltage, such as 208 V or 240 V, in the disconnect in some cases.

The previous voltage measurements are made at the line-side terminals of the disconnect. You will need to make a second set of measurements at the *load-side terminals*, located at the bottom of the disconnect. If you have voltage at the line-side but not at the load-side terminals, the disconnect may have a blown fuse. In some cases you can determine that the fuse is blown if you can measure full voltage across the top and bottom of a fuse. The reason you can read full voltage across the fuse is that it has an open, and you are reading voltage that backfeeds from the bottom of the fuse through a load such as a transformer winding to the other supply voltage line. If you suspect one or both of the fuses are faulty, you can remove them and check them for continuity.

7.6
Voltage and Current in AC Circuits

Voltage waveforms and current waveforms are in phase in an AC circuit that has only resistance in it. This means that both the waveforms start at the same point in time. If the load in the circuit is a resistance heating element, the voltage waveform and current waveform are in phase with each other. If a capacitor or inductor is used in an AC circuit such as a motor circuit, the voltage waveform and current waveform are not in phase. Figure 7–8 shows the voltage, identified by the letter v, and current waveform, identified by the letter i. You can see that these waveforms are in phase, because they start at the same point in time and end each of their cy-

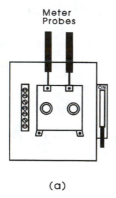

(a)

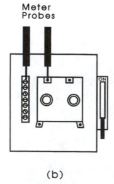

(b)

Figure 7–7   (a) Reading 230 V on the line side of a fusible disconnect; (b) reading 115 V on the line side of a fusible disconnect.

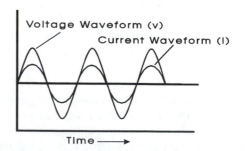

Figure 7–8   Voltage waveform (v) and current waveform (i) are shown in this example.

cles at the same time. The voltage in this circuit is larger than the current, so the voltage waveform is larger than the current waveform.

When the voltage and current waveforms are in phase, there are no losses in the circuit due to inductive reactance or capacitive reactance. If an AC circuit has an inductor in it, such as a motor coil, it has *inductive reactance*, and if the circuit has a capacitor in it, such as a start capacitor for a motor, it has *capacitive reactance*. *Reactance* is an opposition to the AC circuit that is similar to the opposition in a DC circuit caused by resistance. *Inductive reactance* is the opposition caused by the inductor and *capacitive reactance* is the opposition caused by capacitors. If a circuit has inductors (transformer windings or motor windings) or capacitors, there will be a phase shift between the voltage and current waveforms for that circuit. The amount of phase shift can be defined as a number of degrees or it can be defined as the total amount of reactance (opposition) in the circuit. As a technician you need to know that reactance exists in these types of circuits and that the affects of reactance can be calculated, predicted, and used to an advantage. For example, the more phase shift that can be created between the start winding and the run winding of an induction motor, the higher starting torque the motor will have. This is useful for helping larger compressor motors to start. The next sections will help you understand the theory about capacitive reactance and inductive reactance. Even though you may never complete reactance calculations when you are troubleshooting a compressor motor, you must comprehend the theory behind capacitive reactance and inductive reactance to understand why additional capacitance may be required to make a single phase motor start more easily. You will also need a basic understanding of reactance to learn about electronic control circuits that are designed to make pump motors and fan motors run more efficiently.

7.7
Resistance and Capacitance in an AC Circuit

In the previous section we learned that when capacitors or inductors are used in an AC circuit, they will create an opposition that is similar to resistance. The opposition caused by a capacitor is called capacitive reactance, and the opposition caused by an inductor is called inductive reactance. The combined opposition caused by capacitive reactance, inductive reactance, and resistance in an AC circuit is called *impedance*. The main difference between capacitive reactance, inductive reactance, and resistance is that a phase shift

occurs between the voltage waveform and current waveforms. Figure 7–9a shows a diagram of a capacitor in an AC circuit with a resistor. Figure 7–9b shows the voltage and current waveforms for this circuit, and Fig. 7–9c shows the vector diagram that is used to calculate the amount of phase shift for this diagram. A vector diagram is a graph derived from a trigonometry calculation used to determine phase angle.

When voltage encounters a capacitor in a circuit, it will take time to charge up the capacitor and then discharge it. This causes the reactance, which becomes an opposition to the voltage waveform. The capacitor does not bother the current waveform. In Fig. 7–9b you can see that the voltage waveform *lags*, or starts later than, the current waveform for this circuit.

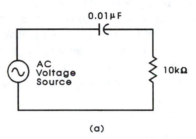

(a)

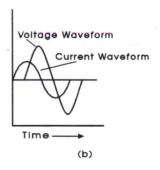

(b)

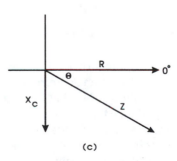

(c)

Figure 7–9 (a) Capacitor and resistor in an AC circuit. (b) waveform of voltage and current for the AC circuit. Notice that the current wave form leads the voltage wave form. (c) Vector diagram that shows the relationship of voltage across the resistor and the voltage across the capacitor.

7.8
Calculating Capacitive Reactance

The total amount of opposition caused by the capacitor is called *capacitive reactance*, X_C. Even though you may never need to calculate capacitive reactance, you should understand the effects of changes in capacitance and frequency on the amount of capacitive reactance. Capacitive reactance can be calculated by the following formula:

$$X_C = \frac{1}{2 \pi FC}$$

where $\pi = 3.14$

F = frequency

C = capacitance in microfarads

From this formula you can see that you must know the value of the capacitor and the frequency of the AC voltage. For example, if a 40-μF capacitor is used in a 60-Hz circuit, the amount of opposition (capacitive reactance) is 66.35 Ω. Notice that because the capacitive reactance is an opposition, its units are ohms. This calculation is as follows:

$$X_C = \frac{1}{2\pi FC} = \frac{1}{2 \times 3.14 \times 60 \times 0.000040} = 66.35 \ \Omega$$

▶ *Example 7–4*

Calculate the capacitive reactance of a 60-Hz AC circuit that has a 90-μF capacitor.

Solution:

$$X_C = \frac{1}{2\pi FC} - \frac{1}{2 \times 3.14 \times 60 \times 0.000090} = 29.49 \ \Omega \ ◀$$

7.9
Calculating the Total Opposition for a Capacitive and Resistive Circuit

The amount of total opposition (impedance) caused by a capacitor and resistor in an AC circuit can be calculated. Figure 7–10c shows the diagram that is used to determine the impedance for this type of circuit. For example, if a circuit has 70 Ω of resistance and 40 Ω due to capacitive reactance, the total impedance must be calculated using the distance formula, because the voltage and current in the circuit are out of phase. The formula for calculating impedance of this circuit is

$$Z = \sqrt{R^2 + X_c^2}$$

$$Z = \sqrt{70^2 + 40^2} \qquad Z = 80.62 \ \Omega$$

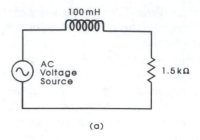

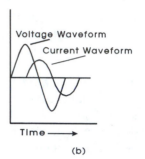

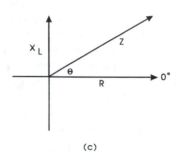

Figure 7–10 (a) An inductor and resistor in an AC circuit. (b) Waveforms for the voltage and current in the inductive and resistive circuit. (c) Vector diagram that shows voltage across the inductor leading the voltage across the resistor.

7.10
Resistance and Inductance in an AC Circuit

When inductors (coils of wire) are used in an AC circuit, they will create an opposition that is similar to resistance. The opposition caused by an inductor is called *inductive reactance*. The main difference between capacitive reactance, inductive reactance, and resistance is that a phase shift between the voltage and current waveforms occurs. Figure 7–10a shows a diagram of an inductor in an AC circuit with a resistor. Figure 7–10b shows the voltage and current waveforms for this circuit, and Fig. 7–10c shows the vector diagram that is used to calculate the amount of phase shift for the circuit.

When current encounters an inductor in a circuit, it will take time to charge up the inductor and then discharge it. This causes the reactance, which becomes an opposition to the current waveform. The inductor does not bother the voltage waveform. In Fig. 7–10b you can see that the current waveform *lags*, or starts later than, the voltage waveform.

7.11
Calculating Inductive Reactance

The total amount of opposition caused by the inductor is called *inductive reactance*, X_L. As in the case of capacitive reactance, inductive reactance can be calculated, and you should remember that even though you do not calculate reactance in the field when you are troubleshooting, it is important that you understand the effect that changing the size of inductor or changing frequency has on the total amount of inductive reactance. Inductive reactance can be calculated by the following formula:

$$X_L = 2\pi FL$$

where π $= 3.14$

F = frequency

L = inductance in (H)enries

From this formula you can see that you must know the value of the inductor and the frequency of the AC voltage. For example if an 80-H inductor is used in a 60-Hz circuit, the amount of opposition (inductive reactance) is 30,144 Ω (30.144 kΩ). Notice that because the inductive reactance is an opposition, its units are ohms. This calculation is as follows:

$$X_L = 2\pi FL = 2 \times 3.14 \times 60 \text{ Hz} \times 80 \text{ H} = 30,144 \ \Omega$$

▶ *Example 7–5*

Calculate the inductive reactance of a 60-Hz AC circuit that has a 20-H inductor.

Solution:

$$X_L = 2\pi FL = 2 \times 3.14 \times 60 \times 20 = 7,536 \ \Omega \quad ◀$$

7.12
Calculating the Total Opposition for an Inductive and Resistive Circuit

The amount of total opposition (impedance) caused by an inductor and resistor in an AC circuit can be calculated. Figure 7–10c shows the diagram that is used to determine the impedance for this type of circuit. For example, if a circuit has 70 Ω of resistance and 50 Ω due to inductive reactance, the total impedance must be cal-

culated with a vector diagram, because the voltage and current in this circuit are out of phase. The formula for calculating impedance of this circuit is

$$Z = \sqrt{R^2 + X_L^2}$$
$$Z = \sqrt{70^2 + 50^2}$$
$$Z = 86.02 \ \Omega$$

7.13
True Power and Apparent Power in an AC Circuit

In a DC circuit you could simply multiply voltage times the current and determine the total power of the circuit. In an AC circuit you must account for the current caused by any resistors and calculate it separately from the power caused by resistors and capacitors or resistors and inductors. The reason for this is that when current is caused by a resistor, it is called *true power* (TP), and current caused by capacitive reactance and inductive reactance is called *apparent power* (AP). Apparent power does not take into account the phase shift caused by the capacitor or inductor. The power that occurs when current flows through a resistor is called true power (TP) because the resistor does not cause a phase shift between the voltage waveform and the current waveform.

The main point to remember is that true power can be used to determine the heating potential in an AC circuit. Thus if you have 1,000 W due to true power, you can determine that you can get 1,000 W of heating power from this circuit.

7.14
Calculating the Power Factor

The amount of true power and the amount of apparent power in an AC circuit combine to become a ratio called the *power factor* (PF). The formula for power factor is

$$PF = \frac{TP}{AP}$$

▶ *Example 7–6*

Calculate the power factor for a circuit that has 600 VA of apparent power and 500 W of true power. Note that when volts are multiplied by amperes in a reactive circuit, the units for apparent power are VA, which stands for volt amperes.

Solution:

$$PF = \frac{TP}{AP} = \frac{500 \text{ W}}{600 \text{ VA}} = 0.83, \quad \text{or} \quad 83\% \quad ◀$$

The true power in a circuit is always smaller than the apparent power, so the power factor is always less than 1. In a pure resistance circuit, the true power and the apparent power are the same, so the power factor is 1. When the power factor becomes too low, the power company adds a penalty to the electric bill that increases the bill substantially. When a factory has an electric bill of more than $20,000 per month, this penalty becomes important. In these types of applications, the power factor can be corrected by adding extra capacitance to an inductive circuit (having large or multiple motors) or by adding extra inductance to a capacitive circuit. Commercial power factor–correction systems are available for these applications.

Questions for This Chapter

1. Draw a diagram of an AC sine wave and identify its peak voltage and peak-to-peak voltage levels.

2. Explain the term *frequency* as it refers to AC electricity.

3. Discuss the difference between a load center and a fused disconnect.

4. Explain how you would test for voltage in a fusible disconnect on the line-side and load-side terminals.

5. Trace the path of a generated AC voltage from the generating station where it is produced to the point where you would connect it to a machine in a factory. Be sure to identify all the important parts of the system between the generation station and the disconnect on the machine.

True or False

1. Impedance is the total opposition to an AC circuit caused by inductive reactance, capacitive reactance, and resistance.

2. The units of impedance are ohms.

3. A load center is the point where incoming voltage is connected in a residence and where circuit breakers are mounted.

4. RMS voltage is the voltage your VOM voltmeter measures.

5. If the RMS voltage in an AC circuit is 110 V, the peak voltage is also 110 V.

Multiple Choice

1. If the size of a capacitor in an AC circuit increases from 10 μF to 20 μF and the frequency stays the same, the capacitive reactance will _____ .
   a. increase.
   b. decrease.
   c. remain the same.

2. When a capacitor is added to an AC circuit, the voltage wave form will _____ the current waveform, which provides additional torque to a motor if it is connected to the circuit.
   a. lag
   b. lead
   c. remain in phase with

3. If a circuit has two large compressor motors and a low power factor, the power factor can be raised by _____ to make the true power equal to the apparent power, which will avoid a penalty on the electric bill.
   a. adding capacitance
   b. removing capacitance
   c. adding inductance

4. The frequency of AC voltage is defined as _____ .
   a. the period of the voltage.
   b. the number of cycles per second.
   c. the peak voltage of the voltage.

5. The number of cycles per second in an AC voltage is called the _____ of the voltage.
   a. frequency
   b. period
   c. impedance

Problems

1. Calculate the RMS voltage for an AC sine wave that has 80 V peak.

2. Calculate the peak voltage for an AC sine wave that has 208 V RMS.

3. Calculate the period of a 60-cycle AC sine wave.

4. Calculate the frequency of an AC sine wave that has a period of 16 ms.

5. Calculate the impedance of an AC circuit that has 40 Ω of resistance and 30 Ω of capacitive reactance.

Chapter 8

Transformers and Three-Phase and Single-Phase Voltage

Objectives

After reading this chapter you will be able to:

1. Explain the operation of a transformer.
2. Discuss the difference between step-up and step-down transformers.
3. Explain how 440, 220, and 110 VAC are developed from a delta-connected transformer.
4. Explain how 440, 208, and 125 VAC are developed from a wye-connected transformer.

8.0
Overview of Transformers

Transformers are needed to step up or step down voltage levels. For example, when voltage is generated it needs to be stepped up to several hundred thousand volts so its current will be lower when it is transmitted. Because the current is lower, the size of wire used to transmit the power can be smaller, so the power can be transmitted over longer distances. When the voltage arrives at a city, it needs to be stepped down to approximately 40,000 V so that it is less dangerous. This level of voltage is sufficient to transmit power throughout a city. When the voltage arrives at a factory it must be further stepped down to 480, 240, or 208 V and 120 V for all the power circuits in the factory, including the office. Transformers provide a means of stepping up or stepping down this voltage. Step-up transformers are used in some industrial applications, such as air cleaners, to increase the line voltage (120 VAC) to several thousand volts. Note that different parts of the country use slightly different amounts of voltage, such as 440, 220, 208, or 110 V or other similar variations. These other voltage values are determined by the amount of the supply voltage from the power company and the different variations of transformers used. For this text we will try to standardize on 480, 240, 208, and 120 V to avoid confusion.

Another type of transformer is also used in factory electrical systems to step down 240 V AC to 120 V. The 120 VAC is used as control voltage for all the controls in the machine systems so that higher voltages are not used in start and stop buttons or other controls that people touch when they are setting them. The lower voltage provides a degree of safety against electrical shock, and it also allows the controls to last longer. This type of transformer is called a *control transformer*, and it is shown in Fig. 8–1.

As a technician you will encounter different voltages, such as 440, 220, 208, and 120 VAC, while you are working on electrical systems in factories. The equipment you are working with will require a specific amount of single-phase or three-phase voltage, and you will be responsible for going to the disconnect or load center where supply voltage is provided and making the proper connections to provide the correct voltage. When the proper amount of voltage is not present, you will need to know what transformers are required to step the voltage up or down to the proper level. This chapter explains how transformers operate and how the various levels of voltage are derived from connecting transformers to provide the proper amount of three-phase and single-phase voltage. The chapter also explains how to take voltage readings at the transformer and disconnect to troubleshoot the loss of a phase.

Figure 8–1   An example of a control transformer that is used in factory equipment.

(Courtesy of Honeywell's Micro Switch Division)

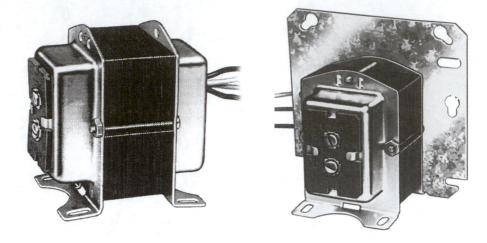

8.1
Operation of a Transformer and Basic Magnetic Theory

The transformer consists of two windings (coils of wire) that are wrapped around a laminated steel core. The winding where voltage is supplied to a winding transformer is called the *primary winding*. Line 1 (L1) and neutral are connected to the primary terminals. The winding where voltages come out of the transformer is called the *secondary winding*. The terminals for the secondary winding of a 24-V transformer are identified by the letters R and C. The letter R stands for *red*, which is the color of wire that is connected to it, and the letter C stands for *common*. Figure 8–2 shows a diagram of a typical transformer with the primary coil and secondary coil identified. From this figure you can see that the windings are also identified. If the transformer's secondary voltage is 110 V, then the secondary terminals will be identified as X_1 and X_2.

A transformer works on a principle called *induction*. Induction occurs when current flows through the primary winding and creates a magnetic field. The magnetic field produces flux lines that emanate from the wire in the coil as current flows through it. When this current is interrupted or stopped, the flux lines collapse, and the action of the collapsing flux lines causes them to pass through the winding of the secondary coil that is placed adjacent to the primary coil. You should remember that the AC sine wave continually starts at 0 V and increases to a peak value and then returns to 0 V and repeats the wave form in the negative direction. The process of increasing to peak and returning to zero provides a means of creating flux lines in the wire as current passes through it and then allowing the flux lines to collapse when the voltage returns to zero. Because the AC voltage follows this pattern naturally 60 times a second, it makes the perfect type of voltage to operate the transformer.

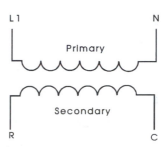

Figure 8–2   Diagram of a typical control transformer. The coil where incoming voltage is applied is called the primary, and the other coil is called the secondary.

(Courtesy of Honeywell's Micro Switch Division)

When the AC voltage returns to zero during each half cycle, the flux lines that were created in the primary winding begin to collapse and start to cross the wire that forms the secondary coil of the transformer. When the flux lines from the primary collapse and cross the wire in the coil of the secondary winding, the electrons in the secondary winding begin to move, which creates a current. Because the current in the secondary winding begins to flow without any physical connection to the primary winding, the current in the secondary winding is called an *induced current*.

The following provides a more technical explanation of the relationship between the AC voltage and the windings of the transformer: the magnetic field in the primary winding of the transformer builds up when AC voltage is applied during the first half-cycle of the sine wave (0° to 180°). When the sine voltage reaches its peak at 90°, the voltage has peaked in the positive direction and it begins to return to 0 V by moving from the 90° point to the 180° point. When the voltage reaches the 180° point, the voltage is at 0 V, and it creates the interruption of current flow. When the sine-wave voltage is 0 V, the flux lines that have been built up in the pri-

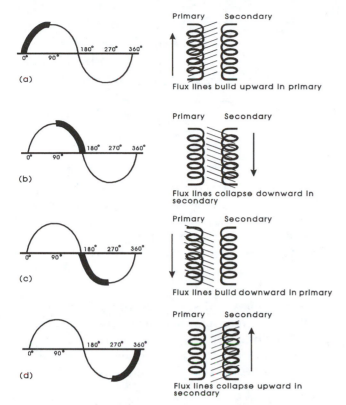

Figure 8–3 (a) AC voltage builds to a positive peak as the sine wave moves from 0 to 90°. Flux lines are shown building in the primary coil during this time. (b) AC voltage drops off from peak at 90° to 180°. During this time the flux lines in the primary coil begin to collapse and cut across the coils of the secondary. (c) AC voltage builds to a negative peak as the sine wave moves from 180° to 270° and the flux lines build in the primary coil again. (d) When voltage decreases from the negative peak back to zero as the sine wave moves from 270° to 360°, the flux lines collapse.

mary winding collapse and cross the secondary coils, which creates a current flow in the secondary winding.

The sine wave continues from 180° to 360°, and the transformer winding is energized with the negative half-cycle of the sine wave. When the sine wave is between 180° and 270°, the flux lines are building again, and when the sine wave reaches the 360° point the sine wave returns to 0 V, which again interrupts current and causes flux lines to cross the secondary winding. This means the transformer primary energizes a magnetic field and collapses it once in the positive direction and again in the negative direction during each sine wave. Because the sine wave in the secondary is not created until the voltage in the primary moves from 0° to 180°, the sine wave in the secondary is *out of phase* by 180° to the sine wave in the primary that created it. The voltage in the secondary winding is called *induced voltage* because it is created by induction.

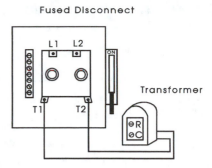

Figure 8–4 A control transformer connected to L1 and L2 terminals in a fusible disconnect. The primary voltage is 230 V, and the secondary is 120 V.

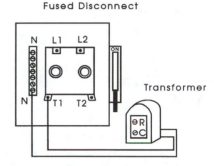

Figure 8–5 A control transformer connected to L1 and L2 in a disconnect for 480 V to test the transformer. The primary voltage is 480 VAC and the secondary voltage is 120 VAC.

The induced voltage is developed even though there is total isolation between the primary and secondary coils of the transformer. Figure 8–3 shows the four stages of voltage building up in the transformer and the flux lines building and collapsing at each point as the sine wave flows through the primary coil.

8.2
Connecting a Transformer to a Disconnect for Testing

An easy way to test a control transformer is to connect it directly to a disconnect and apply power to the transformer primary circuit. When voltage is applied to the primary circuit, voltage should be available at the secondary. The amount of voltage at the secondary will depend on the rating of the transformer. Figure 8–4 shows a transformer connected to L1 and L2, which is the source for 240 V at the disconnect. The secondary voltage available at this transformer at terminals X1 and X2 is approximately 120 V. Figure 8–5 shows a similar transformer connected to L1 and L2 in a disconnect to provide 480 V to the

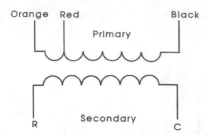

Figure 8–6 A control transformer that is made to operate with either 208 VAC or 230 VAC primary voltage and provide 24 VAC secondary voltage.

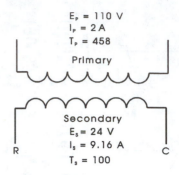

Figure 8–7 Diagram of a transformer that shows the primary voltage (V_p), primary current (I_p), and the number of turns in the primary winding (T_p). This diagram also shows the secondary voltage (V_s), secondary current (I_s), and the number of turns in the secondary (T_s).

primary circuit. Because this transformer is rated as 480/120, the secondary voltage is also 120 V.

8.3
Transformer Rated for 230 V and 208 V Primary

If the equipment is used in the office part of a factory application, such as heating or air conditioning, the supply voltage may be 230 or 208 V, depending on the transformer connections used by the utility company that supplies the power. The secondary voltage for the thermostat will be stepped down to 24 VAC. Because the equipment manufacturers do not know where their equipment will be used, they may provide a control transformer whose primary side can be wired to either 230 V or 208 V and whose secondary side will produce 24 V. This is accomplished by providing a second connection that is *tapped* in the primary winding. Figure 8–6 shows an example of this type of control transformer.

8.4
Transformer Voltage, Current, and Turns Ratios

The amount of voltage a transformer will produce at its secondary winding for a given amount of voltage supplied to its primary is determined by the ratio of the number of turns in the primary winding compared to the number of turns in the secondary. This ratio is called the *turns ratio*. The amount of primary current and secondary current in a transformer also depends on the turns ratio.

Figure 8–7 shows a diagram that indicates the primary voltage (E_p), the primary current (I_p), the number of turns in the primary winding (T_p), the secondary voltage (E_s), the secondary current (I_s), and the number of turns in the secondary winding (T_s).

The primary and secondary voltage, primary and secondary current, and the turns ratio can all be calculated from formulas.

The turns ratio for a transformer is calculated from the following formulas:

$$\text{Turns ratio} = \frac{T_s}{T_p} \text{ or } \frac{V_s}{V_p} \text{ or } \frac{I_p}{I_s}$$

The ratio of primary voltage, secondary voltage, primary turns, and secondary turns is

$$\frac{E_p}{E_s} = \frac{T_p}{T_s}$$

From this ratio you can calculate the secondary voltage with the formula

$$E_s = \frac{E_p \times T_s}{T_p}$$

The ratio of primary voltage, secondary voltage, primary current, and secondary current is

$$\frac{E_p}{E_s} = \frac{I_s}{I_p}$$

(Notice that the ratio of voltage to current is an inverse ratio.) Using this ratio you can calculate the secondary current:

$$I_s = \frac{E_p \times I_p}{E_s}$$

If you know the primary voltage is 110 V, the primary current is 2 A, the primary turns are 458, and the secondary turns are 100, you can easily calculate the

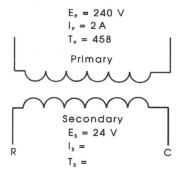

E_p = 240 V
I_p = 2 A
T_p = 458

Primary

Secondary
E_s = 24 V
I_s =
T_s =

R   C

Figure 8–8 Diagram for Example 8–1.

secondary voltage and the secondary current using these formulas:

$$E_s = \frac{E_p \times T_s}{T_p}$$

$$\frac{110\text{ V} \times 110\text{ turns}}{458\text{ turns}} = 24\text{ V}$$

$$I_s = \frac{E_p \times I_p}{E_s}$$

$$\frac{110\text{ V} \times 2\text{ A}}{24\text{ V}} = 9.16\text{ A}$$

▶ *Example 8–1*

Calculate the secondary current and secondary turns of the transformer shown in Fig. 8–8.

Solution:
Find the turns ratio:

$$\frac{E_p}{E_s} = \frac{240\text{ V}}{24\text{ V}} = 10{:}1 \qquad ◀$$

Because the number of turns in the primary is 458, the number of turns in the secondary will be 45.8.

The secondary current can be calculated from the following formula:

$$I_s = \frac{E_p \times I_p}{E_s}$$

$$\frac{240\text{ V} \times 2\text{ A}}{24\text{ V}} = 20\text{ A}$$

8.5
Step-up and Step-down Transformers

If the secondary voltage is larger than the primary voltage, the transformer is known as a *step-up transformer.* If the secondary voltage is smaller than the primary

voltage, the transformer is known as a *step-down transformer.* The control transformer that was presented in the previous sections is an example of a step-down transformer because the secondary voltage is smaller than the primary voltage. Step-up transformers are used to boost the secondary voltage, which also has the affect of lowering the secondary current so that more power can be transferred on smaller-size wire. Some industry applications where a step-up transformer is used include electrostatic air cleaners and the ignition system of an oil furnace. The oil furnace uses an ignition transformer that is a step-up transformer to increase the 120 VAC used to power the furnace to approximately 16,000 V. The high voltage is supplied to the igniter points so that the high voltage can jump the gap between the ignition points, creating a spark that ignites the fuel oil as it comes out of the nozzle.

8.6
VA Ratings for Transformers

The VA (voltampere) rating for a transformer is calculated by multiplying the primary voltage and the primary current or the secondary voltage and secondary current. In the example in Fig. 8–8 you can see that the primary voltage is 110 V and the primary current is 2 A. The VA rating for this transformer is 220 VA. The VA rating indicates how much power the transformer can provide. As a technician you can say that the primary VA is equal to the secondary VA. If you were a circuit designer, you would need to be more precise; you would find that the secondary VA is slightly less than the primary VA because of transformer losses. It is important that the VA rating of the transformer be large enough for an application. If the VA rating is too small, the transformer will be damaged and fail prematurely. Any time you need to replace a transformer you must be sure the voltage ratings match and that the VA size of the replacement transformer is equal or larger than that of the original transformer.

8.7
The 120-VAC Control Transformer

The control transformer provides 120 VAC in the secondary for use in factory-control systems. You should remember that utility companies in different areas of the country provide voltage that is between 110 and 120 V. This means that one part of the country may refer to their lower voltage as 110, whereas another part calls their voltage 115 or 120 V. For this section the voltage will be referred to as 110 VAC. The 110-VAC secondary voltage is necessary because some of the control devices require 110 VAC, and the larger loads in the system,

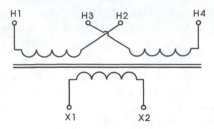

Figure 8–9   Diagram of a control transformer whose secondary voltage is 110 V. The primary windings of this transformer are identified as H1, H2, H3, and H4. The secondary of this transformer is identified as X1 and X2.

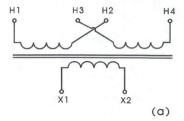

(a)

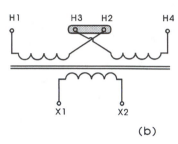

(b)

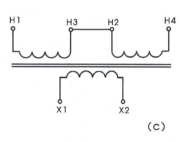

(c)

Figure 8–10   (a) Electrical diagram of the control transformer. Notice that the H2 and H3 terminals are located so that a jumper can be placed on them to connect them in series or parallel. (b) Diagram showing the jumper in place connecting the two primary windings in series. (c) The equivalent electrical diagram that shows the two windings connected in series.

such as compressors and fan motors, require 240 VAC. The control transformer allows the larger loads in a system, such as compressor motors, pump motors, and fan motors, to be supplied with 240 V while providing the control circuit with 110 VAC.

The 240/110-VAC type of control transformer operates on a similar principle as the 110/24-VAC transformer. The diagram for this transformer is also provided in Fig. 8–9. From this diagram you can see that the primary windings of this transformer are identified with the letter *H* and the secondary windings are identified with the letter *X*. The primary winding for this transformer is constructed in two equal sections (windings). The terminals of the first winding are identified as H1 and H2. The terminals of the second winding are identified as H3 and H4. The primary winding is broken into two separate windings so that the transformer primary side can be powered with 480 V or 240 V. If these two sections are connected in parallel with each other, the primary side of the transformer will be powered with 480 V, and if the two sections are connected in series with each other, the transformer will be powered with 240 V.

8.8
Wiring the Control Transformer for 480 VAC Primary Volts

The control transformer can also be powered with 480 VAC by connecting its two primary windings in series. Figure 8–10 shows three diagrams of the control transformer. Figure 8–10(a) shows the transformer as you would see it on a wiring diagram without any jumpers connected. Notice that the H2 and H3 terminals are positioned so that they can be connected in either series or parallel. In this application the jumper is attached between terminals H2 and H3 to connect them in series. Figure 8–10(b) shows the physical location of the jumper when it is connected across these terminals. Figure 8–10(c) shows the equivalent electrical diagram. The two windings are connected in series because each

winding is rated for 240 V, and the primary voltage is 480 VAC. Because the windings are connected in series, each of them will receive 240 V of the total 480 V.

8.9
Wiring the Control Transformer for 240 VAC Primary Volts

The control transformer can also be connected to operate with 240 VAC primary volts and produce 110 VAC on the secondary terminals. Figure 8–11 shows a set of three diagrams for the same control transformer that is connected for 240 VAC primary. Figure 8–11(a) shows an electrical diagram of the control transformer as you would see it in a circuit diagram without any jumpers. The reason this diagram is shown is that some equip-

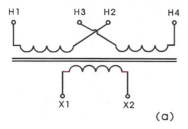

(a)

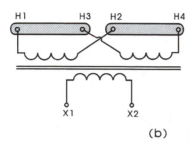

(b)

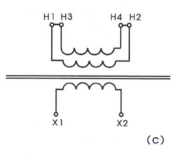

(c)

Figure 8–11 (a) Electrical diagram of a control transformer. (b) Diagram that shows the jumpers in place for the transformer to have 240 VAC primary. (c) Equivalent diagram that shows the two primary windings connected in parallel.

ment manufacturers will show the control transformer in the electrical diagram for the system since they are not sure whether the system voltage will be 480 or 240. This diagram indicates that the control transformer is to be connected by the technician when the unit is installed and the supply voltage can be verified. Figure 8–11(b) shows the diagram with two jumpers in place to connect the two primary windings in parallel with each other. This diagram will help you connect the jumpers to the correct terminals so that the windings are connected in parallel. Figure 8–11(c) shows the equivalent circuit of the two windings in parallel as you would see them in the system's electrical diagram provided by the equipment manufacturer. Some times it is difficult to see that the connections of the two jumpers actually connect the two windings in parallel, so the diagram in Fig. 8–11(c) shows how the windings look when they are in parallel.

As a technician you will be responsible for making the proper jumper connections based on the primary voltage that the system will have. If the jumpers on the

transformer are connected for 480 VAC primary when the unit is shipped and the unit is connected to 240 V, you will need to make the changes in the jumper so that the secondary winding of the transformer provides exactly 110 VAC.

8.10
Troubleshooting a Transformer

As a technician you will need to troubleshoot a variety of transformers. If the transformer you are testing is not connected to power, you can test each of the windings with an ohmmeter for continuity. Each of the coils should have some amount of resistance that indicates the amount of resistance in the wire in each coil. If you measure infinite resistance (∞), it indicates the winding has an open, and if you measure 0 Ω, it indicates one of the windings is shorted.

Another way that you can test a transformer is by applying power to the primary winding. It is important that you provide the continuity test first to detect any shorts before you apply power. If the transformer windings are not shorted, you can apply any amount of AC voltage that is equal or less than the primary rating of the transformer. If the transformer is operational, a voltage will be present at the secondary terminals of the transformer. If no voltage is present at the secondary, be sure to check all the connections to ensure they are correct. If no voltage is present at the secondary, the transformer is defective and should be replaced. It is important to remember that the transformer has a primary winding and a secondary winding that are placed in close proximity to each other; when AC voltage is applied to one winding, induction will cause voltage to be available at the other winding.

The transformer may have a problem where its secondary voltage is higher or lower than its rating. If this occurs, it is possible that the amount of primary voltage is incorrect or that the transformer jumpers are not connected properly. For example if the primary voltage is 208 V instead of 240 V, the secondary voltage will be less than 110 V, which will cause a problem. Be sure that the secondary is the proper amount before you use the transformer in a circuit.

8.11
Nature of Three-Phase Voltage

Industrial equipment installed in factories may require three-phase AC voltage or single-phase voltage. Three-phase voltage is usually indicated on equipment data plates by the symbol 3ϕ voltage. Three-phase voltage is generated as three separate AC sine waves. Figure 8–12 shows an example of the three sine waves. From this figure you can see that the sine waves are identified as phase A, phase B, and phase C. All the phases are separated by 120° from each other. The 120° phase separation is caused by the generator windings being 120° out of phase with each other. Three-phase voltage is generated and transmitted because it is more efficient to produce than single phase since the generator can have three separate windings.

8.12
Why Three-Phase Voltage Is Generated

Three-phase voltage is more useful than single-phase voltage for larger equipment and systems because it provides more power than single-phase voltage for an equal-size system. If a 10-hp motor is required as the hydraulic pump motor, a single-phase motor would need to be larger than the 10-hp three-phase motor. The three-phase motor can be smaller because it has three sets of windings and it receives three equal sources of voltage (L1, L2, and L3), whereas a single-phase motor receives only two sources of power. Because the power is shared by three circuits instead of two, the wire sizes, fuse sizes, and switch sizes are also smaller for a three-phase system.

For example, if a 10-hp motor is connected to a single-phase 230-V power source, each line (L1 and L2) would need to be capable of providing 50 A. A three-phase 10-hp motor connected to a three-phase power distribution would need only 28 A from each wire. This means that the three-phase system could use much smaller wire, which would be lighter and less expensive.

8.13
Three-Phase Transformers

Figure 8–13 shows a picture of a typical three-phase transformer you will encounter on the job. The transformer may be mounted near the equipment, or it may be located in a transformer vault (a special room where transformers are mounted). In some cases the transformers are mounted on a utility pole just outside of the commercial site. Figure 8–14 shows the three-phase transformer with its cover removed so you can see three separate transformer windings. The three-phase transformer operates exactly like a single-phase transformer in that AC voltage is applied to the primary side of the transformer and induction causes voltage to be created in the secondary winding. In the case of the three-phase transformer, the 120° phase shift of the three-phase voltage applied to the primary windings will be maintained in the secondary side of the transformer.

When you are working on an industrial system, you may need to make connections between the transformer and the disconnect box or between the trans-

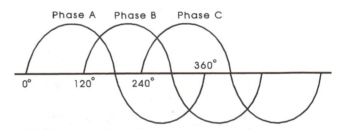

Figure 8–12 Sine waves for three-phase electricity. Notice that phase A is 120° out of phase with phase B, and phase B is 120° out of phase with phase C.

Figure 8–13 A typical three-phase transformer with its cover in place.
(Courtesy of Acme Electric Corp.–Power Distribution Products Division)

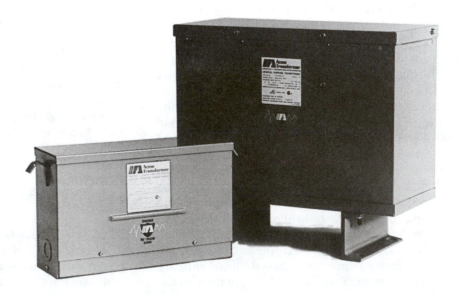

former and a load center. You may also need to make a voltage test at the terminals of the transformer. These connections will be made at the terminal strip of the transformer, as shown in Fig. 8–14. The three-phase transformer is essentially three single-phase transformers that are connected together in a *wye configuration* or *delta configuration.* The picture in Fig. 8–14 shows the three separate transformers. In most cases, you will not be requested to make the physical connections at the transformer, but you must understand the amount of voltage that will be available when a transformer is connected in a wye configuration or in a delta configuration. Because the primary voltage may be rather high (1300–1700 VAC), the electrical technicians for the electric utility company or technicians from a high-voltage service company will make the primary voltage connections for the transformer and connect the secondary to a disconnect box or a load center. You will be expected to make the connections for the equipment at the disconnect or at the load center.

8.14
The Wye-Connected Three-Phase Transformer

Figure 8–15 shows a diagram of a wye-connected three-phase transformer. Notice that the physical shape of the transformer windings looks like the letter *Y*, which gives

this type of configuration its name. Also notice that this diagram shows only the secondary coils of the transformer. It is traditional to show only the primary side connections or only the secondary side connections when discussing a transformer power-distribution system, because showing both the primary and secondary connections in the same diagram tends to become confusing.

The amount of voltage measured at the L1–L2, L2–L3, and L3–L1 terminals on the secondary winding will be 208 V for the wye-connected transformer if its turns ratio is set for low voltage. If the turns ratio of the transformers is set for high voltage, the amount of voltage between each terminals will be 480 V. The voltage indicated between each of the windings for the

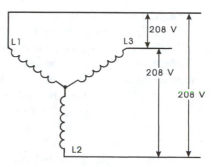

Figure 8–15 A diagram that shows the secondary windings of a three-phase transformer connected in a wye configuration. The voltage available at L1–L2, L2–L3, and L3–L1 is 208 V for this transformer.

Figure 8–14 A typical three-phase transformer with its cover removed so that you can see the three independent transformers and connection terminals.

(Courtesy of Acme Electric Corp.–Power Distribution Products Division)

transformer shown in Fig. 8–15 is 208 V, which indicates the turns ratio for this transformer will provide the lower voltage. If 480 V is needed, a transformer with a higher turns ratio must be used. The primary and secondary voltages for each transformer are provided on its data plate and can be specified when the transformer is purchased and installed.

8.15
The Delta-Connected Three-Phase Transformer

Figure 8–16 shows a diagram of a delta-connected three-phase transformer. You should notice that the shape of the transformer windings in this diagram look like a triangle (Δ). This shape is the Greek letter *D*, which is named *delta*. This diagram also shows only the secondary side of the three-phase transformers.

The amount of voltage measured at L1–L2, L2–L3, and L3–L1 is 230 V for the delta-connected transformer if it is wired for its lower voltage. If the delta-connected transformer is wired for its higher voltage, the voltage between L1–L2, L2–L3, and L3–L1 is 480 V. If the delta-connected transformer is wired for its lower voltage, it is very easy to differentiate it from a wye-connected transformer. If the delta-connected transformer and wye-connected transformers are connected for their higher voltage, you cannot tell them apart, because both will provide 480 V between their terminals.

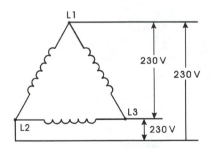

Figure 8–16 Electrical diagram of the secondary windings of a transformer connected in a delta configuration. Notice that the voltage between L1–L2, L2–L3, and L3–L1 is 230 V.

It is important to understand at this point that you will basically purchase the equipment so that its voltage requirements match the voltage supplied by the transformers at the commercial or industrial site. This means that you do not need to determine if the transformer is connected as in a delta configuration or a wye configuration. You will need only to measure the amount of voltage at the secondary of the transformer; if it is 480 V, you will need to ensure that the system is rated for 480 V. If the secondary voltage is 230 V, the system must be rated for 230, and if it is 208 V, the equipment must be rated for 208.

If you need to know if the transformer windings are connected as delta or wye, you can check the physical connections or you can make an additional voltage measurement between each line and the neutral terminal of the transformer if one is provided. The next section explains these measurements.

8.16
Delta- and Wye-Connected Transformers with a Neutral Terminal

Figure 8–17 shows the diagrams of a wye-connected transformer and delta-connected transformer, each with a neutral terminal. The neutral terminal on the wye-connected transformer is at the point where each of the three ends of each individual winding are connected together. This point is called the *wye point*.

Notice that the neutral point for the delta-connected transformer is actually the midpoint of the secondary winding connected between L1 and L2. This point is essentially the center tap of one of the transformer windings. Traditionally it will be the winding that is connected between L1 and L2 if a neutral connection is used with the three-phase transformer.

Figure 8–18a shows the amount of voltage that you would measure between terminals L1–L2 of a wye-connected transformer and between terminals L1–N and L2–N. Notice that the voltage between L1–L2 is 208 V, so this transformer is wired for the lower voltage. The voltage between L1–N and L2–N is shown as

Figure 8–17 Electrical diagram of a wye-connected transformer with a neutral point and a delta-connected transformer with a neutral point.

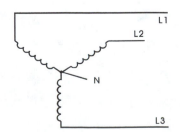

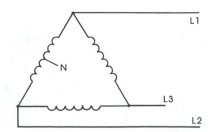

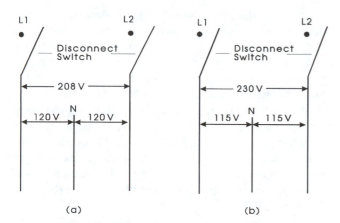

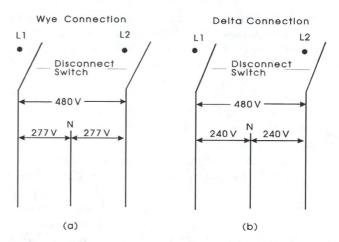

Figure 8–18 (a) Voltage from L1–L2 for a wye-connected transformer is 208 V and from L1–N or L2–N is 120 V. (b) Voltage from L1–L2 for a delta-connected transformer is 230 V and from L1–N or L2–N is 115 V.

Figure 8–19 (a) Voltage from L1–L2 for a wye-connected transformer is 480 V and that from L1–N or L2–N is 277 V. (b) Voltage from L1–L2 for a delta-connected transformer is 480 and that from L1–N or L2–N is 240 V.

120 V. Figure 8–18b shows the amount of voltage that you would measure between terminals L1–L2 and between L1–N and L2–N for a delta-connected transformer. The voltage between L1–L2 is 230 V, so this transformer is connected for its lower voltage. The voltage between L1–N and L2–N is shown as 115 V. Because the neutral point for the delta-connected transformer is exactly halfway on the transformer winding, the voltages L1–N and L2–N will always be exactly half of the voltage between L1 and L2.

This is the main difference between a wye-connected transformer and a delta-connected transformer. The voltage between L1–N and L2–N for any delta-connected transformer is always exactly half that of L1–L2, but the L1–N or L2–N voltage for a wye-connected transformer will always be more than half the voltage L1–L2. The exact amount of voltage L1–N can be calculated by dividing the voltage between L1–L2 by 1.73, which is the square root of 3 ($\sqrt{3}$ = 1.73). The square root of 3 is used because of the relationship between the phase shift of the three phases. Notice that 208 divided by 1.73 is 120 V.

The same relationship of voltage between L1–L2 and L1–N exists when the transformers are wired for their higher voltage. Figure 8–19a shows a wye-connected voltage between L1–L2 is 480 V and that between L1–N is 277 V. The 277 V can be calculated by dividing 480 by 1.73. The 277 V that comes from L1–N or L2–N is generally used for fluorescent lighting systems in commercial buildings. Because the supply voltage originates from a three-phase transformer, the lighting system in a commercial or industrial building will also use L3–N, so that voltage is used from all three legs of the transformer. This voltage is referred to as single-phase voltage because only one line of the three-phase transformer is used for each circuit. For example L1–N is a single-phase circuit.

It is also important to understand at this time that L1–L2, L2–L3, or L3–L1, can each be used to supply complete power for a machine's electrical system. Even though this type of power supply uses two legs of the transformer, it is still called single-phase power supply because only one phase of voltage is used at any instant in time. For example, if the system is powered with voltage from L1–L2, during any given half-cycle of AC voltage, the power source for the system would come from L1, and then during the next half cycle it would come from L2. The source of power would continue to oscillate between L1 and L2, but only one phase is in use during any instant in time.

Figure 8–19b shows the higher voltage for the delta system between L1–L2 is also 480 V, but this time the voltage between L1–N or L2–N is 240 V. Again the L1–N or L2–N voltage is exactly half of the supply voltage, because the neutral point on the transformer is the center tap of one of the transformers. It is important to understand that it is very easy to distinguish between a wye-connected power source and a delta-connected power source by measuring the voltage L1–L2 and L1–N. If the L1–N voltage is half of the L1–L2 voltage, the system is a delta-connected system. If the L1–N voltage is more than half, it is a wye-connected system. You should remember that the L1–N wye voltage can always be calculated by dividing the L1–L2 voltage by 1.73.

8.17
The High Leg Delta System

When a three-phase transformer system is used for the power source for an electrical system, it may have the neutral tap. It is important to remember that a three-phase system does not need to have a neutral to operate correctly. The neutral is added only if the lower voltage (115 or 120 V) is needed for some part of the system. In general, equipment manufacturers make all the components in a three-phase system a higher voltage, or they may supply a small transformer inside the equipment's power panel to drop the higher voltage to the necessary voltage level. For example, if the equipment needs 208 V three-phase for a pump motor, a control transformer can be provided in the power panel of the equipment to drop the 208 V between L1–L2 to 110 V. The 110 V is used to provide power for the control circuit and relay coils. Because the primary side of the control transformer can be powered by L1–L2 (208 V), a neutral is not needed in this system. If the system has small pump motors, they may require 120 V for power, a neutral tap is needed; these components will be connected between L1–N, L2–N, or L3–N.

If the transformer is connected as a delta transformer, it is important to understand that a different voltage becomes available between L3–N. Figure 8–20 shows the diagram for this voltage. From this diagram you can see that the secondary winding of this three-phase transformer is connected as a delta transformer. The voltage from L1–N and L2–N is 115 V, so the voltage between L1–L2, L2–L3, or L3–L1 is 230 V.

The different voltage occurs between L3–N. Because the winding between L1–L2 has the center tap, it stands to reason that the voltage between L1–N or L2–N will be exactly half of the voltage L1–L2. Because the center tap is not between L3 and L1, the voltage for L3–N must come from one complete phase (230 V) and half of the next phase (115 V). This voltage uses two phases, so the 230 V and the 115 V are out of phase, and the resultant voltage from them is 208. This voltage is not the same as the voltage of 208 V that occurs between L1–L2 or L2–L3 or L3–L1 of the wye-connected transformer, because the neutral point is not at a midpoint.

Because the 208 V between L3–N comes from two phases, it will cause the transformer to overheat if it is used to power any components that require 208 V. For this reason the voltage is called the *high leg delta voltage*, to indicate that it is derived from L3–N of a delta-connected transformer and that it should not be used to power 208-V components. It is very important to understand that the L3 leg of the transformer poses no problem when it is used with L1–L3 or L2–L3 as part of the three-phase system or the 230-V single-phase system. The only problem occurs when the L3 terminal is used in conjunction with the neutral, which creates the L3–N voltage of 208 V.

The L3 terminal in any power distribution box for a delta-wired system should always be marked with an orange wire to identify it as the high leg delta. In some areas of the country, the high leg delta is also called the *wild leg*. The high leg delta voltage of 208 V may occur between L1–N if the neutral point is produced by a center tap of the L2–L3 winding, or it may occur between L2–N if the neutral point is produced by a center tap of the L1–L3 winding.

8.18
Three-Phase Voltage on Site

The source for three-phase voltage is an electrical panel or a load center. As you know, the voltage actually comes from the three-phase transformers, but as stated before, as a maintenance technician you will not be expected to make connections right at the transformers. Instead technicians from the utility company will connect the wires between the transformers and the load centers or main disconnects. The load center may also be called a *circuit breaker panel*. Figure 8–21 shows an

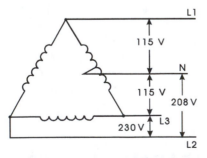

Figure 8–20 A high leg delta voltage occurs between L3–N because the center tap that produces the neutral is the center tap of the L1–L2 winding.

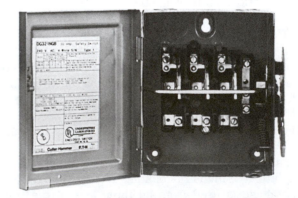

Figure 8–21 A three-phase fusible disconnect that is used to provide three-phase voltage to equipment in commercial and industrial applications.

(Courtesy of Cutler-Hammer, Inc.)

example of a three-phase fusible disconnect, and Fig. 8–22 shows an example of a three-phase load center. In the picture of the three-phase disconnect, you can see the three fuses. Incoming voltage is connected to the line-side terminals at the top of the disconnect. The line-side terminals are identified as L1, L2, and L3. The disconnect has a handle on the right side that is used to turn its switch on and off. If the disconnect switch is depressed at the top, the switch will close and the disconnect will have voltage at the load-side terminals at the bottom of the disconnect. The load-side terminals are identified as T1, T2, and T3. It is important to understand that the line side terminals will always have power even when the disconnect switch is open (in the off position).

The three-phase load center is similar to the circuit breaker panel for a single-phase system, except it has three individual circuits. In smaller shops that have only a few pieces of equipment, the equipment may be connected directly to the circuit breakers in the load center. This is also generally the case for loads such as exhaust fans and lighting, because they need to be individually protected. The load center is specifically designed to mount three-phase circuit breakers or each individual circuit that will be used. Thus if you have four exhaust fans mounted in the ceiling and they are connected to the load center, each one will be connected to its own individual circuit breaker. When you are installing a new fan or servicing existing equipment, you must locate the load center and circuit breakers for the circuit you intend to use, or you must locate the fusible disconnect if one is installed near the unit location.

In larger factories, the power is distributed through bus ducts throughout all areas of the factory. The bus duct has a number of openings that allow the technician to connect bus box disconnects directly onto the bus. These disconnects allow power to be tapped from the three-phase supply as close to the machine as possible. Most bus ducts are suspended approximately 8 to 15 ft above the floor, and service to the disconnect on the machine is made by using flexible cable or fixed conduit.

Figure 8–22 A three-phase load center that contains the circuit breakers for the electrical system. (Courtesy of Cutler-Hammer, Inc.)

8.19
Installing Wiring in a Three-Phase Disconnect

When you are ready to connect the wires between the disconnect and the equipment you are installing, you will need to ensure that the source of the three-phase voltage is turned off and a padlock is used to lock out the system. The padlock should also have a tag attached to it that has your picture and name on it to identify you as the technician who has locked out the system. The lock-out and tag-out procedure ensures that you can safely work on the system without anyone accidentally turning the power on when you are still working in the panel.

Figure 8–23 shows a diagram of the locations in the disconnect where you will connect the equipment wires. The utility company will have the supply voltage wires previously installed at the three terminals at the top of the disconnect box. These terminals are the line-side terminals, and they are identified as L1, L2, and L3. You will make your connections at the bottom terminals, which

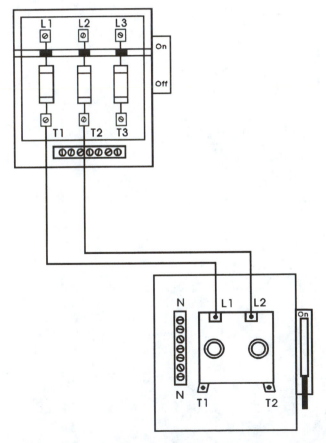

Figure 8–23   A single-phase disconnect is connected to the load-side terminals of a three-phase disconnect to provide a single-phase power source.

are identified as T1, T2, and T3. These terminals are the *load-side terminals*. The other end of each wire will be connected in the power panel for the equipment. This process is called *connecting the field wiring*.

When you need to make connections for field wiring in a circuit breaker panel, you will connect to its *main breaker* at the top of the panel. The individual circuit breakers will receive power any time the main breaker is in the *on* position. You need to add a three-phase circuit breaker for each piece of equipment for which you are providing power. Again, you should notice that the wire from the circuit breaker is connected at the incoming power terminals in the equipment.

8.20
Testing for a Bad Fuse in a Disconnect

At times you will be working on a system that receives its power from a three-phase disconnect. The system will not run, and you may suspect one or more of the fuses are bad. You can check for a blown fuse by measuring the voltage or by removing the fuses and testing them for continuity with an ohmmeter. If you select to check the fuses with a voltage test, you will need to test the incoming voltage at L1, L2, and L3 to determine that the supply voltage is good. Check the voltage between terminals L1–L2, L2–L3, and L3–L1. The voltage at each of these sets of terminals should be the rated voltage for the system, such as 480 V. If you do not get the same voltage reading at each set of terminals, the incoming power has a problem, and you will need to determine where the voltage has been interrupted.

The next test should be at the bottom terminals of the disconnect at T1–T2, T2–T3, and T3–T1. Each of these readings should be the same as the voltage readings at the line side of the disconnect. If any of the voltage readings are lower than the line-side readings or 0 V, you have an open fuse. At this point the simplest way to find the bad fuse is to turn the disconnect off and remove all the fuses and test them for continuity. Be sure to use a fuse puller when you are removing or installing fuses in the disconnect. It is also important to remember that the line-side terminals of the disconnect are "hot" (powered) even though the disconnect switch is open.

The reason the load-side voltage may be lower than the line-side voltage rather than 0 V is because voltage may feed back through two of the windings in any three-phase motor or three-phase transformer that is connected to the load side of the disconnect. It is difficult to determine the exact amount of voltage that is dropped when voltage feeds back through other windings, so you must test for an open fuse if the line-side voltage is not the same as the load-side voltage at each set of terminals where you take a reading. If you find an open fuse, you will need to replace it with one that has the same current rating and same voltage rating.

Remember that you can also measure for voltage across each fuse; if you measure full voltage across any fuse, the fuse is open and the voltage is actually *back-feed voltage*.

8.21
Single-Phase Voltage from a Three-Phase Supply

At times you will need to provide a single-phase source of voltage for a small motor, pump, or shop tools such as grinders or drill presses. The single-phase voltage can be either 480 V, 208 V, or 230 V L1–L2. In Fig. 8–23 you can see the connections between the load-side terminals of a three-phase disconnect and the line-side terminals of a single-phase disconnect. If the single-phase circuit requires a neutral, a neutral wire can be connected between the neutral terminal of the three-phase disconnect and the single-phase disconnect and between the three-phase disconnect and the transformer.

This type of circuit may also be used in applications inside a factory or commercial location where three-phase voltage is supplied to the system and you need to provide a single-phase voltage source to power a window-type air conditioner or a small refrigerator. Once you understand that you can get single-phase voltage from the three-phase system, you will begin to see how the AC power-distribution system works in industrial, commercial, and residential applications.

The remaining chapters of this book will continue to refer to the supply voltage as you would find it connected to the equipment at the line-side terminals. As a technician you will need to be aware of where this voltage comes from and how it is distributed so that you can trace the circuits if no voltage is present at the unit.

8.22
Bus Duct and Bus Disconnect Boxes

In most factories, the power is distributed throughout the factory by a series of bus ducts that are suspended in the ceiling. In some case the bus duct is mounted on the wall or in the floor. The bus box is a disconnect switch that is designed specifically to mount on the bus duct and provide a disconnect and fuse combination. Because the bus duct has multiple bus boxes, one or more can be turned off while others remain powered, so that you can turn the power off to the machine you are working on while allowing power to be connected to a machine right beside you that you are not working on. The front of the bus box is a door that can be opened to be tested or serviced. A safety latch is provided to keep you from opening the door while power is turned on. You can use a screwdriver to override the latch to check the fuses in the box while power is supplied. It is recommended that you turn off the disconnect switch on the box any time you are working on the system. The only time that it is permissible to override the latch is when you need to take a voltage reading across the fuses while power is applied. Figure 8–24 shows a typical bus duct and bus box.

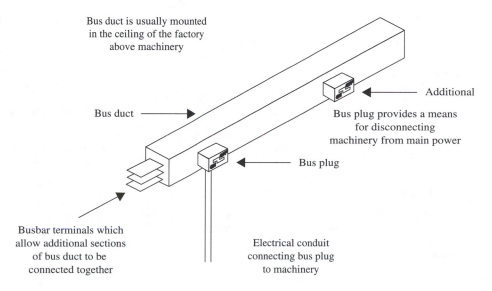

Figure 8–24 Typical bus duct with bus boxes shown connected in the bus.

Bus duct is usually mounted in the ceiling of the factory above machinery

Bus duct

Busbar terminals which allow additional sections of bus duct to be connected together

Additional

Bus plug provides a means for disconnecting machinery from main power

Bus plug

Electrical conduit connecting bus plug to machinery

Figure 8–25   A duplex receptacle that provides 110 V for tools and equipment.

8.23
Wiring a Duplex Receptacle as a Utility Outlet

In most factory equipment it is necessary to provide one or more duplex receptacles so that you can plug in 110-V service equipment, such as service lamps or soldering irons, or hand power tools, such as electric drills. Figure 8–25a shows a duplex receptacle. When you hold a duplex receptacle in your hand you should notice that it has gold-colored screws on one side and silver-colored screws on the other side. The figure shows that the L1 wire should be connected to a gold-colored screw and the neutral wire should be connected to the silver-colored screw. The ground wire should be connected to the green screw on the receptacle.

The openings in the face of the receptacle are different sizes. The larger opening on the left side of the receptacle face is connected to the neutral terminal, and the smaller opening on the right side of the receptacle face is connected to the L1 terminal. The round opening at the top is connected to the ground terminal. Because the openings in the face of the receptacle are different sizes, any plugs that are connected to it have to be connected correctly. If a tool such as an electric drill is made of plastic, its plug will be manufactured so that it can be plugged into the receptacle either way. If the electric drill has a metal case, the plug will have a ground terminal and the plug will only connect to the receptacle only one way.

When you are ready to connect L1 and N wires to the receptacle and disconnect, be sure the disconnect switch is open. It is also important to test the voltage from L1–N at the line side of the disconnect to ensure that you have 115 V instead of the 208 V of the high leg delta voltage. When all the connections have been made, you can close the main disconnect switch and test the two terminals of the receptacle with your voltmeter to ensure that 115 V are present.

Questions for This Chapter

1. Explain how inductance is used in the operation of a transformer.

2. Provide an example of a step-up transformer in an industrial system.

3. Discuss how it is possible to have a control transformer whose primary winding can be connected to 220 or 208 V and provide 110 V at its secondary.

4. Explain how you would test for 240 and 110 VAC in a fusible disconnect on the line-side and load-side terminals.

5. Explain why a control transformer is required in some electrical equipment.

True or False

1. If voltage L1–L2 is 240 V and voltage L1–N is 120 V, the transformer is connected as a delta transformer.

2. The secondary voltage is larger than the primary voltage in a step-up transformer.

3. Voltage in a transformer moves from the primary winding to the secondary winding by conductance.

4. The VA rating for a transformer is determined by multiplying the voltage at the secondary winding by the current flow in the secondary winding.

5. The 208 V from L3 to N for a high leg delta transformer can be used for motor loads that require 208 V.

Multiple Choice

1. The voltage at the secondary terminals of a _____ transformer will be larger than the applied voltage to the primary terminals.

   a. step-up

   b. step-down

   c. isolation

2. A transformer that is wired as a delta transformer has _____ .

   a. 440, 208, and 125 V.

b. 440, 220, and 110 V.

c. 440, 277, and 125 V.

3. A transformer that is wired as a wye transformer has
_____ .

   a. 440, 208, and 125 V.

   b. 440, 220, and 110 V.

   c. 480, 240, and 120 V.

4. If a control transformer has a 2:1 turns ratio and it has 220 V applied to its primary winding, the voltage at the secondary winding will be _____ .

   a. 440 V.

   b. 220 V because the secondary voltage is strictly controlled to equal the primary voltage.

   c. 120 V.

5. The pumps and motors in a system are connected to the _____ of a fused disconnect.

   a. line-side terminals

   b. load-side terminals

   c. power company terminals

Problems

1. Calculate the secondary voltage of a transformer that has a 4:1 turns ratio and 110 VAC applied to the primary terminals.

2. Calculate the VA of a 120-V control transformer that requires 5 A.

3. Determine the turns ratio of a transformer that has 4,000 turns in its primary windings and 800 turns in its secondary winding.

4. If the transformer in Problem 3 has 440 V applied to its primary winding, how many volts will be measured at its secondary?

5. Calculate the amount of secondary amps for a control transformer that is rated for 800 VA and 110 V at the secondary.

Chapter 9

Relays, Contactors, and Solenoids

Objectives

After reading this chapter you will be able to:

1. Identify the contacts and coil in a relay and contactor.
2. Explain the operation of a relay and contactor.
3. Identify the components in a control circuit.
4. Select the proper size of contactor from a NEMA table.
5. Identify the basic parts of a solenoid and explain their functions.

9.0
Overview of Relays and Contactors in the Control Circuit

The *control circuit* in any industrial electrical system is the heart of the system. It consists of the control devices, such as start-stop switches, limit switches, or pressure switches that determine when to energize or deenergize each system. Each control circuit uses relays or contactors to control the various motors, such as pump motors and fan motors. This chapter introduces the main components that are used in the control circuits of these systems and their theory of operation. The components in the control circuit are energized, and in turn they energize components such as compressors and motors in the *load circuit*. Figure 9–1 shows an example of a typical control circuit for a system. Notice that it is the bottom part of the circuit; the load circuit is the top part of the electrical diagram.

9.1
The Control Transformer

The control transformer was introduced in Chapter 8. In that chapter you learned that the transformer converts line voltage to lower voltage power for the control circuit. The lower voltage is used in each case to provide power for the control circuit.

In the diagram in Fig. 9–1, the control circuit is shown as the bottom part of the diagram, and it starts at the secondary side of the transformer, where it is identified by the letters L1 and N. The terminal that is identified as the L1 terminal is considered the source of power; it is called the *hot side of the circuit*. The right side of the diagram, identified by the transformer terminal N, is considered to be the *common side of the circuit*. It is important to understand that because the voltage in the transformer is AC voltage, both sides of the circuit have the same potential during every other half-cycle. The concepts of a hot side and a common side are defined for humans to make it easier to understand the operation of the control circuit.

9.2
The Theory and Operation of a Relay

A *relay* is a magnetically controlled switch that is the main control component in a typical electrical system. Figure 9–2 shows a picture of a typical relay, and Fig. 9–3 shows a cutaway diagram of a typical relay, which

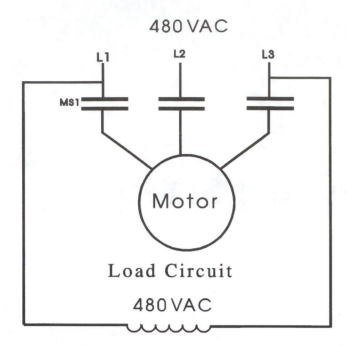

480 VAC

L1   L2   L3

MS1

Motor

Load Circuit

480 VAC

L1   N

110 VAC
Control Circuit

Motor
Starter
Coil

Stop   Start

1   3   4   2

MS1

Figure 9–1 A typical control circuit at the bottom of the diagram. This start-stop button controls a three-phase motor.

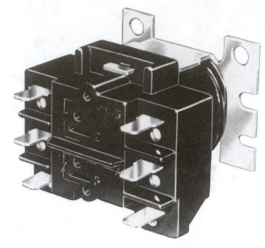

Figure 9–2 A typical relay.
(Courtesy of Honeywell's Micro Switch Division)

consists of a *coil* and a number of *sets of contacts*. The coil becomes an electromagnet when it is energized, and its magnetic field causes each set of normally open contacts to close and each set of normally closed contacts to open. The contacts are basically a switch that is operated by magnetic force. The part of the relay that moves and causes the contacts to move is called the *armature*. Power is applied to the coil of the relay, and the magnetic flux causes the armature to move, which in turn causes the contacts to change position. The coil is part of the control circuit, and the contacts are part of the load circuit.

Figure 9–4 shows the electrical symbol for the coil and contacts of a relay. The armature is not shown in the electrical symbols because it is considered part of

the contacts. When you encounter a relay in the control panel of a system, you must relate the physical parts with the electrical symbols that are used to identify the components. You must also learn to envision its operation as two separate pieces, the coil and contacts, even though they are mounted near each other and operate almost simultaneously. It is important to understand that the coil must be energized first, and a split second later the magnetic field that is built up in the coil will cause the contacts to move.

9.3
Types of Armature Assemblies for Relays

You will encounter many different types and name brands of relays that look like they all have different theories of operation. In reality the operation of all relays can be categorized into four basic methods, because one of four basic armature assemblies is used in all relays. Figure 9–5 shows an example of these four basic armature assemblies. A few relays will use hybrid armatures that involve some features of more than one type of armature assembly.

Figure 9–5a shows an example of a *horizontal-action armature* for a relay. In this example a set of stationary contacts is mounted in the horizontal position. A set of movable contacts is mounted on the armature assembly directly across from the stationary set. The armature is mounted inside the coil in such a way that when the coil is energized, its magnetic field will cause the mass of the armature to center in the middle of the coil. Because the armature is slightly offset when it is placed in the coil, it will move to the left to center itself

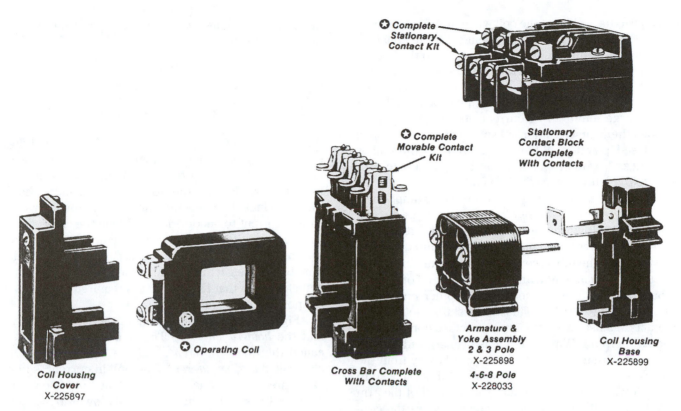

Figure 9–3   A cutaway diagram of a relay. Notice that the coil and contacts are mounted so that the magnetic field of the coil will make the contacts move.

(Courtesy of Rockwell Automation's Allen-Bradley Business)

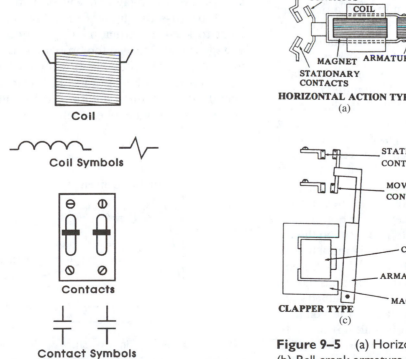

Figure 9–4   Electrical symbols and diagram for the coil and contacts of a relay.

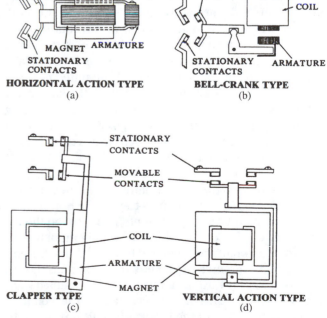

Figure 9–5   (a) Horizontal-action armature assembly. (b) Bell-crank armature assembly. (c) Clapper armature assembly. (d) Vertical-action armature assembly.

(Courtesy of Square D Co.)

directly in the middle of the coil when the magnetic field is developed in the coil. The movement of the armature to the left will shift the movable contacts to the left until they come into contact with the stationary set of contact. When the movable contacts and stationary contacts touch each other, they complete a circuit that allows large amounts of current and voltage to pass through them, just as a normal switch does when it is in the closed position. When the current to the coil is deenergized, a small spring causes the armature to shift back to the left to its original position.

The diagram in Fig. 9–5b shows an example of a *bell-crank armature assembly.* In this example the coil is mounted above the armature, and when it is energized it produces a magnetic field that pulls the armature upward. The movable contacts are connected to an arm that is bent at a right angle. When the end of the arm is pulled upward by the magnet, the other end of the arm is moved to the left. This action shifts the movable contacts against the stationary contacts to complete the electrical circuit. When the coil is deenergized, gravity causes the armature to drop downward away from the coil. This movement causes one end of the arm also to move downward, which makes the left end of the arm shift back to the left. This movement causes the contacts to move to the open position again.

The diagram in Fig. 9–5c shows an example of a *clapper armature assembly.* The armature is a large arm on the right side of the coil. The armature (arm) has a pin through the bottom to act as an axis. The movable contacts are mounted at the top of the armature on the left side. When the coil is energized, it creates a magnetic field that pulls the armature to the left toward the coil. This action causes a small amount of travel at the bottom of the arm and a large amount of travel at the top of the arm, because the bottom of the arm is held in place at the axis. The movable contacts are mounted near the top of the arm, so they will move to the left a significant distance until they come into contact with stationary contacts. When the coil is deenergized, a small spring pulls the armature to the left, which causes the contacts to return to their open position.

Figure 9–5d shows an example of a *vertical-action armature assembly.* The armature is mounted on a bracket that is shaped like the letter *C.* The bottom part of the C-shaped bracket is mounted directly on the armature, and a set of movable contacts is mounted directly on the top of the C-shaped bracket. When the coil is energized, it creates a magnetic field that pulls the entire C bracket upward until the armature is pulled tight against the coil. This movement causes the contacts mounted on the top of the bracket to shift upward until they touch the stationary contacts. When the coil is deenergized, gravity causes the complete bracket to drop downward and move away from the coil, which makes the movable contacts move away from the stationary contacts.

9.4
Pull-in and Hold-in Current

When voltage is first applied to a coil of a relay, it draws excessive current. This occurs because the coil of wire presents resistance to the circuit only when current first starts to flow. As the flow of current increases in the coil, inductive reactance begins to build, which causes current to become lower. When the current is at its maximum, it creates a strong magnetic field around the coil, which causes the armature to move. When the armature moves, it makes the induction in the magnetic coil change so that less current is required to maintain the position of the armature.

Figure 9–6 shows a diagram of the *pull-in current* and the *hold-in current.* The pull-in current is also called the *inrush current,* and the hold-in current is also called *seal-in current.* The pull-in current is typically three to five times as large as the hold-in current. At times the supply voltage is slightly lower in the summer. This condition is called a brownout, and it may cause a problem with activating relays. If the voltage is too low, it may not be able to supply sufficient current to pull the armature into place when the relay is first energized because the pull-in current is too small.

If the system is running when the brownout condition occurs, the relay coil will probably remain energized, because the amount of hold-in current is sufficient to keep the armature in place even though the voltage is low. Because the brownout condition occurs when it is very hot outside, it will most likely affect air conditioners or other systems that cycle on and off frequently through this condition. If the air conditioner ever gets the room cool enough so it can cycle off dur-

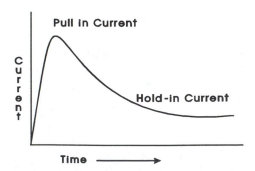

Figure 9–6 A diagram that shows pull-in and hold-in current for relay coils. Notice that the pull-in current is approximately three to five times as large as the hold-in current.

ing a brownout condition, it will probably not energize again because of the low voltage and the large amount of pull-in current required.

9.5
Normally Open and Normally Closed Contacts

A relay can have normally open or normally closed contacts. The electrical symbol in Fig. 9–7 shows examples of normally open and normally closed contacts. It is important to understand that the word *normal* for contacts indicates the position the contacts are in when no voltage is applied to the coil. The contacts can be held in their normal position by a spring or by gravity. The contacts will move from their normal position to their energized position when power is applied to the coil.

ADDING or CONVERTING CONTACT CARTRIDGES HAVING "SWINGAROUND" TERMINALS

General Instructions (Specific cases below.)

1.1 Adding a contact cartridge:

As received, accessory cartridges are in the normally open mode with terminal screws adjacent to N.O. symbols. If normally closed mode is desired, convert contact as indicated in Step 1.2 below. When cartridges are inserted, the terminal screws must face the front. The clear cover may face either side. **Do not install more than 8 N.C. contacts per relay.** When installing one cartridge, locate it at an inner pole position. When installing 2 cartridges, locate both in inner or outer (balanced) positions.

1.2 Converting a contact to its alternate mode (N.O. ⇌ N.C.):

Withdraw an assembled cartridge for replacement or conversion by inserting the blade of a suitably-sized screwdriver under a terminal screw pressure plate. Slide cartridge out. See Figure 2. Back the terminal screws out of the cylindrical nuts a sufficient amount (approximately 2 turns for a fully-tightened screw) to permit rotation of each screw and nut assembly to its alternate position. See Figure 3.

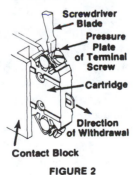

FIGURE 2

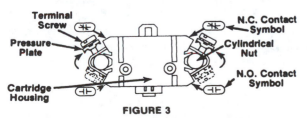

FIGURE 3

Figure 9–7 Example of changing normally open contacts to normally closed contacts in the field by turning them over. Normally closed contacts can also be converted to normally open contacts in the same manner.

(Courtesy of Rockwell Automation's Allen-Bradley Business)

Some types of contacts can be changed or converted from normally open to normally closed in the field. Other types are manufactured in such a way that they cannot be changed. Figure 9–7 shows examples of converting normally open contacts to normally closed contacts while the relay is installed. The contacts can be converted in the field by the technician by simply removing them from the relay and turning them upside down. When normally open contacts are inverted, they become normally closed, and when normally closed contacts are inverted, they become normally open. This means that as a technician in the field, you can change the contacts in a relay to get the exact number of normally open or normally closed contacts needed for the application.

9.6
Ratings for Relay Contacts and Relay Coils

When you must change a relay that is worn or broken, you need to ensure that the coil of the new relay matches the voltage of the control circuit exactly. This means that if the voltage for the control circuit is 110 VAC, the coil must be rated for 110 VAC. If the control voltage is 220 VAC, the coil must be rated for 220 VAC. The voltage rating for a relay coil is stamped directly on the coil. If the coil is rated for 24 VAC, it will be color-coded black. If the coil is rated for 110 VAC, it should be color-coded red or have a red-colored stamp on the coil. If the relay coil is rated for 208 or 230 VAC, it will be color-coded green or identified with a green stamp or green printing on the coil. DC coils are color-coded blue. It is important to understand that the current rating for a relay coil is seldom listed on the component. If it is important to know the current rating for the coil, you can look for it in the catalog or on the specification sheet that is shipped with the new relay. If you change a relay you must also make sure that the rating for the contacts meets or exceeds the current rating and the voltage rating of the load to which it will be connected. For example, if the contacts of the relay are used to energize a 240-VAC fan motor that draws 3 A, the contacts must be rated for at least 240 V and 3 A. It is permissible to have the contacts in this example rated for more voltage and current, such as 600 V and 10 A.

Contact ratings are grouped by voltage and by current. The voltage ratings are generally broken into two groups, 300 V and 600 V. This means that if you are using the contacts to control 240 or 208 VAC, you would use contacts that have a 300-V rating. If you are using contacts to control a 480-VAC motor, you would need to use contacts with a 600-V rating.

The current ratings of contacts are listed in amperes or horsepower. The current rating or horsepower rating must exceed the amount of current the relay is controlling. This means that if the relay is controlling 12 A, the contacts need to be rated for more than 12 A of current. The current rating of the contacts and the voltage rating for the contacts are printed directly on the contacts or on the side of the relay.

9.7
Identifying Relays by the Arrangement of Their Contacts

Some types of contact arrangements for relays have become standardized so that they are easier to recognize when they are ordered for replacement or when you are trying to troubleshoot them. The diagrams in Fig. 9–8 show examples of some of the standard types of relay arrangements. Figure 9–8a shows a relay with a set of normally open contacts. This type of relay could also have a single set of normally closed contacts instead of normally open contacts. Because this relay has only one contact and it can only close or open, this type of relay is called a *single-pole, single-throw (SPST) relay*. The word *pole* in this identification refers to the number of contacts, and the word *throw* refers to the number of terminals to which the input contacts can be switched. Because the contact in this relay has one input and it can be switched only to a single output terminal, it is said to have a single throw.

Figure 9–8b shows a relay with two sets of normally open contacts. Because this relay has two sets of single contacts, it is called a *double-pole, single-throw relay (DPST)*. The double-pole part of the name comes from the fact that the relay has two individual sets of normally open contacts, and the single-throw part of the name comes from the fact that each contact has only one output terminal. When the coil is energized, both sets of contacts move from their normally open position to the normally closed position.

Figure 9–8c shows a relay with a set of normally open and a set of normally closed contacts that are connected on the left side. The point where this connection is made is called the *common point*, and it is identified with the letter *C*. When the relay coil is energized, the normally open part of the contacts will close, and the normally closed part of the contact will open. Because these contacts basically have a common point as the input terminal and two output terminals, one normally open (NO) and one normally closed (NC), it is called a *single-pole, double-throw relay (SPDT)*. The most important part of this relay is that the contacts have two terminals on the output side, so it is called a double-throw relay. The single-pole, double-throw relay is used

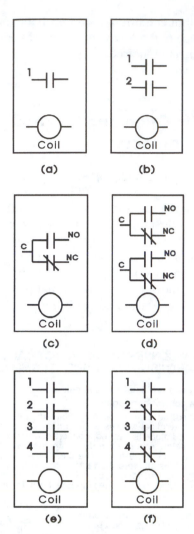

Figure 9–8 (a) A relay with a single set of normally open contacts. This type of relay is called a single-pole, single-throw (SPST) relay. (b) A relay with two individual sets of contacts. This relay is called a double-pole, single-throw (DPST) relay. (c) A relay with two sets of contacts that are connected at one side at a point called the common (C). The output terminals are identified as normally open (NO) and normally closed (NC). This type of relay is called single-pole, double-throw (SPDT). (d) A relay with two sets of SPDT contacts. This relay is called a double-pole, double-throw (DPDT) relay. (e) A relay with multiple sets of normally open contacts. (f) A relay with a combination of normally open and normally closed contacts.

where two exclusive conditions exist and you do not ever want them both to occur at the same time. For example, if this relay is controlling the cooling (air conditioning) and the heating (furnace) for the cooling and heating system to the offices of a factory, you would never want them both to be on at the same time. By connecting the air-conditioning system to the normally open terminal on the right side of the relay and the furnace to the normally closed terminal, you create the

Figure 9–9 Variety of relays used in industrial electrical applications.
(Courtesy of Honeywell's Micro Switch Division)

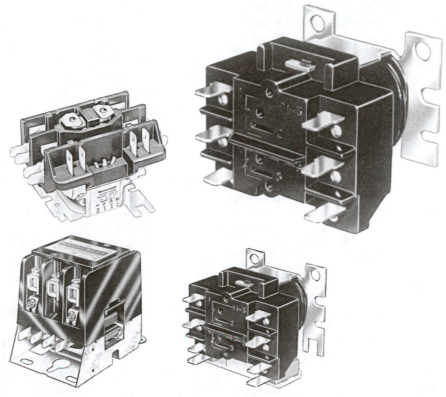

conditions so that the furnace and air conditioning system cannot be on at the same time.

The relay in Fig. 9–8d has two sets of single-pole, double-throw contacts, so it is called a *double-pole, double-throw relay (DPDT)*. In this case the word *double throw* is used because two sets of normally open/normally closed contacts are provided. Each set has a common point on the left side (input side) and a terminal that is connected to the normally open (NO) set and a terminal that is connected to the normally closed (NC) set on the right side. This type of relay is used where exclusion is needed and the loads of 208 VAC or 230 VAC need power from both L1 and L2. In this type of application, L1 is connected to the common terminal (C) of one set of contacts, and L2 is connected to the common terminal (C) of the other set. This causes L1 and L2 to be switched the same way in both conditions.

Figure 9–8e shows multiple sets of normally open contacts. This type of relay can have any number of sets of normally open contacts. The additional sets of contacts can be added to original contacts in some types of relays. If the original relay is manufactured with this provision, you can purchase the additional contacts and add them to the original relay by placing them on top of the original relay and tightening the mounting screws to make the additional contacts operate with the relay armature. The contacts for this type of relay can all be normally closed if the application requires it. The main feature of this type of relay is that it can have any number of contacts.

Figure 9–8f shows a relay with multiple sets of individual normally open and normally closed contacts. The combination of normally open and normally closed contacts can be any mixture of the two. This type of relay is similar to the one shown in Fig. 9–8e, except in this type of relay the contacts can be any combination of normally open or normally closed sets. In most cases, the contacts in this type of relay are convertible in the field, and as a technician, you can add sets of contacts and change them from normally open to normally closed, or vice versa, as needed. In most original installations, the relays are provided in the original equipment, and you will need only to identify them for installation and troubleshooting purposes. Later, if the equipment is modified or if additional machinery such as electronic air cleaners, humidifiers, or other similar equipment is added, you may need to locate additional contacts on a relay to connect the add-on equipment so that it will operate correctly with the original system.

9.8
Examples of Relays Used in Industrial Electrical Systems

Relays are used in a variety of applications in industrial electrical systems. For example, one of the most common relays is the fan relay used in heating systems. Another common relay is the sequencing relays used in some electric furnaces. Figure 9–9 shows examples of several types of relays that you will find in these systems. In some newer systems, the relays use plug-in ter-

Figure 9–10 (a) A typical current relay. (b) An electrical diagram of a current relay.

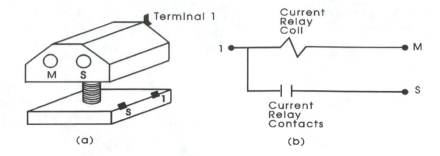

(a)

(b)

minals so that the relay can be removed and replaced without changing any wiring. The relay is simply plugged into a socket-type base. You will begin to become familiar with relays and recognize them in the control panels of the equipment you work on.

9.9
Current Relays and Potential Relays for Starting Single-Phase Compressors

Current relays and potential relays are special relays that are designed to energize the start winding of single-phase compressor motors during starting and then immediately deenergize as soon as the motor is running. Single-phase air-conditioning and refrigeration compressor motors have a special problem in that they are motors that are sealed inside a metal case, yet they need a device to operate as a start switch to switch the start winding of the motor into the circuit during starting and then disconnect the winding during the run phase. Because the motor uses lubrication, an explosion is possible if a switch is inside the sealed case. This means that starting device or relay must be mounted outside the compressor.

The current relay has a low-impedance coil that takes advantage of pull-in and hold-in current to energize a set of normally open contacts. The coil for the current relay is connected in series with the run winding of a single-phase motor, and the contacts of the relay are connected in series with the start winding of the motor. In some cases a capacitor is also connected in series with the contacts and start winding. When voltage is applied to the motor, current can flow only through the run winding, which causes very large current to be drawn. This large current is large enough to exceed the pull-in current level of the contacts, which makes them close. When the current relay contacts close, voltage is applied to the start winding of the motor, which provides additional power that allows the motor to start and continue running. After the motor has started and begins to run, the large initial current that is flowing through the run

winding and current relay coil diminishes to a minimal level, which is below the hold-in current level. When this occurs, the current relay coil is not able to keep its contacts closed, so they open and deenergize the start winding of the motor. This action is exactly what the single-phase motor needs, because its start winding can be energized only a few seconds while the motor is starting. Figure 9–10 shows the current relay and electrical diagram. The current relay is also available as an electronic-type relay.

The *potential relay* uses a high-impedance coil with a set of normally closed contacts to energize a start capacitor for a capacitor-start, capacitor-run (CSCR) single-phase compressor motor. Figure 9–11 shows a picture and diagram of a typical potential relay. When power is applied to the compressor motor, current flows through the run winding and through the normally closed contacts of the potential relay to the start winding. When the motor begins to run, its windings create a back voltage called back EMF. The back EMF is large enough to cause the coil of the potential relay to energize and pull its contacts open, which deenergizes the start winding. When main power is turned off to the motor, it stops running, and the normally closed contacts of the potential relay return to their normally closed position so the relay is ready to start the motor again.

9.10
The Difference between a Relay and a Contactor

A contactor is similar to a relay in that it has a coil and a number of contacts. The main difference is that the contactor is larger and its contacts can carry more current. A *relay* is generally defined as magnetically controlled contacts that carry current less than 15 A. A *contactor* is defined as having contacts that are rated for 15 A or more. Some manufacturers do not follow the 15-A rating, so sometimes you will find a relay that has a current rating for its contacts in excess of 15 A, and you may also find a contactor with contact ratings less than 15 A. In general the main difference is that a contactor

(a)

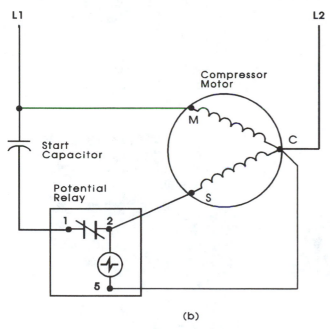

(b)

Figure 9–11 (a) A potential relay used to start single-phase compressors. (b) A potential relay connected to a CSCR compressor motor.

(Courtesy of Supco Sealed Unit Parts Co. Inc.)

is specifically designed so its contacts can carry larger amounts of current, up to 2,250 A. Contactors are rated by the National Electrical Manufacturers Association (NEMA), and their sizes range from a size 00 to a size 9.

An example of a place in which a contactor is used instead of a relay is the compressor motor of an air-conditioning system. In this type of system the motor is controlled by a contactor instead of a relay because the amount of current the compressor motor draws is usu-

ally in excess of 15 A. When you look at the compressor contactor, you will see that it looks like a very large relay. Because its contacts can carry current larger than 15 A, it will technically be called a contactor instead of a relay.

9.11
NEMA Ratings for Contactors

Figure 9–12 shows a table of NEMA ratings for contactors. This table shows that a size 00 contactor is rated to safely carry up to 9 A for a continuous load. You should also notice that the contacts are rated for up to 600 V. You should remember that the current rating for contacts depends on the size of the contacts, and the voltage rating of the contacts depends on whether the relay is manufactured so that arcs do not jump between different terminals. This means that contacts rated for a higher voltage have more plastic or insulating material between the sets of contacts so arcs do not jump from the contacts to other parts of the relay or other sets of contacts. This table also identifies the load as a maximum horsepower rating. Thus, you may identify the load from its current rating or its horsepower rating and select the proper contactor size to safely control the load.

Notice from this table that the next-larger size of contactor is a size 0. This contactor is rated for up to 18 A on a continuous basis. The largest contactor is the size 9 contactor; its current rating is 2,250 A.

Industrial electrical systems usually use contactors up to a size 4. One application for contactors is to control electric heating elements for electric furnaces. These contactors are called *heating contactors*. Because heating elements may draw up to 100 A, the size 4 contactor can easily carry this current. If larger heating elements are used, a larger contactor is used. Figure 9–13 shows two examples of contactors, a size 3 and a size 5. The size 3 and size 5 contactors are approximately 5 and 12 inches tall, respectively.

9.12
The Function of the Control Circuit

As you have seen in this section the function of the control circuit in industrial electrical systems is to control the conditions that cause the equipment to become energized or deenergized. The components in the control circuit of typical industrial electrical systems are generally low-voltage (110 VAC) devices. The relay coil is located in the control section of the electrical system and the relay contacts are located in the load section, and they are typically connected in series with the load it is controlling. It is important to

Figure 9–12   NEMA ratings for contactors.

(Courtesy of Rockwell Automation's Allen-Bradley Business)

NEMA Size	Continuous Ampere Rating	Maximum Horsepower Rating Full Load Current Must Not Exceed "Continuous Ampere Rating"			
		Motor Voltage			
				50 Hz	
		200V	230V	380V-415V	460V-575V
		3 Ø • 4 Power Poles • 600V AC Maxi			
00	9	1-1/2	1-1/2	2	2
0	18	3	3	5	5
1	27	7-1/2	7-1/2	10	10
2	45	10	15	25	25
3	90	25	30	50	50
4	135	40	50	75	100
5	270	75	100	150	200
6	540	150	200	300	400
7 ❷	810	-	300	600	600
8 ❷	1215	-	450	900	900
9	2250	-	800	1600	1600

Figure 9–13   Examples of NEMA size 3 through size 5 contactors. The size 3 is approximately 5 inches tall, and the size 5 is approximately 12 inches tall.

(Courtesy of Rockwell Automation's Allen-Bradley Business)

(a)

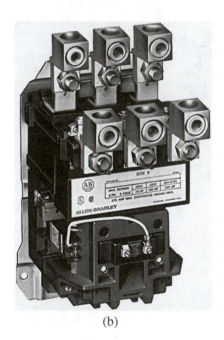

(b)

remember that the function of the control circuit is to energize the relay coil so power can be sent through the relay contacts to the load. The action in the control circuit must always occur before the action in the load circuit.

If you must troubleshoot an electrical system for an industrial electrical system, you should make two checks. First you should see if the loads are energized and operating correctly. If the loads are not energized, you should suspect the control circuit is not energized, move on to the second test, and make checks in the control circuit. As you become more familiar with the control circuit and load circuit, you will find it is almost automatic to check the control circuit to ensure that the relay coil is energized so its contacts are energized. The next sections of this book present in-

Figure 9–14 Example of a solenoid valve in the closed position (de-energized) and the open position (energized).

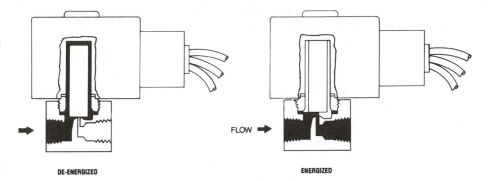

DE-ENERGIZED   FLOW ➡   ENERGIZED

depth material concerning the operation of typical loads, such as compressor motors and fan motors.

9.13
Solenoids Used in Industrial Systems

A *solenoid* is a *magnetically controlled valve*. You learned in earlier chapters about the power of current flowing through a coil of wire and causing a strong magnetic field to develop. The strong magnetic field in the solenoid valve is used to make the valve plunger move from the open to the closed position or from the closed to the open position. The valve is part of the plumbing system, such as an inlet valve for a water system. Solenoid valves allow the plumbing system to be interfaced with the electrical system, which means the plumbing system can be automated.

9.14
Basic Parts of a Solenoid Valve

Figure 9–14 shows a typical solenoid valve in the closed position and in the open position. From this figure you can see that the solenoid valve is a plumbing-type valve that has a movable plunger used to close off the valve. The valve either can be designed so that the plunger is held in the open position by spring pressure or it can be designed to be held in the closed position by spring pressure. When electric current is applied to the coil in the valve, the strong magnetic field causes the plunger to move upward. If the valve is the type where spring pressure is causing the plunger to stay in the closed position, the magnetic field will cause the plunger to open the valve passage. If the valve is the type where spring pressure is causing the plunger to stay in the open position, the magnetic field will cause the plunger to close the valve passage.

The coil for the solenoid is exactly like the coils that you learned about in the chapter for relays and contac-

tors. Because the coil is basically a long wire that is coiled on a plastic spool, it will have two ends. You can test the coil for continuity, and you should notice that each type of solenoid coil has some amount of resistance that will vary from approximately $50\ \Omega$ to $1,500\ \Omega$. Remember that the exact amount of resistance does not matter; because the coil is made from one long piece of wire, it is doubtful that the amount of resistance will change. The main problem a coil will have is an open somewhere along its wires. If the coil is open, its resistance will be infinite.

9.15
Troubleshooting a Solenoid Valve

At times you will be working on a system that has one or more solenoid valves and you will suspect the solenoid valve is not operating correctly. The first test you should make when troubleshooting a solenoid valve is to read its data plate and determine the amount of voltage the valve should have; then measure the amount of voltage that is available across the two wires that are connected to the coil. It is important to remember that the coil of any solenoid valve must have the correct amount and type of voltage applied to its two wires if it is to conduct sufficient current to cause the plunger to move. This means that if the valve is rated for 120 VAC, you should measure 120 VAC across the two coil wires. If the voltage is missing or is an incorrect amount, the coil will not operate properly. If the right amount of voltage is not present, you will need to use troubleshooting procedures to locate the open in the circuit that is causing the loss of voltage before it reaches the coil.

If the proper amount of voltage is present and the coil will not activate, you must make several additional tests. You can test to see if the coil has developed a magnetic field by placing a metal screwdriver near the coil to see if it is attracted by the magnetic field when voltage is applied. If the screwdriver is attracted to the coil by the magnetic field but the valve is not opening or closing, the valve seat may be frozen in place; in that case the valve will need to be replaced.

The other problem that the coil may have if the correct amount of voltage is present at the coil wires is that the coil may have an open in its wire. You can test the coil wire with a continuity test. Be sure to turn off all voltage to the solenoid and disconnect the coil wires so that the coil is isolated from back feeding to the remainder of the circuit. If the continuity test indicates infinite (∞) resistance at the highest resistance range, you can suspect the coil has an open and should be replaced. You should notice that the coil can be replaced without unsoldering the valve.

Questions for This Chapter

1. Identify the main parts of a relay and explain their operation.

2. Explain the operation of the four types of armature assemblies presented in this chapter.

3. Explain the operation of a current relay and a potential relay.

4. Discuss the differences between a contactor and a relay.

5. Identify the main parts of a solenoid and explain its operation.

True or False

1. Plug-in relays are used in industrial electrical systems because they are easy to change out when they become faulty.

2. The main difference between a relay and a contactor is the contactor usually has more contacts.

3. A solenoid is similar to a relay in that it has a coil that becomes magnetic and moves the plunger of a valve.

4. It is possible to change normally open contacts to normally closed contacts in the field for some relays.

5. For a relay to operate correctly, its contacts must close first to provide current to its coil.

Multiple Choice

1. The normally closed contacts of a relay _____ .
   a. pass current when the coil is not energized.
   b. pass current when the coil is energized.
   c. pass current at all times, whether the coil is energized or not energized.

2. A current relay is _____ .
   a. a special relay designed to control only current.
   b. a relay that is designed to be used as a relay or a contactor.
   c. a relay that is designed to start single-phase compressors.

3. A solenoid is _____ .
   a. a valve that is controlled (opened and closed) by gravity and springs.
   b. a reversing relay that is used to start heat pumps.
   c. a valve that is controlled (opened and closed) by springs and a magnetic field produced by a coil.

4. A contactor is _____ .
   a. a special type of solenoid used to start a pump motor.
   b. similar to a large relay in that it has contacts and a coil, but its contacts are rated for more than 15 A.
   c. a relay that has both normally open and normally closed contacts.

5. An SPDT relay has _____ .
   a. one set of contacts that connect to two output terminal points.
   b. two sets of contacts that connect to one terminal point.
   c. one set of contacts and one output terminal point.

Problems

1. Draw a wave form that shows pull-in and hold-in current for a relay coil.

2. Use the NEMA table provided in Fig. 9–12 to select a NEMA starter size that can control 45 continuous amps.

3. Draw an example of single-pole, double-throw (SPDT) contacts and explain their operation.

4. Draw an example of a double-pole, single-throw (DPST) contact and explain its operation.

5. Draw an example of a double-pole, double-throw (DPDT) contact.

◀ **Chapter 10** ▶

Motor Starters and Over-Current Controls

Objectives

After reading this chapter you will be able to:

1. Explain the operation of motor starter.
2. Identify the main parts of a motor starter.
3. Identify the main parts of an overload and explain their operation.
4. Explain the operation of single-element and dual-element fuses.
5. Identify single-pole, two-pole, and three-pole circuit breakers and explain their operation.

10.0
Overview of Motor Starters

Motor starters are magnetically controlled contacts that are usually used in industrial applications. The motor starter is a larger version of a relay or contactor, and it is used to control larger motors. The motor starter also has *over-current protection* for motors built into it, but relays and contactors do not. This over-current protection is called an *overload*, and it is sized to trip if the amount of current drawn by the motor exceeds the designated limit.

10.1
Why Motor Starters Are Used in Industrial Electrical Systems

In Chapter 9 you learned about relays and contactors. Relays and contactors are designed to close or open their contacts and provide current to small loads in the sys-

tem. If a contactor is used to control a motor, such as a hydraulic pump, conveyor motor, or pneumatic compressor motor, it is protected against over-current by fuses or internal overloads that are built into the winding.

In most industrial systems the motors are large and expensive. These motors must be protected by fuses or circuit breakers in the disconnect for short-circuit protection and by the overloads in motor starters to protect against slow over-currents. If an open motor has a problem such as the loss of lubrication or if the motor is overloaded, it will draw extra current, which will damage the motor if the overload is allowed to continue for any length of time. The overloads in the motor starter sense the excess current and trip the motor starter coil so that its contacts open and stop all current flow to the motor until a technician manually resets the overloads. A special contactor that has integral over-current protection built in is called a *motor starter*. This chapter explains the basic parts and operation of a motor starter.

10.2
The Basic Parts of a Motor Starter

Figure 10–1 shows a typical motor starter with all the parts identified. From this picture you can see that it is a three-pole starter used in three-phase circuits. The incoming voltage is connected on the top at the terminals identified as L1, L2, and L3, and the motor leads are connected to the bottom terminals identified as T1, T2, and T3. The three major sets of contacts are located in the top part of the motor starter, and the overload assembly is mounted in the lower part of the motor starter. The coil is located in the middle of the motor

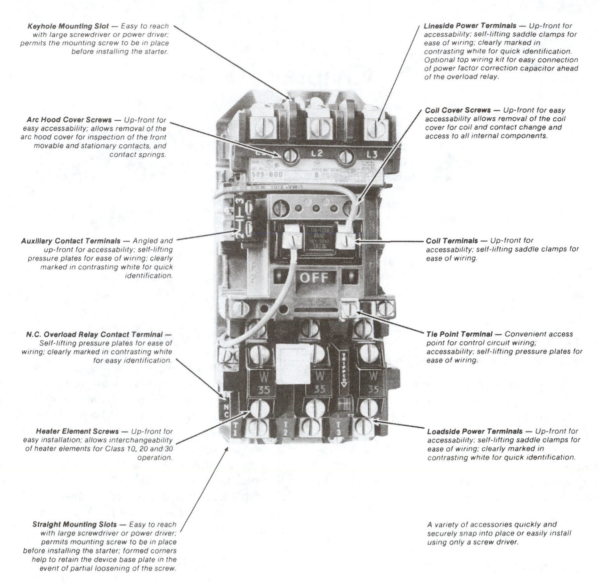

Keyhole Mounting Slot — Easy to reach with large screwdriver or power driver; permits the mounting screw to be in place before installing the starter.

Arc Hood Cover Screws — Up-front for easy accessibility; allows removal of the arc hood cover for inspection of the front movable and stationary contacts, and contact springs.

Auxiliary Contact Terminals — Angled and up-front for accessability; self-lifting pressure plates for ease of wiring; clearly marked in contrasting white for quick identification.

N.C. Overload Relay Contact Terminal — Self-lifting pressure plates for ease of wiring; clearly marked in contrasting white for easy identification.

Heater Element Screws — Up-front for easy installation; allows interchangeability of heater elements for Class 10, 20 and 30 operation.

Straight Mounting Slots — Easy to reach with large screwdriver or power driver; permits mounting screw to be in place before installing the starter; formed corners help to retain the device base plate in the event of partial loosening of the screw.

Lineside Power Terminals — Up-front for accessibility; self-lifting saddle clamps for ease of wiring; clearly marked in contrasting white for quick identification. Optional top wiring kit for easy connection of power factor correction capacitor ahead of the overload relay.

Coil Cover Screws — Up-front for easy accessibility allows removal of the coil cover for coil and contact change and access to all internal components.

Coil Terminals — Up-front for accessibility; self-lifting saddle clamps for ease of wiring.

Tie Point Terminal — Convenient access point for control circuit wiring; accessability; self-lifting pressure plates for ease of wiring.

Loadside Power Terminals — Up-front for accessability; self-lifting saddle clamps for ease of wiring; clearly marked in contrasting white for quick identification.

A variety of accessories quickly and securely snap into place or easily install using only a screw driver.

Figure 10–1 A typical motor starter with all its parts identified.

(Courtesy of Rockwell Automation's Allen-Bradley Business)

starter, and it has an indicator that shows the word *ON* when the coil is energized and *OFF* when the coil is deenergized.

Figure 10–2 shows a wiring diagram of a motor starter connected to a three-phase motor at the top of the diagram and a ladder diagram of just the control circuit, which consists of the start push button, the stop push button, the motor-starter coil, and the motor-starter overloads, in the bottom diagram. It is easy to see the parts of the motor starter in the wiring diagram because they are all in the shaded area. You can see that the coil circuit uses smaller wire than the contact circuit. The motor starter has three sets of contacts that have a heater in series with it. This ensures that all the current that flows to the motor must pass through a heater. If the

current flowing to the motor is normal, the heater does not provide sufficient heat to cause the overload to trip. If the motor draws excess current, the current flowing through the heater will cause it to create excess heat that will trip the normally closed overload contacts. Because the normally closed overload contacts are connected in series with the motor-starter coil, current to the motor-starter coil will be interrupted when the overload contacts open. A reset button on the motor starter must be manually reset to set the overload contacts back to their normally closed position.

The control circuit is shown as a ladder diagram at the bottom of this figure. In this diagram you can see that the motor-starter coil is energized when the start push button is depressed.

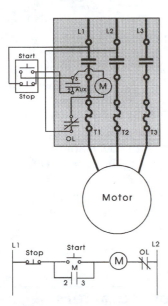

Figure 10–2 Electrical diagram of a motor starter and control circuit.

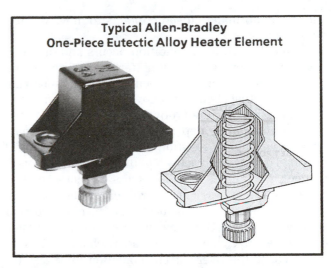

Figure 10–3 A typical heater assembly. Notice the ratchet that protrudes from the bottom of the heater.
(Courtesy of Rockwell Automation's Allen-Bradley Business)

The motor starter also has one or more additional sets of normally open contacts called *auxiliary contacts*. The auxiliary contacts are connected in parallel with the start push button. These contacts serve as a seal-in circuit after the motor-starter coil is energized. The start push button is a momentary-type switch, which means that it is spring-loaded in the normally open position. When the start push button is depressed, current flows from L1 through the normally closed stop push button contacts and through the start push button contacts to the motor starter coil. This current causes the coil to become magnetized so that it pulls in the three major sets of load contacts and the auxiliary set of contacts. When the auxiliary contacts close, they create a parallel path around the start push button contacts so that current still flows around the start push button contacts to the coil when the push button is released. Because the stop push button is connected in series in this circuit, the current to the coil is deenergized, and all the contacts drop out when the stop push button is depressed.

10.3
The Operation of the Overload

The overload for a motor starter consists of two parts. The heater is the element that is connected in series with the motor, and all the motor current passes through it. The heater is actually a heating element that converts electrical current to heat. The second part of the overload device is the trip mechanism and overload contacts. The trip mechanism is sensitive to heat, and if it detects excess heat from the heater, it will trip and

cause the normally closed overload contacts to open. Because the motor-starter coil is connected in series with the normally closed overload contacts, all current to the coil will be interrupted, and the coil will become deenergized when the overload contacts are tripped to their open position. When the coil becomes deenergized, the motor-starter contacts will return to their open position and all current to the motor will be interrupted. When the overload contacts open, they remain open until the overload is reset manually. This ensures that the overloaded motor stops running and cools down until someone comes to investigate the problem and resets the overloads.

Figure 10–3 shows a typical heating element for a motor-starter overload device, which is called the *heater*. In the cutaway diagram for this figure you can see that the heater is actually a heating element that converts electrical current into heat as it passes through the heating element. You should also notice the knob protruding from the bottom of the heater. This knob has a shaft that is held in position inside the heater and teeth machined into the part that protrudes from the heater. The teeth are called the *ratchet mechanism*.

Figure 10–4 shows the heating element mounted into the trip mechanism. The trip mechanism consists of the ratchet from the heater and a paw. The ratchet is actually the knob that protrudes from the bottom of the heater that has teeth in it. The paw has spring pressure, which tries to rotate the ratchet. Because the heater holds the shaft of the ratchet tight with solder, the paw cannot move. When the heater becomes overheated, it melts the solder that holds the ratchet in place and allows it to spin freely. When the ratchet spins, it allows the paw to move past it, which in turn allows the normally closed contacts to move to their open position.

After the overload condition occurs and the overload contacts open, the motor starter deenergizes and the motor stops running. When the motor stops running, the heating element is allowed to cool down. After the heating element cools down for several seconds, the reset button can be depressed, which moves the paw back to its original position, and the normally closed overload contacts move back to their closed position. If the motor continues to draw excess current when it is restarted, the excess current will cause the heater to trip the overload mechanism again. If the motor current is within specification, the heaters will not produce enough heat to cause the overload mechanism to trip.

Because the over-current condition must last for several minutes for the overload mechanism to trip, the overloads allow the motor to draw high locked-rotor amperage (LRA) during the few seconds the motor is trying to start without tripping. If the motor has an over-current condition while it is running, the overloads allow the condition to last several minutes to see if it clears up on its own before the motor is deenergized. If the problem continues, the overloads sense the over-current and trip to protect the motor. Figure 10–4 shows the overload contacts in their normally closed position and in their tripped, or open, condition.

10.4
Exploded View of a Motor Starter

It may be easier to see all the parts of the motor starter in an exploded view. Figure 10–5 shows an exploded view of the motor starter. At the far left in this picture

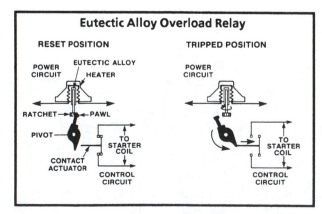

Figure 10–4 Diagram of the ratchet and paw mechanism that trips and allows the normally closed overload contacts to open.

(Courtesy of Rockwell Automation's Allen-Bradley Business)

you can see that the main contacts of the motor starter are much larger than those in a traditional relay. The coil is shown in the middle of the picture. The coil has two square holes in it, which allow the magnetic yoke to be mounted through it. The magnetic yoke and coil are mounted in the movable contact carrier. When the coil is energized, it pulls the magnetic yoke upward, which causes the movable contacts to move upward until they make contact with the stationary contacts that are shown at the far left. The overload mechanism is shown at the far right in this figure. It is mounted at the lower part of the motor starter, and all current that flows through the contacts must also flow through the overload mechanism.

10.5
Sizing Motor Starters

At times you will need to select the proper-size motor starter for an application. The size of motor starters is determined by the National Electrical Manufacturers Association (NEMA). The ratings refer to the amount of current the motor starter contact can safely handle. The sizes are shown in the table provided in Fig. 10–6, and you can see the smallest-size starter is a size 00, which is rated for 9 A, which is sufficient for a 2-hp three-phase motor connected to 480 V or for a 1-hp single-phase motor connected to 240 V. You can see that the size 1 motor starter is rated for 27 A, which is sufficient for a 10-hp, three-phase motor connected to 480 V or a 3-hp single-phase motor connected to 240 V. A size 00 motor starter is about 4 in. high, and a size 2 motor starter is approximately 8 in. high. You will typically use up to a size 3 or 4 motor starter to protect compressor motors for commercial air-conditioning and refrigeration systems. The size 4 starter will protect motors up to 100-hp. It is important to understand that the overload heaters for the motor starter can also be purchased for a specific current rating. This means that each motor starter can have a heater that is rated specifically to the amount of current drawn by the motor that is connected to it.

10.6
Solid-State Motor Protectors

Solid-state motor protectors provide a feature similar to motor starters in that they protect motors against problems. For example, when the motor is supplied with three-phase voltage, the solid-state motor protector provides protection against the loss of any phase of voltage, low voltage on any phase, high voltage on any

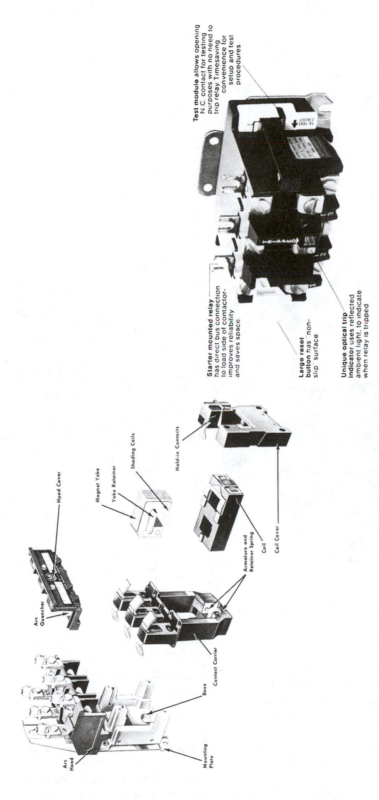

Figure 10-5 Exploded view of a motor starter. The stationary contacts are shown at the far left, and the movable contacts are shown to the right of the stationary contacts. The coil and magnetic yoke are shown in the middle of the picture. The overload mechanism is shown at the far right side of the picture.

(Courtesy of Rockwell Automation's Allen-Bradley Business)

Figure 10–6 The NEMA sizes for motor starters. Notice the smallest motor starter is a size 00, and the largest is a size 9.
(Courtesy of Rockwell Automation's Allen-Bradley Business)

Ratings for NEMA Full Voltage Starters					
		Maximum Horsepower Rating *Full Load Current Must Not Exceed* *"Continuous Ampere Rating"*			
		Motor Voltage			
				50 Hz	
NEMA Size	*Continuous Ampere Rating*	*200V*	*230V*	*380V-415V*	*460V-575V*
00	9	1-1/2	1-1/2	2	2
0	18	3	3	5	5
1	27	7-1/2	7-1/2	10	10
2	45	10	15	25	25
3	90	25	30	50	50
4	135	40	50	75	100
5	270	75	100	150	200
6	540	150	200	300	400
7	810	–	300	600	600
8	1215	–	450	900	900
9	2250	–	800	1600	1600

Figure 10–7 A solid-state motor protector. This device protects the motor against low voltage, loss of phase, and short cycling.
(Courtesy of SymCom, Inc.)

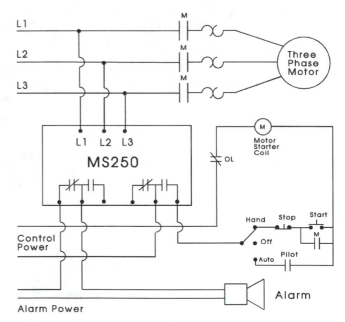

Figure 10–8 Electrical diagram of a motor saver that is connected to a coil of the motor starter or contactor.
(Courtesy of SymCom, Inc.)

phase, reversal of any phase, unbalanced voltage between any phase, and short cycling. In some systems, such as hydraulic pumps, air compressors, and refrigeration compressors, short cycling is a common problem that occurs when a motor is turned off and turned on again after a few seconds before the pressure in the system has a chance to equalize. This can occur if someone sets the controls incorrectly or if power is interrupted due to a power outage such as one that occurs during a lightening storm and then immediately comes back on. If the motor is allowed to try and start before its pres-

sure is equalized, it will draw large locked-rotor current and possibly not start.

Figure 10–7 shows a picture of a solid-state motor-protection device, and Fig. 10–8 shows a diagram of the location of a motor protection device in a compressor-starting circuit. In the diagram you can see that the motor saver has a set of normally open contacts that are

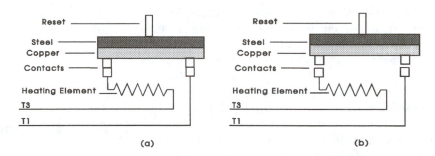

Figure 10–9 (a) Electrical diagram of thermal overload with contacts open. (b) Electrical diagram of thermal overload with contacts open.

connected in series with the motor-starter coil. The motor starter samples the voltage at L1, L2, and L3 from the three-phase voltage supplied to the compressor. If the supply voltage has a problem (low voltage, phase loss, etc.), the motor-saver circuit sensing the problem opens its contacts and causes current to the motor starter to be interrupted. A second circuit provides contacts for an audible alarm. When the motor saver trips, the alarm indicates the problem exists.

10.7
Over-Current Controls

Over-current controls include thermal overloads used in motor starters, thermal overloads used internally in motor windings, circuit breakers, and fuses. In this section you will learn about thermal overloads, solid-state overloads, circuit breakers, and fuses used to protect compressors and other types of motors from drawing too much current and overheating. You will find that the heat that damages motors can come from over-currents or from other conditions such as insufficient air flow that cause the motor to overheat when the current is within normal limits.

10.8
Thermal Overloads

Thermal overload is specifically designed to detect excessive heat that comes from over-current or from other physical conditions. A thermal overload is mounted directly on the dome of a compressor to sense the temperature in the compressor, and it is connected in series with the compressor run and start windings to detect excessive current. Fig. 10–9 contains a series of diagrams that shows how the thermal overload operates. In part (a) of this figure, you can see the physical shape of the overload and the location of all the parts inside the overload. Part (b) of this figure shows the contacts when the overload senses too much current and the overload contacts are open.

Figure 10–10 Plug-type and cartridge-type fuses.
(Courtesy of Cooper Industries, Bussman Division)

All the current flowing through the compressor motor windings flows in terminal C1 at the bottom of the diagram, through the heating element, and through the contacts and out terminal C2. Because all the current for the motor must flow through the heating element, the overload must be sized for the exact amount of current the motor will draw safely. When the motor draws current in excess of this limit, the heater produces extra heat, which causes the bimetal contacts to warp to the open position. When the overload is allowed to cool, the bimetal will snap the contacts to the closed position.

10.9
Fuses

Fuses perform a similar function as an overload, except the fuse uses an element that is destroyed when the over-current occurs. The fuse provides a thermal sensing element that is capable of carrying current. When the amount of current becomes excessive, the heat that is generated is sensed by the fuse element, which melts when the temperature is high enough.

Fuses are available in a variety of sizes and shapes for different applications. Figure 10–10 shows a plug-type fuse and a cartridge-type fuse. Each fuse is sized for the amount of current it will limit. When the amount of current is exceeded, the fuse link melts and opens the fuse. Figure 10–11 shows examples of a single-element fuse and a dual-element fuse. The single-element fuse provides protection at one level. This type of fuse is generally used for noninductive loads, such as heating elements or lighting applications. The dual-element fuse provides two levels of protection. The first level is called *slow over-current protection*, and it consists of a fusible link that is soldered to a contact point and attached to a spring. When a motor is started, it draws locked-rotor amperage (LRA) for several seconds. This excess current causes heat to build in the fuse, and this heat is absorbed in the slow over-current link. If the motor starts and the current drops to the FLA level, the link will cool off and the fuse will not open. If the LRA current continues for 30 s, the amount of heat generated will cause the solder that holds the slow over-current link to melt. When the solder melts, the spring pulls the link open and interrupts all current flowing through the fuse. The slow over-current link allows the fuse to sustain over-current conditions for short periods of time, and if the condition clears, the fuse will not open. If the condition continues to exist, the fuse will open.

The second type of element is called the short-circuit element. This element opens immediately when the amount of current exceeds the level of current the link is designed to handle. In the dual-element fuse, the short-circuit link is sized to be approximately five times the rating of the fuse. A short circuit is by definition any current that exceeds the full-load current rating by five times. (Some manufacturers use a factor of 10 times the rating.)

The single-element fuse has only a short-circuit element in it. These types of fuses are generally not used in circuits to start motors because the motor draws LRA. If a single-element fuse is used to protect a motor circuit, it must be sized large enough to allow the motor to

start, and then it is generally too large to protect the motor against an over-current condition of 20%, which will eventually damage the motor if it is allowed to occur for several hours.

10.10
Fused Disconnect Panels

The cartridge-type and screw-base-type fuses are generally mounted in a panel called a *fused disconnect*. The fused disconnect is normally mounted near the equipment it is protecting and it serves two purposes: First, it provides a location to mount the fuses; second, it provides a means of disconnecting the electrical supply voltage to a circuit. A three-phase disconnect is shown in Fig. 10–12. The fuse disconnect has a switch handle

Figure 10–12   A three-phase fused disconnect.
(Courtesy of Cutler-Hammer, Inc.)

Figure 10–11   (a) Single-element fuses in the process of blowing.
(b) Dual-element fuses in the process of blowing.
(Courtesy of Cooper Industries, Bussman Division)

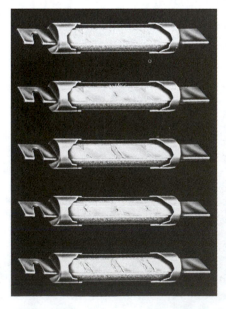
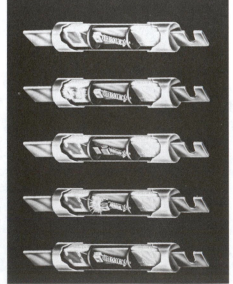

that is used to disconnect the main power from the circuit and the fuses so that you can safely remove and replace or test the fuses.

> **SAFETY NOTICE!**
> It is recommended that you always use plastic fuse pullers to remove and replace fuses from a disconnect to protect you from electrical shock even when the fuse-disconnect switch is in the off position. It is important to remember that even though the switch handle is open, line voltage is still present at the top terminal lugs in the disconnect. You should never use metal pliers or screwdrivers to remove fuses.

10.11
Circuit Breakers and Load Centers

The *circuit breaker* is similar to the thermal overload in that it is an electromechanical device that senses both over-current and excess heat. Some circuit breakers also sense magnetic forces. Circuit breakers are mounted in electrical panels called *load centers*. The load center can be designed for three-phase circuits or for single-phase circuits. The single-phase panel provides 240 VAC and 120 VAC circuits. Figure 10–13 shows a typical load center without any circuit breakers mounted in it.

The circuit breakers are manufactured in three basic configurations for single-phase, 120-VAC applications that require one supply wire; 240-VAC, single-phase applications that require two supply wires; and three-phase applications that require three wires. The three-phase circuit breakers can be mounted only in a load center that is specifically manufactured for three phase circuits. The two-pole and single-pole breakers can be mounted in a single-phase or three-phase load center. Figure 10–14 shows examples of circuit breakers.

The operation of the circuit breaker is similar to the thermal overload in that it senses excessive current that will trip its circuit after a specific amount of time. The main difference between the circuit breaker and the thermal overload is that the circuit breaker is mounted in the load center to protect both the circuit wires as well as the load. The circuit breakers are sized to total current rating for the wire and all the loads that are connected to the wire. In some cases this means that the circuit breaker is sized for the current flowing to several motors that are connected on the circuit. For instance, if a system has a hydraulic pump that draws 20 A and two conveyor motors that each draw 5 A, the circuit breaker would be rated for 30 A. A problem can occur if the hydraulic pump is turned off and the conveyors are running. If one of the conveyors has an overload that doubles its current to 10 A, the circuit breaker will not be able to detect 10 A in the conveyor motor as an overload because the total circuit current is below the 30-A total. In these circuits it may be necessary to use a circuit breaker to protect the entire circuit and overloads at each motor to protect them individually from over heating. This means that the main job of the circuit breaker is to protect the circuit against short-circuit conditions and protect the entire circuit against over-current conditions rather than protecting individual motors against over-current. This is why many circuits have a combination of circuit breakers and overload devices.

Figure 10–13 A load center that is used to mount circuit breakers. The load center is sometimes called a circuit breaker panel.

(Courtesy of Cutler-Hammer, Inc.)

Figure 10–14 Single-pole, two-pole, and three-pole circuit breakers for load centers.

(Courtesy of Cutler-Hammer, Inc.)

Questions for This Chapter

1. Explain the differences and similarities between a motor starter and a contactor.

2. An overload on a motor starter has contacts and a heater. Describe the function of each.

3. Identify the parts of thermal overload used to protect a compressor and explain each part's operation.

4. Discuss the differences between a single-element fuse and a dual-element fuse.

5. Explain the difference between slow over-current and short-circuit current.

True or False

1. The heater is the part of the motor-starter overload that opens and interrupts current flow.

2. The overload contacts on a motor starter are normally open and interrupt current when an overload condition is detected.

3. The main job of the circuit breaker is to protect the circuit against short circuits and the entire circuit against over-current rather than to protect individual motors against over-current.

4. A fused disconnect provides a location for mounting fuses for the system and also provides a means of disconnecting main power from the circuit.

5. A thermal overload device may be mounted internally in a motor such as an air-conditioning system's sealed compressor to detect heat inside the compressor shell.

Multiple Choice

1. A motor starter _____ .
   a. is just a larger version of a relay and it has only a coil and contacts.
   b. has a coil and contacts like a relay or contactor and also has overloads to provide over-current protection.

   c. is more like a circuit breaker than a relay in that it can detect excessive heat buildup on the outside of the motor through its heaters.

2. A dual-element fuse _____ .
   a. can detect slow over-current and large short-circuit currents.
   b. can be used two times before it must be replaced.
   c. has two elements so it can be used in single-phase circuits to protect both L1 and L2 supply-voltage lines.

3. The overload device on a motor starter _____ .
   a. has a heater to detect excess current and overload contacts to interrupt the control circuit current.
   b. has a heater to detect excessive current and overload contacts that directly interrupt the large load current to the motor.
   c. has a heater that interrupts control-circuit current and overload contacts that directly interrupt the large load current to the motor.

4. The motor starter has auxiliary contacts that _____ .
   a. are connected in parallel with a start switch to seal in the circuit when it is operating correctly and ensure the coil remains deenergized if the overload contacts trip.
   b. are connected in series with a start switch to seal in the circuit when it is operating correctly and ensure the coil remains deenergized if the overload contacts trip.
   c. can switch current to an auxiliary motor such as a conveyor motor at the same time the main contacts supply current to a hydraulic pump motor.

5. Heaters for motor starters are available in different sizes so that they can _____ .
   a. provide the correct amount of current protection to match the motor connected to the motor starter.
   b. match the voltage rating of the fuses used in the circuit.
   c. match the temperature rating of the ambient air in which the motor will operate.

Problems

1. Draw a sketch of a motor starter that has its contacts connected to a three-phase motor and its coil controlled by start and stop push buttons. Be sure to show the heaters and the overload contacts in the proper circuits.

2. Draw a sketch of a thermal-type overload and explain the function of each of the parts.

3. Draw a sketch of a single-element fuse and a dual-element fuse and explain how they are different.

4. Draw a sketch of thermal overload device connected to the outside of a compressor and explain how it protects the compressor against overheating.

5. Use the table shown in Fig. 10–6 to select a motor starter for a single-phase, 3-hp motor that draws 27 A at 240 V.

Chapter 11

Single-Phase AC Motors

Objectives

After reading this chapter you will be able to:

1. Explain the theory of operation of a simple induction AC motor.
2. Identify the main parts of an AC motor and explain their function.
3. Identify the parts of the centrifugal switch and explain their operation and function.
4. Explain the data found on a typical motor data plate.

11.0
Overview of Single-Phase AC Induction Motors

The main loads on which you will work in most systems are motors of all types. You will work with both single-phase and three-phase motors while you are on the job. Because these motors are slightly different, you will learn about single-phase motors in this chapter and three-phase motors in the next chapter. Single-phase motors are somewhat smaller than three-phase motors; typically they are less than 2 hp. You typically find single-phase motors in shop power tools such as bench grinders, drill presses, small fans, and smaller pumps and conveyor motors. The motors used in these applications are called *open-type motors* because they have a shaft on one or both ends. Figure 11–1 shows a picture of a single-phase motor. The theory of operation, installation, and troubleshooting for open-type motors that are used to power fans, pumps, and belt-driven loads is

Figure 11–1 Typical AC single-phase motor.
(Courtesy of GE Industrial Systems, Fort Wayne, Indiana)

also explained in this chapter. The material in these sections shows that some motors require special switches and components to start and operate correctly. These devices are explained so that when you must troubleshoot a motor you will fully understand their function. A diagram and picture are provided for each type of motor so that you will learn to identify each type of motor from its physical appearance and from its electrical diagram. As a technician you will be expected to identify, install, troubleshoot, and repair or replace any motors you encounter. This chapter provides a comprehensive overview for all single-phase motors found in electrical systems.

11.1
AC Split-Phase Motor Theory

The AC split-phase motor is widely used in a variety of applications, and it has many of the basic parts of the other types of open motors. This motor has three basic parts, the stationary coils called the *stator*, the rotating shaft, called the *rotor*, and the *end plates*, which house the bearings that allow the rotor shaft to turn easily. Figure 11–2 shows an exploded view of a split-phase motor so that you can see where these parts are located inside the motor. From this diagram you can see that the rotor is supported by the bearings in the end plates of the motor. The rotor is also mounted directly inside the stationary windings so that when voltage is applied to them, their magnetic field can be induced into the rotor.

If the AC split-phase motor requires a *centrifugal switch* to help it get started, it will be mounted inside the end plate. The actuator for the centrifugal switch is

mounted on the end of the rotor's shaft, and it has flyweights that swing out when the motor reaches 75% to 85% full rpm to open the centrifugal switch. The next sections of this chapter show how these parts operate with each other to provide a rotating force to the motor's shaft to turn a fan, conveyor, or pump.

11.2
The Rotor

The rotating part of the motor is called the *rotor*. When voltage is applied to the coils of wire in the stator, they produce a very strong magnetic field. This magnetic field is passed to the rotor by induction, in much the same way as voltage is passed from the primary winding to the secondary winding of a transformer. After the rotor becomes magnetized, it begins to rotate. The speed of the rotor is determined by the number of poles and the frequency of the AC voltage that is applied to the motor. The rotor has a shaft that rotates to do the work of the motor, such as to turn a fan blade or to move the pulley for a conveyor. Figure 11–3 shows a diagram and a picture of a rotor that is used in most single-phase AC motors. As a rule, the rotor of an AC motor does not have any wire in it (only repulsion start or some synchronous AC motors have wire in their rotors, but these motors are not used too often in applications). The rotor is made from pressing laminated steel plates on the frame. The frame of the rotor looks like the wireframe exercise wheel that is used by hamsters. For this reason it has been named a *squirrel cage*, and when it is used in the rotor it is called a squirrel-cage rotor. The squirrel cage is actually the frame for the rotor, and the parts that look like the squirrel cage are actually rotor

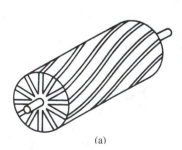

Figure 11–2 Exploded view of a capacitor-start, induction-run motor.

(Courtesy of Leeson Electric Corporation)

(a)

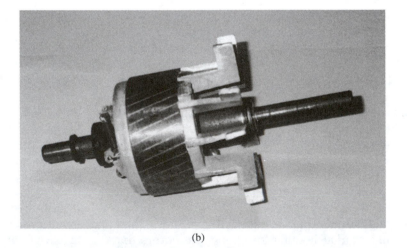

(b)

Figure 11–3 (a) Diagram of a squirrel-cage rotor for an AC motor. The rotor bars will become multiple-bar magnets when the rotor is magnetized. (b) Picture of a squirrel-cage rotor for an AC motor.

bars, which act exactly like multiple-bar magnets. The laminated steel plates that are pressed onto the frame of the squirrel-cage rotor allow the rotor bars to be magnetized by induction when current is passed through the stator. This means that the permanent field in the stator acts like the primary of a transformer, and the rotor acts like the secondary windings of a transformer. The result of this is that the induction motor does not need brushes to pass current to the rotor to get its magnetic field to build. This is also why these types of motors are called *induction motors*. It is also important to understand that the laminated steel sections in the rotor allow it to become magnetized very easily by the AC voltage and then quickly change the polarity of its magnetic field as the AC voltage oscillates from positive to negative as the AC sine wave changes. The polarity of the magnetic field in the rotor of the AC motor changes because the sine wave of the AC voltage changes from positive to negative naturally during each cycle of the sine wave. This causes the magnetic fields in the rotor to continually change polarity and spin as long as AC voltage is applied to the motor.

11.3
Locked-Rotor Amperage and Full-Load Amperage

When voltage is first applied to the motor, the run and start windings will draw a large amount of current, called *locked-rotor amperage* (LRA). This current is called locked-rotor amperage, because the rotor has not started to turn at this time. The amount of current is large because the amount of resistance in the windings is very small and because the rotor has not yet started to rotate and create *counterelectromotive force* (CEMF) voltage. When the rotor starts to spin and comes up to full rpm, the current will return to normal, and it is called *full-load amperage* (FLA). Figure 11–4 shows a

graph of the large LRA current that occurs when the motor is first started, and it also shows the current returning to lower levels (FLA) when the motor is running at full-load speed.

The reason the current returns to its normal value after the rotor is spinning at full speed is because the rotor produces CEMF. When the rotor starts to spin, its magnetic field begins to pass the coils of wire in the run winding and a voltage is generated when the magnetic field passes the coils. This generated voltage is out of phase with the applied voltage, so it is called counter-EMF (CEMF). Counter-EMF is also sometimes called *back EMF*.

The amount of CEMF is equivalent to the speed at which the rotor is turning. If the rotor is turning at 90% of full speed, the CEMF will be approximately 90% of the applied EMF. For example, if the applied voltage is 230 VAC and the rotor is turning at approximately 90% rpm, the CEMF may be 200 VAC. If the impedance of the run winding is 10 Ω, the full-load current is determined by dividing the difference in voltage (230 − 200), which is 30 VAC, by the impedance. This means the FLA current is approximately 3 A.

11.4
Rotor Slip and Torque in an AC Induction Motor

The amount of LRA and FLA that an AC induction motor draws also determines the amount of torque the motor's shaft will have. *Torque* is defined as the amount of *rotating force* that the rotor shaft has. You should remember that this force is needed to pump the piston of the compressor and turn the fan blades for a condenser fan. If the motor has a large amount of LRA during starting, it will also have a lot of starting torque. If the motor draws a large FLA when the motor is running, it will have a lot of running torque.

Figure 11–4 Graph of the locked-rotor amperage (LRA) that occurs when a motor is first started and the full-load amperage (FLA) that occurs when the motor is running at full speed.

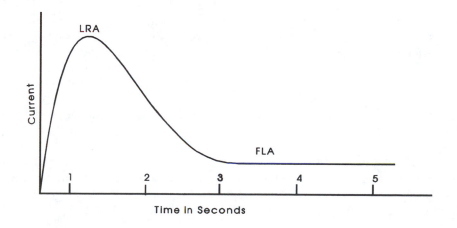

The amount of FLA is determined by the amount of slip a motor has. *Slip* is defined as the *difference* in the *rated speed of a motor* and its *actual running speed.* For example, a four-pole motor is rated for 1,800 rpm, but its actual running speed is approximately 1,725 rpm. The difference of 75 rpm is called slip. At first glance you would think that slip is not good for a motor because the motor is not running as fast as it should be, but you will see that the induction motor relies on slip to create the amount of current necessary to turn the load at the end of the motor shaft. If the amount of slip is minimum and the rotor is spinning at close to the rated speed, the amount of CEMF is large, and the difference between the CEMF and the applied EMF may be as small as 5 VAC. When this occurs, the amount of current in the motor is reduced to less than 0.5 A, and consequently the amount of torque the rotor has is also reduced. It is also important to remember that the rotor gets the current required to build its magnetic fields through induction from the stator windings. The induction can occur only if the rotor is not turning at the same speed as the magnetic field is rotating in the stator.

When the rotor's torque is minimal, the shaft will not be able to turn its load, and it will begin to slow down, which decreases the amount of CEMF. This makes the difference between the CEMF and applied voltage larger, which makes the motor draw more current and produce more torque. All induction motors rely on slip to operate correctly and produce sufficient torque. If it is important that the motor run at its rated speed, a special type of motor called a *synchronous motor* could be used, because it runs at its rated speed with no slip. The synchronous motor is not used very often in industrial applications because it requires DC voltage to be supplied to its field.

It is also possible to use a variable-frequency drive to adjust the frequency of the voltage above 60 Hz to get the motor to run at its rated speed even if it has slip. Variable-frequency drives were introduced into industrial applications in the 1980s. You will learn more about variable-frequency drives in the chapter that covers electronic devices. As a technician you will encounter a variety of variable-frequency drives in newer systems, where they are used to control the speed of conveyors, pumps, and fan motors.

11.5
End Plates

The end plates play a very important part in the operation of the motor. They house the bearings that support the rotor to allow it to spin with a minimum amount of friction, and one end plate houses the end switch that is

Figure 11–5   End plates are shown with a rotor. The bearings are mounted in the end plates, and they support the rotor shaft so that it can spin freely.

used to help the single-phase motor to get started. The bearings in a single-phase motor may be the ball-bearing type or a bushing type. The ball-bearing type is more expensive and is generally used in larger single-phase motors. The bushing is used in smaller and less expensive single-phase motors. Figure 11–5 shows end plates, and Fig. 11–6 shows an example of a ball bearing and a bushing. The ball bearing shown in Fig. 11–6a uses a number of small steel balls, which create a rolling surface between the inner hub and outer hub. The shaft of the motor is pressed securely into the inner hub of the bearing so that the shaft and inner hub of the bearing move together as though they were welded. As the rotor shaft spins, the inner hub rolls on the balls, and the balls make contact with the outer hub. This means that only a small portion of each ball is touching the inner and outer hub at any time, which keeps friction to a minimum. The ball bearing for smaller motors is lubricated with grease when it is manufactured, and it usually is sealed for the life of the bearing, which means that motors with these types of ball bearings will not need to be lubricated. Some larger motors use bearings that must be greased periodically. A grease fitting is provided so that the ball bearing can be greased once or twice a year, depending on the number of hours the motor runs.

The bushing shown in Fig. 11–6b is made of a brass or bronze alloy sleeve with a lubrication hole in it. An *oiler pad*, which is also called an *oil wick*, is mounted on the outer side of the bushing, so oil from the pad can find its way through the hole to the inside of the bushing. The oil creates a "bearing" surface between the shaft of the rotor, which spins while the bushing remains stationary. As long as a sufficient amount of oil remains on the inside of the bushing, between it and the rotor shaft, the amount of friction is held to a minimum. If the bushing ever runs dry, the rotor shaft will be allowed to touch the bushing surface and both surfaces will begin to wear severely. It is vitally important to en-

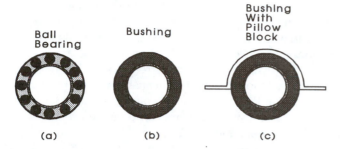

Figure 11–6   (a) An oil-type bearing (bushing) with insulator; (b) sealed-type bronze bushing and sealed ball bearing; (c) pillow block–type bearing with bronze bushing.

sure that the lubrication ports for the bushing type motor are always located on the top side of the motor so that gravity can draw the lubrication through the lubrication pad into the bushing. It is also important to remember that the motor that uses a bushing must be lubricated at least once a year.

11.6
The Stator

The stator is actually made of two separate windings, called the *run winding* and the *start winding*. Each of these windings is subdivided into poles, which represent the poles of a magnet. We use a four-pole motor for our examples in this chapter. This means that the run winding has four poles and the start winding has four poles. Figure 11–7 shows a picture stator so that you can more easily see the coils of wire that are mounted in it. From this picture you can see the four poles of the run winding and the four poles of the start winding. The windings are pressed over pole pieces that also become magnetized when the coils have current flowing in them. The pole pieces are made from laminated steel like the rotor, so the poles can quickly change polarity when AC voltage is applied.

The AC sine wave starts at 0 V and peaks positive, returns to 0 V and peaks negative, and returns to 0 V again. It continues to oscillate from positive to negative and back as long as voltage is applied. This action causes the magnetic field in each winding in the stator to be positive and then negative and continually change. The changing magnetic field actually results in the magnetic field rotating around the poles inside the stator. Because the rotor also has a magnetic field it will begin to spin and follow the rotating magnetic field in the stator. It will be easier to understand the function of the stator if you study the operation of the start winding and run winding separately.

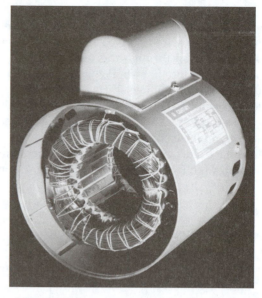

Figure 11–7   Close-up picture of start and run windings in the stator. The run winding has larger wire than the start winding.

(Courtesy of Magnetek)

11.7
The Start Winding

If the AC induction motor only had one winding, the rotor would not tend to rotate when single-phase AC voltage was first applied to the stator windings in the AC induction motor. The rotor needs a small amount of phase shift between the applied voltage and current in the magnetic field created in the stator to begin to spin. There are several ways to provide the phase shift required to start the rotor spinning. One way to provide the phase shift is to use two windings in the stator and physically offset them. A second way to cause a phase shift in the stator is to put more wire in one set of windings than in the other. For this reason, the stator has a start and a run winding, and the start winding is made from smaller wire than the run winding so it can have more turns of wire and thus have more wire.

Figure 11–7 shows a close-up picture of a stator that clearly shows the run and start winding. From this picture you can see that the run winding is made of larger wire and is physically offset from the start winding. Figure 11–8 shows a sketch of how the run winding and start winding are placed in the stator offset from each other. Figure 11–8 also shows an electrical diagram of the start winding and run winding for an eight-lead single-phase AC motor and for a four-lead single-phase motor.

The electrical diagram in Fig. 11–8a shows that the start winding is made of smaller wire and has more turns than the run winding. The start-winding coil has more turns of wire than the run-winding coil, so more feet of wire are used in the start winding. Because the start winding has more wire, it takes the current longer to pass through it than through the run winding. It is important to understand that when voltage is applied to the AC induction motor, voltage starts into the run winding and the start winding at the same time. Because the start winding has more wire and is offset physically from the run winding, a phase shift is created.

This phase shift gives the rotor a sufficient amount of starting torque (rotational force) to get the rotor to start turning and to keep it turning until the motor comes up to full speed. The phase shift occurs between the run winding and the start winding, but it also can be described as the phase shift that occurs between the voltage waveform and the current waveform because inductors (motor windings) cause the current waveform to lag the waveform of voltage. In later sections you will see how capacitors can be added to the start windings to enhance the amount of phase shift so the motor will have more starting torque. Figure 11–9 shows the phase shift between the current in the start winding and that in the run winding.

Because the start winding is made from very small wire so that its coils can have many more turns of wire than the run winding, the start winding will draw a large amount of current when voltage is applied to it. If the start winding were allowed to remain in the circuit continually, this large current would damage it and eventually burn a hole in the windings. For this reason a centrifugal switch is used to disconnect the start winding after the motor's rotor is spinning. The centrifugal switch can determine when the speed of the rotor has reached 75% to 85% rpm and disconnect the start winding from the circuit until it is needed the next time the motor is started.

11.8
The Operation of the Centrifugal Switch

The centrifugal switch is also called the *end switch* in a motor because it is mounted in the end plate. Even though the centrifugal switch is physically mounted in the end plates, it is important to understand that it is

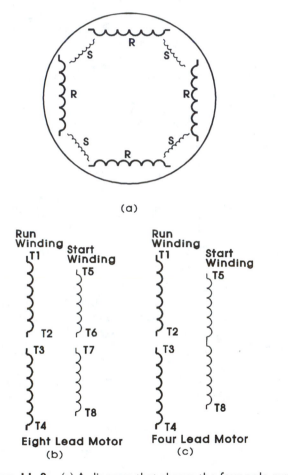

(a)

Run Winding T1   Start Winding T5   Run Winding T1   Start Winding T5

T2   T6   T2

T3   T7   T3

T8   T8

T4   T4

Eight Lead Motor
(b)

Four Lead Motor
(c)

Figure 11–8   (a) A diagram that shows the four-pole run winding and four-pole start winding in the stator of an AC induction motor. Notice that the start winding is physically offset in the stator from the run winding to help provide more of a phase shift and that the run winding is made of larger-gauge wire and has fewer turns than the start winding. (b) A diagram of the run winding and start winding of an eight-lead induction motor. Notice that the two sections of the run windings are identified as T1–T2 and T3–T4, and the two sections of the start winding are identified as T5–T6 and T7–T8. (c) A diagram that shows a four-lead induction motor. The run winding is identified as T1–T4, and the start winding is identified as T5–T8.

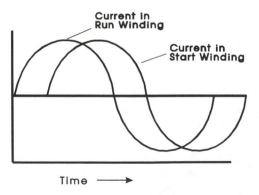

Figure 11–9   A diagram of the phase shift that occurs when voltage is applied to the start and run windings of an AC motor. Because the start winding has more wire than the run winding, a phase shift is created between the magnetic fields in the two windings.

electrically connected to the start winding that is located in the stator. The centrifugal switch is actually part of a larger assembly consisting of an activation mechanism that has flyweights that begin to swing out when the rotor reaches 75% full rpm. The activation mechanism is mounted on the end of the rotor shaft so

Figure 11–10 (a) Centrifugal switch removed from an end plate and the activation mechanism removed from the armature shaft. (b) The centrifugal switch is mounted in the end plate and the end plate is mounted on the end of the motor shaft so that the activation mechanism is touching the centrifugal switch.

it can turn when the rotor shaft is turning. The activation mechanism is mounted on the shaft in such a way that it makes contact with the centrifugal switch and holds the switch contacts in the closed position when the shaft is not rotating.

Figure 11–10 shows a centrifugal switch and the activation mechanism. Figure 11–10a shows a centrifugal switch that is removed from an end plate. Figure 11–10b shows a centrifugal switch mounted in the end plate, with its flyweights at rest as you would find them when the rotor is not turning.

The centrifugal switch is a set of electrical contacts mounted on spring steel. Spring steel is used because it can be flexed so that the spring tension in the steel keeps the contacts in the open position. The activation mechanism that is mounted on the rotor shaft comes into contact with the spring steel of the switch when the end plate is attached to the stator. You should remember from the cutaway diagram in Fig. 11–11b that the rotor shaft goes through the bearing in the end plate, and when the end plate is drawn tight against the stator with screws, the activation mechanism touches the spring steel and causes the switch contacts to press against each other to complete the electrical circuit through the switch.

When voltage is applied to the windings in the motor, current flows through the run winding and through the start winding because the activation mechanism is holding the contacts of the centrifugal switch closed. As the rotor begins to spin, centrifugal force begins to pull the flyweights outward. Because the fly weights are held together with a spring on each side, they will not swing out all the way until the rotor is spinning approximately 75% to 85% of full rpm. The spring tension that holds the flyweights inward will

Figure 11–11 (a) Flyweights and activation device for centrifugal switch. The activation device is mounted on the end of the rotor shaft so that it can come into contact with the centrifugal switch. A spring holds the flyweights against the shaft when the rotor is not turning. (b) When the rotor is spinning at 75% to 85%, the flyweights move outward from the shaft because of centrifugal force. The action of the flyweights causes the activation device to drop away from the switch contacts so that they can move to their open position.

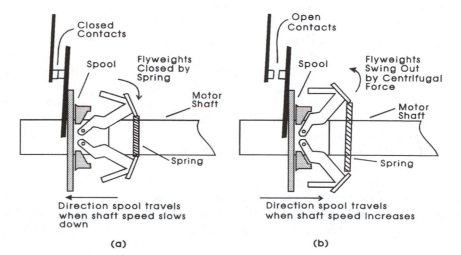

determine the speed at which the flyweights will swing out completely.

When the flyweights swing out completely, they allow the spring tension on the activation mechanism to move the mechanism to a retracted position, which allows the spring tension in the contacts of the centrifugal switch to move the contacts to the open position. When the contacts of the centrifugal switch open, current to the start winding is interrupted, and the motor continues to run with current flowing only through the run winding of the motor.

Figure 11–11a shows a diagram of the activation mechanism when the rotor is not turning. In this part of the diagram you can see that the flyweights are held in position by spring tension. When the flyweights are in this position, the activation mechanism is pressed against the switch contacts, holding them in the closed position so current can flow through them. In Fig. 11–11b the rotor is turning at nearly full speed. The centrifugal force that is created when the rotor shaft is spinning at full speed causes the flyweights to swing out away from the rotor shaft. When the flyweights swing out completely, the activation mechanism is pulled away from the contacts, which allow the tension of the spring steel to open the contacts.

11.9
The Run Winding

The run winding is the part of the stator that remains in the motor circuit at all times. Figure 11–12 shows an electrical diagram for a single-phase motor and you can see that the run winding is easily identified because it uses larger wire and fewer coils in its windings. In most single-phase motors, the run winding is numbered, as shown in the diagram. If the motor is a dual-voltage motor, its run winding is made in two sections, one with terminal ends T1 and T2 and the other with terminal ends T3 and T4. If the motor is wired for high voltage (220 or 208 VAC), the windings will be wired in series with each other.

Figure 11–12b shows the two sections of the run winding connected for low voltage (115 VAC). If the motor is wired for low voltage, the two sections of the run windings are connected in parallel with each other. These two diagrams are used in all electrical system diagrams for installing and troubleshooting motors in industrial applications.

You can test the run winding and start winding with an ohmmeter to determine which one is which. For example, in the diagram each part of the run winding has 4 Ω of resistance and the start winding has 12 Ω.

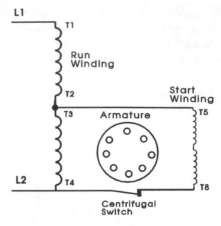

Split-phase open motor
wired for 230 volts
(a)

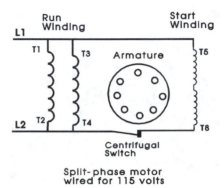

Split-phase motor
wired for 115 volts
(b)

Figure 11–12 (a) Electrical diagram of an AC induction motor wired for high voltage (230 or 208 VAC). The two sections of the run winding are numbered T1 and T2, and T3 and T4 are connected in series. (b) Electrical diagram for an AC induction motor wired for low voltage (115 VAC). The run winding in this diagram is shown in two sections that are connected in parallel.

11.10
Multiple-Speed Motors

The coils of wire in the run winding can be separated into sections called *poles*. The motor has an even number of poles. The speed of an induction motor is determined by the number of poles and the frequency. The speed can be calculated from the following formula:

$$\text{rpm} = \frac{\text{frequency} \times 120}{\text{poles}}$$

A typical motor has two, four, six, or eight poles. The speed of each type of motor is listed in the following table.

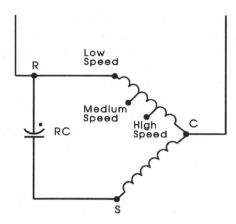

Low Speed = Red Wire
Medium Speed = Blue Wire
High Speed = Black Wire

Figure 11–13 The electrical diagram of a multitapped fan motor. The run winding is tapped so that it can provide three different choices for the number of poles that are used. The black wire is used for the high speed, the blue wire is used for the medium speed, and the red wire is used for the low speed.

No. of poles	rpm
2	3,600
4	1,800
6	1,200
8	900

Motors for typical industrial applications need to be able to operate at a variety of speeds. This means that the motor may be manufactured with a set number of windings to provide a set speed, or the motor may be reconnected in the field to give the motor a different speed.

Another way to change the speed of a motor is to tap the run winding so that more or less inductive reactance is used. This means that if you have a very long run winding, you can place taps on it just like the multiple-tapped transformer. Each tap uses less of the total run winding, and consequently the change in inductive reactance allows the speed of the motor to change. The amount of change in rpm for each tap is approximately 15% to 20%. Figure 11–13 shows an example of a motor with a multitapped run winding. From the diagram you can see the taps are in the run winding, and you can see the color code for the wires. If you want the motor to run at high speed, you use the black wire for the run winding, and if you want the motor to run at medium speed, you use the blue wire; the red wire allows the motor to run at low speed. Generally the multiple speeds are used for fan motors, where the high speed is

used for cooling and the low speed is used for heating. The medium speed is used where a slightly lower speed for air conditioning or a higher speed than the lowest speed for heating is needed. It is important to remember that as the tap is changed for this motor, it will use less of the winding to run faster, but it will also lose power. In some motors the common terminal is connected to the black, red, or blue wire instead of the run terminal.

11.11
Motor Data Plates

The motor's *data plate* lists all the pertinent data concerning the motor's operational characteristics. It is sometimes called the *name plate*. Figure 11–14 shows an example of a data plate for a typical AC motor. The data plate contains information about the ID (identification number), FR (frame and motor design), motor type, phase, horsepower rating, rpm, volts, amps, frequency, service factor (SF), duty cycle (time), insulation class, ambient temperature rise, NEMA design, and code. Each of these features is discussed in detail in the next sections.

11.11.1
Identification Number (ID)

The identification number (ID) for a motor is basically a model and serial number. The model number includes information about the type of motor, and the serial number is a unique number that indicates where and when the motor was manufactured. These numbers are important if a motor is returned for warranty repairs or if an exact replacement is specified as a model for model exchange.

11.11.2
Frame Type

Every motor has been manufactured to specifications that are identified as the *frame size*. These data include

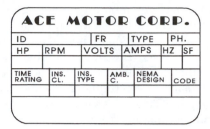

Figure 11–14 Data plate for a typical AC motor.

the distance between mounting holes in the base of the motor, the height of the shaft, and other critical data about physical dimensions. These data are then given a number such as 56. The frame number indicates that any motor that has the same frame number will have the same dimensions, even though it may be made by a different company. This allows users to stock motors from more than one manufacturer or replace a motor with any other with the same frame information and be sure that it will be an exact replacement.

11.11.3
Motor Type

The motor-type category on the data plate refers to the type of ventilation the motor uses. One type is the *open* type, which provides flow-through ventilation from the fan mounted on the end of the rotor. In some motors that are rated for variable-speed duty (used with variable-frequency drives), the fan is a separate motor that is built into the end of the rotor. The fan motor is connected directly across the supply voltage, so it will maintain a constant speed to provide constant cooling regardless of the motor speed.

Another type of motor is the *enclosed* type. The enclosed motor is not air-cooled with a fan; instead it is manufactured to allow heat to dissipate quickly to and from the inside of the motor outward to the frame. In most cases the frame has fins built into it on the outside to provide more area for cooling air to reach.

11.11.4
Phase

The phase of the motor is indicated as single-phase or three-phase. The number may be given as 1 or 3, or it may be written out as one or three.

11.11.5
Horsepower (hp) Rating

The horsepower (hp) rating is given as a fractional horsepower (a number less than 1.0) or a horsepower greater than 1. Fractional horsepower values may be given as fractions $\left(\frac{1}{2}\right)$ or decimals (0.5).

11.11.6
Speed (rpm)

The motor speed is indicated as in terms of rpm. This is the rated speed for the motor and does not account for slip. The actual speed of the motor will be less because of slippage. As you know, the speed of the motor is determined by the number of poles and the frequency of

the AC voltage. Typical speed for a two-pole motor is 3,600 rpm, for a four-pole motor it is 1,800 rpm, and for an eight-pole motor it is 1,200 rpm. The actual speed of a motor rated for a 3,600-rpm motor is approximately 3,450; for the 1,800-rpm motor it is approximately 1,750, and for the 1,200-rpm motor, it is approximately 1,150. The actual amount of slip is indicated by the *motor-design letter*. (The motor-design letter is explained in Section 11.11.14.)

11.11.7
Voltage Rating

Voltage ratings for single-phase motors are typically 115 V, 208 V, or 230 V. Three-phase motors have typical voltage ratings of 208, 240, 440, 460, 480, and 550 V. Other voltages may be specified for special types of motors. You should always ensure that the power supply voltage rating matches the voltage rating of the motor. If the rating and the power supply voltage do not match, the motor will overheat and be damaged.

11.11.8
Amps Rating

The amps rating is the amount of full-load current (FLA) the motor should draw when it is under load. This rating helps the designer calculate the proper wire size, fuse size, and heater size in motor starters. The supply wiring for the motor circuit should always be larger than the amps rating of the motor. The NEC (National Electric Code) provides information to help you determine the exact fuse size and heat sizes for each motor application.

11.11.9
Frequency

The frequency of a motor is listed in hertz (Hz). A typical frequency rating for motors in the United States is 60 Hz. Motors manufactured for use in some parts of Canada and all of Europe and Asia are rated for 50 Hz. You must be sure that the frequency of the motor matches the frequency of the power supplied to the motor.

11.11.10
Service Factor (SF)

The *service factor* (SF) is a rating that indicates how much a motor can be safely overloaded. For example a motor that has an SF of 1.15 can be safely overloaded by 15%. Thus, if the motor is rated for 1 hp, it can actually carry 1.15 hp safely. To determine the overload capability of a motor, you multiply the rated hp by the service factor. The motor is capable of being overloaded because it is designed to be able to dissipate large amounts of heat.

11.11.11
Duty Cycle

The *duty cycle* of a motor is the amount of time the motor can be operated out of every hour. If the motor's duty cycle is listed as continuous, it means the motor can be run 24 h a day and does not need to be turned off to cool down. If the duty cycle is rated for 20 min, it means the motor can be safely operated for 20 min before it must be shut down to be allowed to cool. A motor with this rating should be shut down for 40 min of every hour of operation to be allowed to cool.

Another way to specify the duty cycle of a motor is to use the *motor rating*. The motor rating on a data plate refers to the type of duty the motor is rated for. The types of duty include continuous duty, intermittent duty, and heavy duty, which includes jogging and plugging duty. Continuous duty includes applications where the motor is started and allowed to operate for hours at a time. Intermittent duty includes operations where the motor is started and stopped frequently. This type of application allows the motor to heat up because it will draw LRA more often than will a motor rated for continuous duty.

Motors that are rated for jogging and plugging are built to withstand very large amounts of heat that build up when the motor draws large LRA during starting and stopping. Because the motor can be reversed when it is running in the forward direction for plugging applications, it builds up excessive amounts of heat. Motors with this rating must be able to get rid of heat as much as possible to withstand the heavy-duty applications.

11.11.12
Insulation Class

The *insulation class* of a motor is a letter rating that indicates the amount of temperature rise the insulation of the motor wire can withstand. The numbers in the insulation class are listed in degrees Celsius (°C). The table in Fig. 11–15 shows typical insulation classes for motors. The insulation class and other temperature-related features of a motor will help determine the temperature rise the motor can withstand.

11.11.13
Ambient Temperature Rise

The ambient temperature rise is also called the Celsius rise. It is the amount of temperature rise the motor can

Class	Temperature Rise, °C
A	105
B	130
F	155
H	180

Figure 11–15 Insulation class for motors. This table indicates the amount of temperature for which the wire's insulation is rated.

withstand during normal operation. This value is listed in degrees Celsius. A typical open motor can withstand a rise of 40°C (104°F), and an enclosed motor can withstand a 50°C (122°F) rise. This means the motor should not be exposed to environments where the temperature is 104°F above the ambient. If the ambient is considered to be 72°F, the motor is limited to a temperature of 176°F. Another classification that helps determine the amount of temperature a motor can withstand is the type of insulation the motor winding has. The classes of insulation used with motors and the amount of temperature these classes can handle are listed in Section 11.11.14.

11.11.14
NEMA Design

The National Electric Manufacturers Association (NEMA) provides design ratings that may also be listed as motor design. The motor design is identified on the data plate by the letter *A*, *B*, *C*, or *D*. This designation is determined by the type of wire, insulation, and rotor used in the motor and is not affected by the way the motor might be connected in the field.

Type A motors have low rotor-circuit resistance and have approximate slip of 5% to 10% at full load. These motors have low starting torque with a very high locked-rotor amperage (LRA). This type of motor tends to reach full speed rather rapidly.

Type B motors have low to medium starting torque and usually have slip of less than 5% at full load. These motors are generally used in fans, blowers, and centrifugal pump applications.

Type C motors have a very high starting torque per ampere rating. This means that they are capable of starting when the full load is applied for applications such as conveyors and crushers and reciprocating compressors such as air-conditioning and refrigeration compressors. These motors are rated to have slip of less than 5%.

Type D motors have a high starting torque with a low LRA rating. This type of motor has a rotor made of

NEMA Code Letter	Locked-Rotor kVA per hp
A	0–3.15
B	3.15–3.55
C	3.55–4.00
D	4.00–4.50
E	4.50–5.00
F	5.00–5.60
G	5.60–6.30
H	6.30–7.10
J	7.10–8.00
K	8.00–9.00
L	9.00–10.00
M	10.0–11.2
N	11.2–12.5
P	12.5–14.0
R	14.0–16.0
S	16.0–18.0
T	18.0–20.0
U	20.0–22.4
V	22.4 and up

Figure 11–16 Table with locked-rotor amperage (LRA) ratings. The ratings are listed as the amount of kVA per horsepower.

(Courtesy of NEMA, National Electrical Manufacturers Association)

brass rather than copper segments. It is rated for slip of 10% at full load. Normally, this type of motor will require a larger frame to produce the same amount of horsepower as a type A, B, or C motor. These motors are generally used for applications with a rapid decrease of shaft acceleration, such as a punch press that has a large flywheel.

These standards are set by NEMA, and a motor must meet all the requirements of the standard to be marked as a type A, B, C, or D motor. This allows motors made by several manufacturers to be compared on an equal basis according to application.

11.11.15
NEMA Code Letters

NEMA code letters use letters of the alphabet to represent the amount of locked-rotor amperage (LRA) in kVA per horsepower that a motor will draw when it is started. These letters are listed in Fig. 11–16. From the table you can see that letters at the beginning of the alphabet indicate low LRA ratings, and letters toward the end of the alphabet indicate higher LRA ratings. It is im-

portant to remember that the number in the table is not the amount of LRA the motor will draw, but rather it is the number that must be multiplied by the horsepower rating of the motor.

11.12
Wiring Split-Phase Motors for a Change of Rotation and Change of Voltage

The wiring diagram for the run and start windings of the split phase motor were shown in Fig. 11–12. The terminals in the run and start windings can be reconnected in the field to change the rotation of the motor or to allow the motor to operate at a different voltage. Because the split-phase motor and the capacitor-start motor are very similar, these connections are presented in Section 11.16 after you learn more about capacitor-start, induction-run motors.

11.13
Overview of Capacitor-Start, Induction-Run Motors

The *capacitor-start, induction-run* (CSIR) motor is basically a split-phase induction motor that adds a capacitor with the start winding to create a larger-phase shift between the start and run windings to start the motor. The capacitor-start, induction-run motor is generally used in applications where the load is connected directly to the shaft of the motor. For example, a capacitor-start, induction-run motor can be used to turn direct-drive fans and larger pump loads. Some capacitor-start, induction-run motors are used to drive belt-driven compressors. The torque for a capacitor-start, induction-run motor is larger than for the split-phase-type motor. A typical capacitor-start, induction-run motor is shown in Fig. 11–17.

11.14
Theory of Operation for a Capacitor-Start, Induction-Run Motor

The capacitor-start, induction-run motor operates in a manner very similar to the AC split-phase motor described in the previous chapter. It has a rotor and a stator with a run winding and a start winding. A start capacitor is connected in series with the start winding to create a larger-phase shift than the split-phase motor

Figure 11–17 A capacitor start induction run motor used for HVAC and refrigeration applications such as direct drive fan motors and pump motors.
(Courtesy of GE Industrial Systems, Fort Wayne, Indiana)

Figure 11–18 The start capacitor is mounted in a black plastic case that has a round, cylindrical shape.
(Courtesy of North American Capacitor Company)

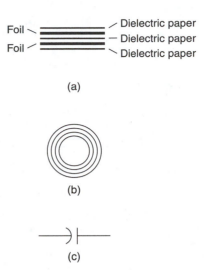

Figure 11–19 (a) Two sheets of conducting foil that are separated by sheets of dielectric paper. The dielectric paper acts as an insulator between the conducting foil. The two sheets of foil are called the plates. (b) The layers of conducting foil and dielectric paper are rolled up so that they will fit into the round, black plastic case. (c) The electrical symbol for the capacitor shows the two conducting plates separated by the insulator.

has. The start capacitor is generally rated from 50 to 90 μF (microfarads). The capacitance and voltage rating for the capacitor are listed on its case. Figure 11–18 shows a picture of a start capacitor. The start capacitor is so named because it is always connected in series with the start winding in the capacitor-start, induction-run motor to provide a larger phase shift, which will also create more starting torque. The start capacitor is also what gives the capacitor-start, induction-run motor its name.

The start capacitor has several physical features that make it easy to recognize when you are servicing equipment. The easiest way to recognize a start capacitor is that it is generally mounted in a black plastic case. A second characteristic is that the black plastic case for the start capacitor is generally round in shape. From the

picture of the start capacitor in Fig. 11–18, you can see that the start capacitor is mounted in a round, black plastic case.

The start capacitor is made of two sheets of conducting foil, called *plates*, that are separated (sandwiched) with three sheets of insulating paper called the *dielectric*. Figure 11–19a shows an example of the foil sheets and dielectric sheets before they are rolled up. Figure 11–19b shows a diagram of the sheets rolled up so that they will fit into the round, plastic case. A lead wire is soldered onto each piece of foil and brought out to the top of the capacitor to be used as the capacitor's terminals. Figure 11–19c shows the electrical symbol for the capacitor. Notice that the symbol shows the capacitor's conducting plates separated by the dielectric so current cannot pass directly from one plate to the other.

When voltage is applied to the capacitor in a circuit, current will flow as far as the plate on the left side of the capacitor. Because the capacitor plates are separated by dielectric paper, current cannot flow directly through the capacitor to the other plate, so the voltage must reverse and move backward through the circuit until it reaches the other capacitor plate. This process is referred to as *charging the capacitor and discharging the capacitor*. If the voltage that causes the current to flow is AC voltage, the AC sine wave will continually charge and discharge the capacitor, which keeps current continually flowing through the components in the circuit.

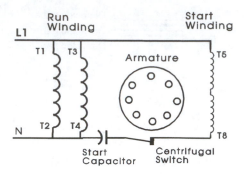

Figure 11–20 Electrical diagram of a capacitor-start, induction-run (CSIR) motor. Notice the start capacitor (SC) is connected in series with the centrifugal switch and start winding.

Because it takes time for the voltage to charge up the first plate and then reverse itself and discharge the plate, the capacitor causes a phase shift of up to 90° between the voltage waveform and the current waveform in the circuit. It is important to remember that the capacitor will always cause the voltage waveform to lag the current waveform.

11.15
Applying Voltage to the Start Winding and Capacitor

Figure 11–20 shows an electrical diagram of a capacitor-start, induction-run motor so that you can see that the start capacitor (SC) is connected in series with the centrifugal switch and the start winding. From this diagram you can see that when voltage is first applied to the leads of the motor, current will flow from L1 through the run windings and back to L2 or N. At the same time, voltage is applied to the first capacitor plate. Because current cannot flow through the capacitor to the other plate, the voltage must reverse itself and move back through the run winding; eventually it finds its way back to the start winding and reaches the other plate of the capacitor. Because the voltage must take the long way around the motor circuit to get to the other plate, it creates a phase shift between the current flowing in the run winding and the current flowing in start winding. This process continues as long as the centrifugal switch remains closed and current flows through the start winding. The capacitor causes a larger phase shift than the split-phase motor can develop without a capacitor. Because the capacitor causes a larger phase shift than the split-phase motor can develop with just the start winding, the capacitor-start, induction-run motor will have more starting torque than the split-phase motor.

When the motor's rpm increases to 75% to 85% of full rpm, the centrifugal switch opens and causes the current to stop flowing through the start winding. After the centrifugal switch opens the start-winding circuit, current will continue to flow through the run winding so the motor will act as an induction motor when it is running. This is why the motor is called a *capacitor-start, induction-run motor*.

11.16
Wiring Split-Phase and Capacitor-Start, Induction-Run Motors for 115 V

The diagram shown in Fig. 11–20 shows the electrical diagram for a capacitor-start, induction-run motor that is connected for 115 V. This connection is also called the *low-voltage connection*. The diagram for a split-phase motor is identical except it does not have a start capacitor connected to its start winding. When you must connect a split-phase motor or capacitor-start, induction-run motor for 115 V, you should connect terminals T1, T3, and T5 together with the supply voltage wire L1, and you should connect terminals T2, T4, and T8 together with the neutral supply voltage wire. You should notice from the diagram that these connections place the two sections of the run winding in parallel with each other and with the start winding. You should also notice that the start capacitor is connected in series with the centrifugal switch. This means that each section of the run winding will have 115 V applied to it, and the start winding will have 115 V.

11.17
Wiring a Capacitor-Start, Induction-Run Motor for 230 V

At times you may need to reconnect the capacitor-start, induction-run motor on which you are working for a higher voltage. Figure 11–21 shows an electrical diagram of a capacitor-start, induction-run motor that is connected for 230 V. This connection is also called the *high-voltage connection*. Again, the diagram for the split-phase motor is similar except it does not have the start capacitor. From this diagram you can see that the two sections of the run winding are connected in series. This connection ensures that 115 V is applied to each winding as a voltage drop. You should understand that the insulation on the wire that is used for the run and start windings is rated for 115 V. The two run windings are connected in series, so you are ensuring that each winding

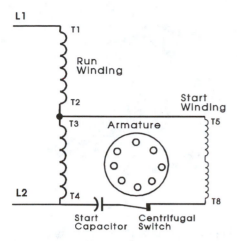

Figure 11–21 Electrical diagram for a capacitor-start, induction-run motor connected for 230 V.

will receive only 115 V, because the 230 V will be dropped equally across each run winding. You should also notice that the capacitor is always connected to the centrifugal switch, so it will always be connected in series with the start winding.

It is important at this time to remember that the motor may be rated for 208 V for its higher voltage rather than 230 V. The diagram for the higher voltage is the same regardless of whether the motor data plate indicates the motor is rated for 230 V or 208 V. The important point to remember is that the motor must match the exact amount of incoming voltage used to supply the motor. This means that you should always measure the exact amount of voltage before you connect a motor.

You should also notice that the start winding is connected so that it receives only 115 V. This means that the start winding is connected so that it is actually in parallel with the lower section of the run winding (T3–T4). One end of the centrifugal switch is connected permanently to one terminal of the start capacitor. The other terminal of the start capacitor is connected to terminal T4. This ensures that it will always be connected across the lower section of the start winding. You can see that if the motor is connected for 230 V or 115 V, one end of the centrifugal switch is always connected through the capacitor to terminal T4.

11.18
Wiring a Split-Phase and Capacitor-Start, Induction-Run Motor for a Change of Rotation

When you are working in the field you will find times when you must change the direction of rotation of the motor's shaft. Identify the rotation of the shaft by stand-

ing behind the motor (at the opposite end from where the shaft comes out of the end plate) and observe the direction of rotation. To change the rotation of the motor shaft, you must change the direction current flows through the start winding. The easiest way to do this is to exchange the T5 and T8 wires.

The capacitor-start, induction-run motor that is connected for high voltage can also be wired so that the direction of its shaft's rotation can be reversed. When the motor is wired for clockwise rotation, terminal T5 of the start winding is connected to terminals T2 and T3 of the run winding, and terminal T8 of the start winding is connected to terminal T4 of the run winding. Figure 11–21b shows the electrical diagram of a high-voltage (230 V) capacitor-start, induction-run motor connected for counterclockwise rotation. In this diagram you can see that terminal T8 of the start winding is connected to terminals T2 and T3 of the run winding, and terminal T5 of the start winding is connected to terminal T4 of the run winding.

11.19
Changing Speeds with a Capacitor-Start, Induction-Run Motor

The split-phase and capacitor-start, induction-run motors are generally not designed to have their speed changed. Some more expensive capacitor-start, induction-run motors have additional sets of poles provided so that their speeds can be changed by changing the number of poles used in each run winding, but typically this type of motor cannot be reconnected for a change of speed. Thus if you need to change the speed of the motor, you would need to replace the motor with one of a different speed. It is also important to understand that if the motor is used to drive a belt-driven fan, the speed of a fan can be altered slightly by changing the size of the pulley on the motor.

11.20
Overview of Permanent Split-Capacitor Motors

The *permanent split-capacitor* (PSC) motor is similar to the split-phase induction motor and capacitor-start, induction-run motor in that it has a start winding and a run winding, but it is different in that it does not use a centrifugal switch to remove power from the start winding when the motor reaches full speed. Instead of a centrifugal switch, the PSC motor uses a run capacitor that is connected in series with the start winding to limit the

Figure 11–22 Typical permanent split-capacitor motors used for applications such as conveyors that are belt-driven or direct-drive blade-type fan motors and pump motors.
(Courtesy of GE Industrial Systems, Fort Wayne, Indiana)

Figure 11–23 A special PSC motor that has a shaft extending out each end of the motor. This type of motor is used in window air conditioners to power the condenser fan and evaporator fan.
(Courtesy of GE Industrial Systems, Fort Wayne, Indiana)

amount of current allowed to flow through the start winding when the motor reaches full speed. The run capacitor is mounted in a metal container so that it can dissipate the heat better, which allows it to remain in the circuit even after the motor is running at full speed. The permanent split-capacitor motor is generally used in applications where the load is connected directly to the shaft of the motor. For example, a permanent split-capacitor motor can be used to turn direct-drive fans, such as squirrel-cage-type fans or blade-type fans, and motors for conveyor applications and pump applications. The torque for a permanent split-capacitor motor is larger than that of a split-phase-type motor and about the same or slightly less than a capacitor-start, induction-run motor. Two typical permanent split-capacitor motors are shown in Fig. 11–22. You should be able to see that the run capacitor is mounted on the top of each motor. Figure 11–23 shows a special PSC motor that has a shaft extending out of both ends of the end plates. This type of PSC motor is used in window air conditioners; a blade-type fan is mounted on one end to operate as the condenser fan, and a squirrel-cage fan is on the other end to operate as the evaporator fan.

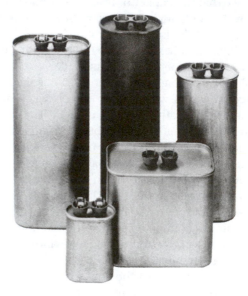

Figure 11–24 Examples of typical run capacitors. Notice that run capacitors are mounted in metal containers that are oval or rectangular in shape.
(Courtesy of North American Capacitor Company)

11.21
Basic Parts and Theory of Operation for a PSC Motor

The basic parts of the PSC motor include the start winding and the run winding, which are mounted in the stator, and the end plate, which supports the rotor that has the motor shaft on one or both ends. The permanent split-capacitor motor operates in a manner very similar to the AC split-phase motor. It has a rotor and a stator with a run winding and a start winding. A run capacitor is connected in series with the start winding to create a larger phase shift than the split-phase motor has. The run capacitor is generally rated from 5 to 60 μF. The capacitance and voltage rating for the capacitor are listed on its case. Figure 11–24 shows a picture of several examples of run capacitors. The run capacitor is so named because it remains in the motor circuit even after the motor is running at full speed.

The run capacitor has several physical features that make it easy to recognize when you are servicing equipment. The easiest way to recognize a start capacitor is that it is generally mounted in a metal container. A second characteristic is that the metal container for the run capacitor is generally oval or rectangular in shape.

The run capacitor is similar to the start capacitor in that two sheets of conducting foil, called *plates*, are separated (sandwiched) by three sheets of insulating paper, called the *dielectric*. The foil plates and dielectric are rolled up and placed in the metal container, and an electrical terminal is connected to each plate.

You should remember that when voltage is applied to the capacitor in a circuit, current will flow as far as the plate on the left side of the capacitor. Because the capacitor plates are separated by dielectric paper, current cannot flow directly through the capacitor to the other plate, so the voltage must reverse and move backward through the circuit until it reaches the other capacitor plate. Recall that this process is referred to as charging the capacitor and discharging the capacitor. If the voltage that causes the current to flow is AC voltage, the AC sine wave will continually charge and discharge the capacitor, which keeps current continually flowing through the components in the circuit.

Because it takes time for the voltage to charge up the first plate and then reverse itself and discharge the plate, the capacitor causes a phase shift of up to 90° between the voltage waveform and the current waveform in the circuit. It is important to remember that the capacitor always causes the voltage waveform to lag the current waveform.

The reason a centrifugal switch is not needed to remove the start winding from the circuit after the motor is running at full speed is that the CEMF that is generated by the rotor is present across the start winding. When this counter-EMF is present, it opposes the applied voltage, and the resulting current in the start winding is determined by the difference of the applied voltage and CEMF. If the motor is running at near full rpm, the CEMF is high, and the amount of current in the start winding is minimal. If the motor encounters a large load and its slip increases because the speed of its shaft slows down, the CEMF decreases and the difference between the applied and CEMF becomes larger, which allows the start winding to draw extra current. This extra current helps provide enough additional torque to get the motor shaft back to its original speed. In this manner, the run capacitor allows the start winding to stay in the circuit and add current when additional torque is required to get the shaft to return to its normal speed. Thus, the PSC motor is able to regulate its speed. This

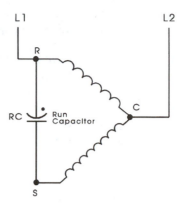

Figure 11–25 Electrical diagram of a permanent split-capacitor (PSC) motor. Notice the run capacitor (RC) is connected in series with the start winding, and this type of motor does not have a centrifugal switch.

is an important feature for the PSC motor if it is used for moving air or pumping cooling water for a chiller.

11.22
Applying Voltage to the Start Winding and Capacitor of a PSC Motor

Figure 11–25 shows an electrical diagram of a typical permanent split-capacitor motor. You can see that the start capacitor (RC) is connected in series with the start winding. When voltage is first applied to the leads of the motor, current flows from L1 through the run windings and back to L2 or N. At the same time, voltage is also applied to the first capacitor plate. Because current cannot flow through the capacitor to the other plate, the voltage must reverse itself and move back through the run winding until it finds its way back to the start winding, where it reaches the other plate of the capacitor. Because the voltage must take the long way around the motor circuit to get to the other plate, it creates a phase shift between the current flowing in the run winding and the current flowing in start winding. In the PSC motor this process continues as long as voltage is applied to the motor windings. Because the run capacitor remains in the start-winding circuit at all times for this motor, it is called a *permanent split-capacitor motor*.

In the PSC motor the initial current flow into the motor windings is rather large because the rotor is not turning at first. This current is called *locked-rotor amperage* (LRA). Because a run capacitor is mounted in series with the start winding, the current cannot flow directly through the capacitor. Instead, the voltage moves as far as the first plate, when L1 voltage is applied, and then stops. At this point one plate of the capacitor has a large positive charge, which creates a potential differential across the plates of the capacitor. This potential differential causes the voltage to move from the first plate back through the run winding and start winding, until the voltage reaches the opposite plate of the capacitor. This causes a large phase shift between the current in the run winding and start winding, which is sufficient to cause the rotor to spin.

When the rotor begins to spin, it begins to generate counter-EMF, which causes the current flow to diminish to a point of normal current flow called full-load amperage (FLA). The actual amount of current the motor draws depends on the difference between the rated speed and the actual speed. This difference is called *slip*. The rotor continues to rotate as long as voltage is applied to the windings.

Questions for This Chapter

1. Identify the main parts of an AC split-phase motor.

2. Explain how you can identify the run winding and start winding in the stator of an AC split-phase motor.

3. Explain how a magnetic field is developed in the rotor of an AC motor.

4. Explain how the squirrel-cage rotor acts like the bar-magnet rotor in an AC motor.

5. Explain why a rotor rotates when a voltage is applied to the stator of an AC motor.

True or False

1. Torque is rotational force.

2. Slip is the difference between the actual speed of a motor and its rated speed.

3. The AC split-phase motor is called an induction motor because it uses a transformer with a capacitor to start.

4. The run winding is physically offset from the start winding in the stator of an AC motor to help it provide sufficient phase shift to start the rotor.

5. The start winding uses larger wire than the run winding.

Multiple Choice

1. The start capacitor is _____ .

   a. connected in series with the run winding of a PSC motor.
   b. connected in series with the start winding of a CSIR motor.
   c. connected in series with the run winding of a CSIR motor.

2. The run capacitor is _____ .
   a. connected in series with the run winding of a PSC motor.
   b. connected in series with the start winding of a PSC motor.
   c. connected in series with the start winding of the CSIR motor.

3. The start winding of a split-phase motor should have _____ the run winding.
   a. less resistance than
   b. more resistance than
   c. the same resistance as

4. The permanent split-capacitor (PSC) motor _____ .
   a. uses a run capacitor and a centrifugal switch.
   b. uses a start capacitor and a centrifugal switch.
   c. uses a run capacitor but does not have a centrifugal switch.

5. The capacitor-start, induction-run (CSIR) motor _____ .
   a. uses a start capacitor that is connected in series with its start winding.
   b. uses a start capacitor that is connected in series with its run winding.
   c. uses a run capacitor that is connected in series with its start winding.

Problems

1. Calculate the speed of a two-pole, four-pole, six-pole, and eight-pole motor.

2. Draw the sketch of a typical data plate (name plate) for an AC motor and identify each of the items listed in Section 11.11.

3. Draw the electrical diagram of a split-phase motor wired for 115 VAC.

4. Draw the electrical diagram of a capacitor-start, induction-run motor that is wired for 230 VAC.

5. Draw the electrical diagram of a PSC motor that is wired for 230 VAC.

◀ Chapter 12 ▶

Three-Phase Motors

Objectives

After reading this chapter you should be able to:

1. Explain the theory of operation of a three-phase motor.
2. Identify the main parts of the three-phase motor.
3. Wire the three-phase motor for high or low voltage.
4. Change the rotation of a three-phase motor.
5. Wire the three-phase motor for a delta or wye configuration.

12.0
Three-Phase Motor Theory

When industrial applications require larger motors (more than 1 hp), three-phase motors are usually used. The three-phase motor can produce extremely large starting and running torque. In this section you learn about how three-phase AC motors have some similarities to single-phase AC motors. Today all maintenance technicians must be able to work on jobs that utilize three-phase motors. You must be able make small changes on three-phase motors in the field so they can operate at a different speed, different voltage, or different rotation. This chapter makes these field changes, installation, and troubleshooting three-phase motors easy to perform.

The nature of three-phase AC voltage is that it has three independent sources of voltage that are 120° apart. Figure 12–1 shows a diagram of three-phase voltage. Notice that this diagram shows three separate sine waves that are 120° apart. This natural phase shift in the voltage provides the necessary shift in the magnetic field

when the voltage is applied to the stationary fields (stator) of the AC motor. This means that three-phase voltage can provide the phase shift required to start a motor naturally, without any capacitors or start windings.

When the three-phase voltage is applied to the stator winding of a motor, the phase shift causes the magnetic field in the stator to actually rotate or move around the stator at the speed of the frequency of the AC voltage. The phase shift in the stator also causes the squirrel-cage rotor in the three-phase motor to become magnetized. When the rotor is magnetized, its field follows the rotating magnetic field in the rotor and causes the rotor to spin.

Because the three-phase voltage has a natural phase shift, it creates a very strong rotating magnetic field in the motor with the same natural phase shift. The strength of the magnetic field creates the strong torque at the shaft of the motor. This provides a means to start the motor under heavy loads such as the high pressure that a compressor must start against. The windings of a three-phase motor may be connected in a number of configurations to provide more or less starting torque and running torque, and the windings can be connected

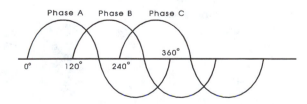

Figure 12–1 Example of three-phase voltage. Notice this voltage consists of three independent sine waves.

for high-voltage (480 VAC) applications and low-voltage (230 VAC) applications. The direction of rotation of the three-phase motor may be reversed by interchanging any two of the three supply voltage wires so that the phase relationship is reversed.

12.1
Basic Parts of a Three-Phase Motor

Figure 12–2 shows a typical three-phase motor. The three-phase motor consists of a stator that has three separate windings mounted out of phase from each other. Figure 12–3 shows an exploded-view picture of the three-phase motor. The windings are equally spaced around the stator, and the wire in each winding is the same gauge size. This means that the three windings in the stator are equal, and if you measure their resistance it should be the same for each winding. The rotor is also shown in this picture. From this picture you can see that the rotor is a squirrel-cage rotor. This squirrel-cage rotor is very similar to the rotors in other AC motors in that it is made of laminated steel plates pressed on the

squirrel-cage frame. The steel plates allow the rotor to become magnetized very easily and to give up its magnetic field very easily. This allows the magnetic bars in the rotor to become magnetized positively and negatively as the rotor spins inside the stator. Because a three-phase motor has three separate voltages, the three-phase motor has three separate and distinct magnetic fields in the rotor at the same time, which gives the rotor optimum torque.

12.2
Theory of Operation for a Three-Phase Motor

The theory of operation for a three-phase motor is very easy to understand. When three-phase voltage is applied to the three windings of the motor, three separate magnetic fields are developed, which immediately begin to rotate through the stator windings. Because the magnetic field rotates around the stator, the three magnetic fields in the rotor of the motor "chase" the magnetic fields and cause the rotor to begin to spin. The speed of rotation of the magnet field is fixed, and it is based on the number of windings and the frequency of the applied voltage. This means that the three-phase motor runs at a constant speed unless the frequency of the applied voltage is varied. In some industrial applications, a variable-frequency drive is used to control the speed of the motor by providing less than 60 Hz at times when the speed of the motor does not have to be maximum.

The natural phase shift in the three-phase voltage and the location of the three-phase windings in the stator of the motor allow the motor to develop the magnetic field in the rotor through induction so no extra components or circuits—such as a centrifugal switch, current relay, potential relay, or capacitors—are needed to provide extra torque from a start winding. As long as voltage is applied to the stator, a magnetic field is provided in the rotor by induction, and it always tries to cause the rotor to spin with the rotating magnetic

Figure 12–2   A typical three-phase motor.
(Courtesy of GE Industrial Systems, Fort Wayne, Indiana)

Figure 12–3   Exploded view of a three-phase motor.
(Courtesy of Reliance Electric)

field in the stator, which allows the rotor to provide torque at its shaft that can be used to turn the shaft to do work.

12.3
Wiring a Simple Three-Phase Motor

The simplest three-phase motor you will encounter has only three wires brought out of its case for external connections. This makes all connections very simple. A diagram of this type of three-phase motor is shown in Figure 12–4. This diagram shows a motor-starter coil that is controlled by a thermostat. The motor terminals are identified as T1, T2, and T3. When you make the field-wiring connections to these terminals, you can connect L1 to T1, L2 to T2, and L3 to T3. Because this type of motor provides only three wires for external connection, you cannot change the connections of the windings to change the speed of the motor or its torque. You can change its direction of rotation by switching L1 to T2 and L2 to T1. At times you may need to reverse the rotation of the three-phase motor for different applications.

12.4
Changing Connections in Three-Phase Motors to Change Torque, Speed, or Voltage Requirements

At times as a technician, you will need to make minor changes to three-phase motors that are used to move conveyor belts, turn fans, pump water and hydraulic fluid, or compress air. These changes include providing more starting torque, operating the motor at different speeds, or changing connections so that the motor can run at a different voltage. Because these motors may cost hundreds of dollars, it is important that you be able to make these changes while the motor is located on its

machine right on the factory floor. These changes will also allow you to use the existing motor rather than order an expensive new motor. After you learn about these field wiring changes, you will become confident in making these changes in the field.

12.5
Wiring a Three-Phase Motor in a Wye Configuration

As you know, the three-phase motor has three equal windings, and it does not need any starting switches or capacitors connected to any of its windings. Because the motor has three equal windings, they can be connected in a wye configuration or a delta configuration. Figure 12–5 shows the three-phase motor windings for a six-lead motor connected in a typical wye configuration. The ends of each lead are numbered so that they can be changed or reconnected for different wiring configurations. The ends of the leads for the first winding are identified as T1 and T4, and the second windings are

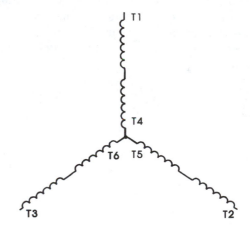

Figure 12–5   A three-phase motor connected in a wye configuration.

Figure 12–4   A simple three-phase motor with only three wires brought out for external connections.

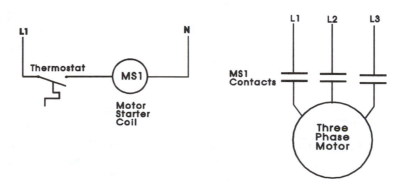

T2 and T5, and the third windings are T3 and T6. You can see that terminals T4, T5, and T6 are all connected together at one point in the center of the motor, which is called the *wye point*. The most important thing to remember about connecting a three-phase motor in a wye configuration is that it will have less starting torque than if the same motor leads are connected in a delta configuration. This means that the wye-connected motor draws less current than the delta-connected motor. These are called "wye" windings because they look like the letter *Y* upside-down.

12.6
Wiring a Three-Phase Motor in a Delta Configuration

The six-lead motor shown in the previous figure can also be connected in a delta configuration. Figure 12–6 shows a six-lead motor connected in a delta configuration. In this diagram you can see that the windings are connected in such a way that the three corners of the delta configuration are identified as T1, T2, and T3. In this configuration the incoming power supply is connected to T1, T2, and T3. You should remember that the delta-connected motor has more starting torque and draws more current than the wye-connected motor. These are called "delta" like the Greek letter Δ.

12.6.1
Wiring a Three-Phase, Six-Wire Open Motor for High Voltage or Low Voltage

All the terminal leads of a three-phase open motor are brought outside the stator so the motor can be reconnected to operate with a high-voltage supply (480 VAC) or a low-voltage supply (230 VAC). The number of leads that are brought out of the stator for a three-phase motor may be 6, 9, or 12. If 6 leads are brought out of the stator, you can use the diagram in Fig. 12–7a to connect the leads for low voltage (240 or 208 V) and Fig. 12–7b to connect the leads for high voltage (480 V). It is important to understand that if the motor is a six-lead, three-phase motor it is wired so it can be connected for high or low voltage. You should notice that the same connections are used for both the delta- and the wye-connected six-lead motor.

12.7
Rewiring a Nine-Lead, Three-Phase, Wye-Connected Motor for a Change of Voltage

A number of three-phase motors have nine leads brought out of the stator. The reason these motors have nine leads is that each of the three windings is broken into two pieces. The two sections that make up the three windings can be connected in a wye configuration or a delta configuration. Figure 12–8 shows the nine leads connected in a wye configuration. From this diagram you can see that the two ends of the first half of the first winding are identified as T1 and T4, and the terminals for the second half of the first winding are identified as T7 and common. The common point is where the three windings are connected together to form the wye point.

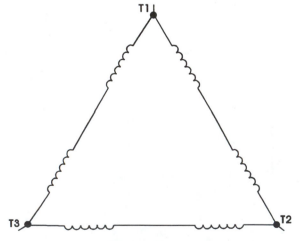

Figure 12–6 A three-phase motor connected in a delta configuration.

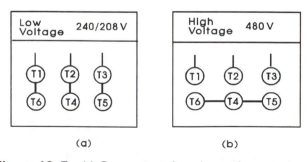

(a)   (b)

Figure 12–7 (a) Connections for a low-voltage, wye-connected and low-voltage, delta-connected motor. (b) Connections for a high-voltage, wye-connected motor and a high-voltage, delta-connected motor.

Figure 12–8   (a) Connections for a nine-lead, three-phase, wye-configured motor wired for high voltage (480 V); (b) Connections for a nine-lead, three-phase, wye-configured motor wired for low voltage (208 or 240 V).

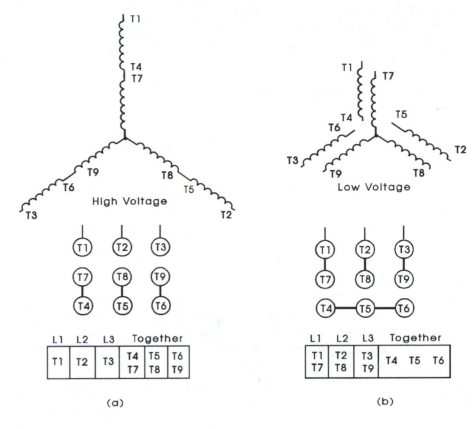

The terminal ends for the first half of the second winding are identified as T2 and T5, and the terminals for the second half of the second winding are identified as T8 and common. The terminal ends for the first half of the third winding are identified as T3 and T6, and the terminals for the second half of the second winding are identified as T9 and common. The diagram in Fig. 12–8 shows the correct way to connect the terminals of the wye-connected motor for high voltage (480 V) and for low voltage (208 or 240 V). It is also important to note that if the three- phase motor brings out nine leads, it is internally connected as a wye or a delta motor. If the motor is connected as a nine-lead, wye-connected motor and you need more starting torque so that you want to change it to delta, you will not be able to make this conversion, because the motor is internally connected so that it can be wired only as a wye-connected motor. Thus if you need a delta-connected motor, you will need to purchase a new nine-lead motor. This is a minor drawback of the three-phase motor, because you very seldom change the motor from delta to wye or wye to delta once the motor is installed.

12.8
Rewiring a Nine-Lead, Three-Phase, Delta-Connected Motor for a Change of Voltage

A nine-lead, three-phase, delta-connected motor can also be rewired for high and low voltage in the field. The nine-lead motor has each winding broken into two sections. Figure 12–9 shows the nine leads connected in a delta configuration. From this diagram you can see that the two ends of the first half of the first winding are identified as T1 and T4, and the terminals for the second half of the first winding are identified as T1 and T9.

The terminal ends for the first half of the second winding are identified as T2 and T5, and the terminals for the second half of the second winding are identified as T2 and T7. The terminal ends for the first half of the third winding are identified as T3 and T6, and the terminals for the second half of the third winding are identified as T3 and T8. The diagram in Fig. 12–9 shows the

Figure 12–9 (a) Connections for a nine-lead, three-phase, delta-configured motor wired for high voltage (480 V); (b) Connections for a nine-lead, three-phase, delta-configured motor wired for low voltage (208 or 240 V).

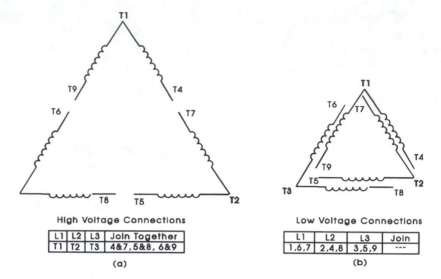

High Voltage Connections

L1	L2	L3	Join Together
T1	T2	T3	4&7,5&8, 6&9

(a)

Low Voltage Connections

L1	L2	L3	Join
1,6,7	2,4,8	3,5,9	---

(b)

correct way to connect the terminals of the delta-connected motor for high voltage (480 V) and for low voltage (208 or 240 V). The important point to remember is that the windings are connected in series for high voltage, and they are connected in parallel for low voltage.

12.9
The 12-Lead Three-Phase Motor

Some larger three-phase motors can be connected as either a wye motor or a delta motor. In order for this change to occur, the motor must have either 6 or 12 leads brought out of the stator. If the motor has 9 leads brought out, the motor cannot be changed from wye to delta in the field. For example, in some applications you must start the motor as a wye-connected motor so it does not draw too much LRA and then redo it as a delta-connected motor after it is running so that it will pro-

vide sufficient running torque. This type of motor is called a *wye-delta motor application*, and the configuration is changed while the motor is running by a wye-delta motor starter. The wye-delta motor starter is a special motor starter that has two independent sets of three-phase contacts connected to it. The motor is connected in wye configuration on the first set of contacts, and it is connected as a delta-configured motor on the second set of contacts. When the motor is started, the coil for the first set of contacts is energized and the motor is connected as a wye-connected motor when voltage is first applied. When the motor reaches full speed, the relay of the second set of contacts is energized and the motor is rewired as a delta-configured motor. The motor for the wye-delta starting must be a 6- or 12-lead three-phase motor. Figure 12–10 shows a 12-lead motor connected as a wye motor, and Figure 12–11 shows a 12-lead motor connected as a delta-connected motor.

12.10
Wiring a Three-Phase Motor for a Change of Rotation

At times you will need to change the direction of rotation for the three-phase motor while it is connected to its application. This is perhaps the most common change you will make to the three-phase motor. As a technician, you will need to change the direction that a pump motor runs when it is turning the wrong way, or you may be requested to change the direction a fan motor or conveyor motor is turning to ensure it is moving in the proper direction. In these cases all that you need to do is exchange any two of the supply leads. For example, you can change L1 and L2 so that L2 is connected to T1 and

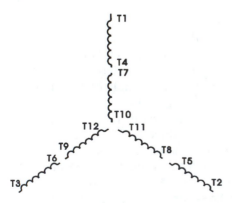

Figure 12–10 A 12-lead, three-phase motor connected in a wye configuration.

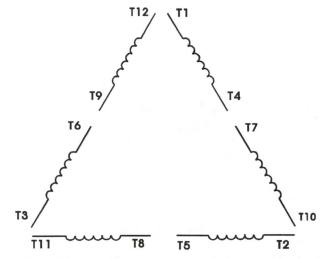

Figure 12–11 A 12-lead, three-phase, motor connected in a delta configuration.

L1 is connected to T2. Figure 12–12a shows the three-phase wye motor connected for clockwise rotation. Notice that L1 is connected to T1 and L2 is connected to T2. In Fig. 12–12b you can see that the compressor is connected to operate in a counterclockwise direction. L1 and L2 are exchanged so that L1 is connected to T2 and L2 is connected to T1. Figure 12–13a and b show similar diagrams for a delta-wired motor.

12.11
Controlling a Three-Phase Motor

The three-phase motor is generally turned on or off with a motor starter. The voltage to the coil of the motor starter is controlled by a stop-start switch. When the contacts in the switches close, voltage is applied to the coil of the motor starter, and it draws current, becomes

Figure 12–12 (a) A three-phase, wye-connected motor connected for clockwise rotation. (b) A three-phase, wye-connected motor connected for counterclockwise rotation. Notice T1 is now connected to L2 and T2 is now connected to L1.

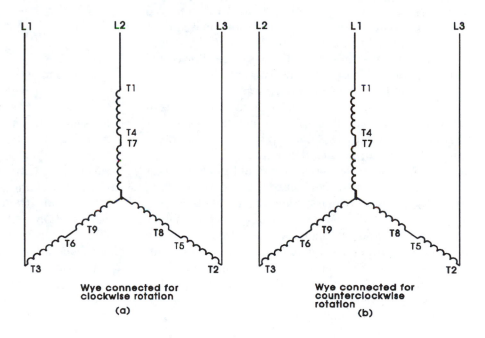

Figure 12–13 (a) A three-phase, delta-connected motor connected for clockwise rotation. (b) A three-phase, delta-connected motor connected for counterclockwise rotation. Notice T1 is now connected to L2 and T2 is now connected to L1.

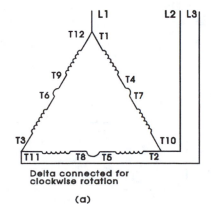

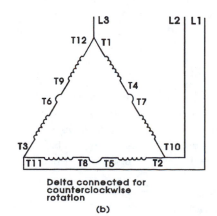

magnetized, and pulls the three sets of normally open contacts to their closed position. When the contacts close, three-phase voltage is applied to the motor and it begins to run. When the stop button is depressed, it deenergizes the relay coil, and the three sets of contacts open and turn the motor off.

12.12
Controlling the Speed of a Three-Phase Motor

The speed of a three-phase motor can be changed by changing the number of poles in the motor windings or by changing the frequency of the voltage supplied to the motor. Prior to the 1980s the only way that the frequency of the supply voltage could be easily changed was by changing the speed of the generator that produced the electricity. Because this was not practical, the preferred method of changing the speed of the three-phase motor was to reconnect the motor in the field so it would have a different number of poles. This basically allowed the fan motor or compressor to have two speeds.

In the 1980s more emphasis was placed on efficiency, so it became important to change the speeds of three-phase fans and three-phase pump motors so that they would run only at the speed necessary to do the job. In the late 1970s and 1980s electronic technology began to produce products that could control larger amounts of voltage and current so that the frequency of the voltage sent to a motor could be changed. Chapter 16 goes into detail about these devices. In this chapter all you need to know is that the variable-frequency drive is specifically designed to change the frequency of voltage that is used to power a three-phase or single-phase motor. Because the frequency is changed, the speed of the motor can be adjusted. For example, if the motor needs to operate at a higher speed, the frequency would be increased above 60 Hz. If the speed needs to be decreased, the frequency would be lowered below 60 Hz. Figure 12–14 shows a picture of a variable-frequency drive, and Figure 12–15 shows a block diagram of a vari-

Figure 12–14 A three-phase, variable-frequency drive that is used to control the speed of three-phase fan motors and three-phase compressor motors.
(Courtesy of Rockwell Automation's Allen-Bradley Business)

able-frequency drive. From the block diagram you can see that the drive is supplied with three-phase voltage. The first section of the drive is the rectifier section. In this section three sets of diodes are used to rectify the AC voltage to DC voltage. The second section of the drive is a filter section. You can see that the voltage waveform is half a sine wave as it enters the filter section, and it changes to filtered pure DC as it leaves the filter section. The final section of the drive is the transistor section, and the transistors can be turned on and off by a triggering circuit so that the output waveform looks similar to the waveform of the original three-phase voltage that supplies this circuit. The main difference between the input voltage and the output voltage is that the frequency of the output voltage can be changed to any value between 1 Hz and 120 Hz. Thus the variable-frequency drive can create frequencies from 1% to 200% of 60 Hz, but the motor speed in practice is generally controlled at between 75% and 125% of the rated motor rpm.

Figure 12–15 A block diagram of a variable-frequency drive. AC three-phase voltage enters the drive and is converted by the rectifiers to half-wave DC. The filter section converts the half-wave DC to pure DC. The transistors convert the pure DC back to variable frequency AC.

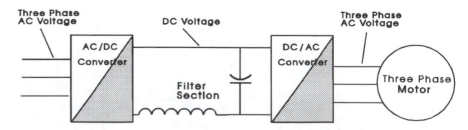

12.13
Troubleshooting the Three-Phase Motor

At times in the field you will be expected to troubleshoot a three-phase motor that will not start or a motor that is malfunctioning. When you are in the field you can test a three-phase motor for voltage and current and to see if each of the windings has continuity. The following sections explain how to make each of these tests. The most frequent test for a three-phase motor is the test of supply voltage at the disconnect or fuse box and for voltage at each of the three motor terminals.

If the three-phase motor fails to run when it is energized, you will need to make two basic tests. The first test involves testing each of the three lines that supply voltage to the motor to ensure that three-phase voltage is flowing through the wires to the terminals of the motor. You can make this test by testing for voltage right at the terminals of the motor. If the voltage supply for this motor is 240 V three-phase, you should measure 240 V between T1–T2, T2–T3, and T3–T1. If you have voltage at two of the three terminals, it indicates that you have lost one phase of the three-phase supply voltage; the most likely problem is a blown fuse. In some cases the motor can continue to run when one of the phases is lost, but the motor will not have sufficient torque to restart, and the overload in the motor starter will be tripped when you find the motor during a troubleshoot-

ing call. Be sure that the voltage test indicates you have the correct amount of voltage at each of the three tests you make at the terminals on the compressor. The voltage tests are similar to the test you learned about to determine if the voltage is present in a three-phase disconnect switch. If the motor is a three-phase hermetically sealed compressor, it may have internal overloads and you may need to allow the compressor to cool down for several hours before the overloads will automatically reset. If the compressor continues to trip its overloads, you will need to have a technician test the system for loss of refrigerant or other problems that cause compressors to overload and overheat.

If the correct amount of voltage is present at all three terminals and the motor will not run, you must suspect that one of the windings in the motor is open and has infinite (∞) resistance. (Remember that the compressor may also have an internal thermal overload that would give a similar symptom. Be sure to allow several hours for the overload to cool down before you make a decision about changing a compressor.) If you suspect the winding is open, you will need to turn the power off to the system, disconnect all the wires from the motor terminals, and test each winding for resistance. It is important to understand that if you leave the motor connected to the circuit, you may read back-feed resistance through a transformer or other motor winding on the system. Figure 12–16 shows the proper location for placing a voltmeter to make these measurements.

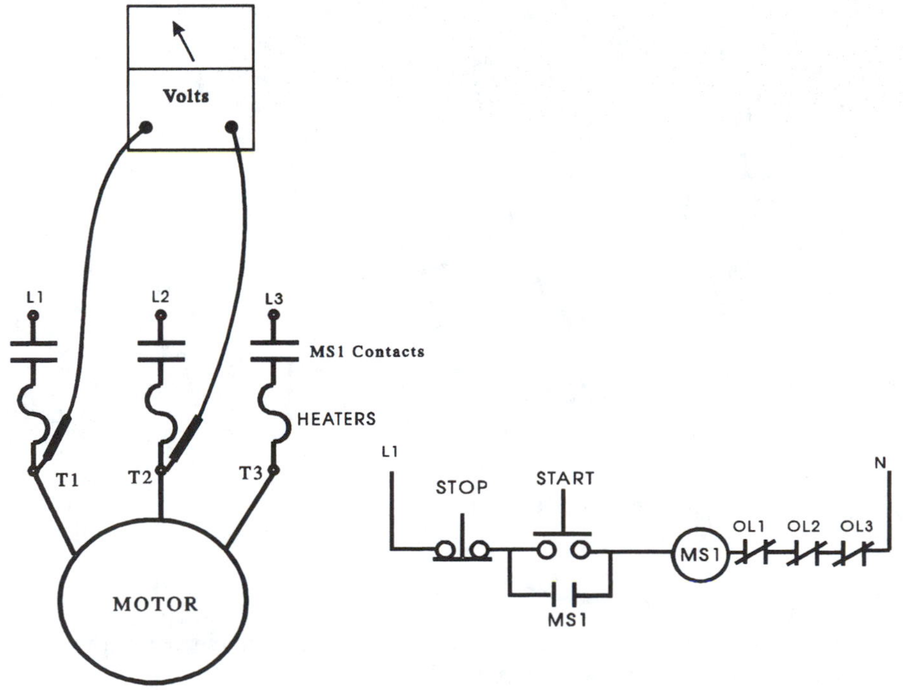

Figure 12–16 The proper location for placing voltmeter terminals to test for three-phase voltage on the three-phase motor.

12.14
Testing the Three-Phase Motor for Continuity

If the voltage test indicates that the proper amount of voltage is present at the terminals, but the motor does not start and hums or does not make any noise when it does not start, you should suspect one or more motor windings have an open, and you must test the motor windings for resistance. Always remember that the voltage to the motor should be turned off for the continuity test and the windings should be isolated if possible. The test for resistance should be between T1–T2, T2–T3, and between T3–T1. If any of these windings are open, the motor must be removed and replaced. As stated before, some compressors have an internal thermal overload built into the windings, and it may open if the internal temperature gets too hot. The motor windings may get too hot if the amount of refrigerant in the system is too low to provide cooling for the motor or if the motor is overloaded. Remember the hermetic motor is cooled only by the extra refrigerant brought back to the motor specifically to cool the windings. If the overload is open, it will not allow current to flow through the windings, and hence the motor will not run. Because the overload may be the problem, it is always a good practice to wait 1 to 2 h for the compressor to cool down and to retest the windings for continuity before you change it out.

It is also important to understand that some refrigeration compressors use the motor windings as a heater when the motor is not running. This is accomplished by connecting capacitors and resistors to allow a small amount of bleed current to flow through the windings to act as a small heater. This small amount of heat keeps the windings warm enough so that the refrigerant oil does not migrate away from the compressor to part of the system that is warmer. A problem may occur when you test the compressor if you do not make sure that all this circuit is isolated from the motor windings when you are testing for an open circuit. Figure 12–17 shows the proper location at which to place the ohmmeter to make the continuity test.

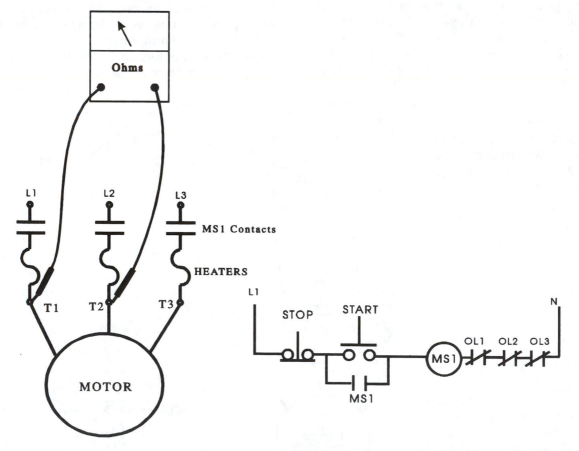

Figure 12–17   The proper locations at which to place ohmmeter terminals to test the three-phase motor windings for continuity to check for an open circuit.

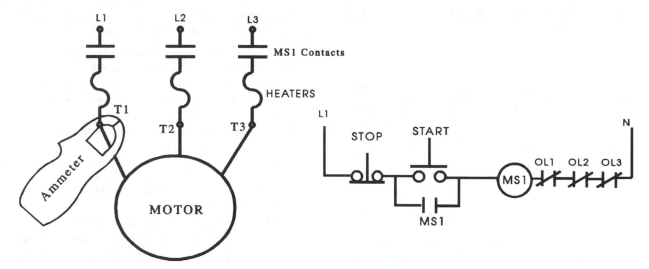

Figure 12–18 The proper locations at which to place a clamp on an ammeter to test the three-phase motor for proper current draw.

12.15
Testing a Three-Phase Motor for Current

If the three-phase motor will start, you must check it for the proper amount of current to ensure that it is not overloaded. Figure 12–18 shows the proper location at which to place the clamp on the ammeter around each of the three-phase wires that supply voltage to the motor. Because the supply wires are connected in series,

the current measurement indicates how much current the motor is drawing through each windings. It is important to understand that the amount of current measured in each of the three wires should be nearly the same. If one of the leads is drawing excess current, it indicates the motor is beginning to fail. If all three measurements are larger than the data plate rating, the motor is being overloaded, and you need to determine the cause of the overload. Sometimes three-phase pump motors draw excess current when the pump is overloaded.

Questions for This Chapter

1. Explain why a three-phase motor can start without using any start relays or capacitors.

2. Identify the main parts of a three-phase open motor.

3. Explain why you would need to reconnect the terminals of a three-phase motor to change its direction of rotation.

4. Explain why you would need to reconnect the terminals of a three-phase motor for a change of voltage.

5. The three-phase motor you are testing for continuity is a sealed compressor, and your test indicates an open circuit. Explain why you should wait 2 h or more before you make the decision to change the compressor.

True or False

1. The direction of rotation of a three-phase motor can be reversed by exchanging any two of its three terminals.

2. One of the windings in a three-phase motor is smaller than the other two, so a phase shift can be created to help the motor get started.

3. The three-phase motor has more starting torque than an equal-size single-phase motor.

4. It is possible to change the internal winding connections on a nine-lead, three-phase motor so that it can operate on 480 VAC or 240 VAC.

5. It is possible to change the connections on a three-phase motor so it will operate as a wye or a delta motor.

Multiple Choice

1. If you change motor terminal T1 to L2 of the supply voltage and terminal T2 to L1 of the supply voltage, a three-phase motor will _____.

a. not run correctly because it must be wired L1 to T1, L2 to T2, and L3 to T3.

b. change the direction of its rotation.

c. be able to run on 240 VAC instead of 480 VAC.

2. A three-phase motor does not require any current relays, potential relays, or centrifugal switches because _____.

a. it has three equal windings instead of a start windings.

b. it is a low-torque motor, and these relays would cause its torque to become too large.

c. these switches are mainly used to control for over-current conditions.

3. The phase shift that is required to create torque for a three phase motor is _____.

a. created by adding capacitors to the motor windings.

b. found naturally in the three-phase voltage used to provide power for the motor.

c. created by making each of the windings in the three-phase motor slightly different.

4. The reason you need to know how to reconnect a three-phase motor for a change of voltage is _____.

a. because the voltage rating of the motor that you have in stock is rated 480 VAC and the motor that it is replacing is rated for 240 VAC.

b. because the motor may need more speed.

c. because some three-phase motors need DC voltage instead of AC voltage.

5. If the three-phase motor runs but draws excessive current, you should suspect _____.

a. a faulty centrifugal switch or current relay.

b. the motor is running the wrong direction and you should exchange any two leads.

c. one of the motor windings is open, or one phase of the three-phase voltage is not supplying voltage.

Problems

1. Draw the sketch of a wye-connected, three-phase motor.

2. Draw the sketch of a delta-connected, three-phase motor.

3. Draw the connections that you would use to connect a wye-wired motor for low voltage and for high voltage.

4. Draw the connections that you would use to connect a delta-wired motor for low voltage and for high voltage.

5. Draw the electrical diagram of a nine-lead, three-phase motor connected for wye configuration and for delta configuration and explain why each is used.

Chapter 13

Lockout, Tag-out

Objectives

After reading this chapter you will be able to:

1. Identify the lockout, tag-out procedure.
2. Identify sources of electrical energy.
3. Identify sources of other types of energy.
4. Identify sources of potential energy and kinetic energy on a machine.

13.0
Overview of Lockout, Tag-out Safety

A wide variety of safety precautions are used in factories today to ensure that workers are not endangered or injured while they are working. One program in use is called *lockout, tag-out*. This program is designed to ensure that all power to the machine is deenergized and locked in the off position prior to working on the system; and tags are then attached to the disconnect switches to indicate who is working on the machine.

The lockout, tag-out program is mandated in all factories by the Occupational Safety and Health Administration (OSHA). The exact standard is OSHA Standard 1910.147. This program explains techniques that are to be used to ensure, before a technician works on a machine, that uncontrolled release of energy cannot occur. The program is designed to prevent injury that may be caused by uncontrolled machine motion, electrical shock, electrical burns, or injuries from coming into contact with hazardous materials.

The National Institute for Occupational Safety and Health (NIOSH) has maintained data that show a wide number of factory accidents that resulted in serious injury or death. These accidents include personnel being crushed between moving parts on a machine, such as the clamps on a molding machine, or having arms and legs caught in moving conveyor equipment and amputation of arms or legs from blades. These accidents occurred because coworkers did not pay attention when workers were inside equipment and machines or did not understand the dangers of working around machinery that still has power applied to it.

The lockout, tag-out procedure is devised to allow maintenance and other personnel to work safely on and around machinery where unexpected start-up of the equipment or the unexpected release of stored hydraulic, pneumatic, or other types of energy may cause injury or death. The procedures should be used anytime equipment is being set up, inspected, or adjusted or anytime when maintenance is being conducted.

These procedures are particularly important when guards must be removed to set up a machine or change molds and dies. During this time workers are especially vulnerable to conditions where body parts such as arms and legs are exposed to moving parts. In these conditions workers may sustain serious injuries from a crushing accident or they may have their clothing snagged, which can pull them deeper into the moving part of the machinery.

The procedure should also cover the disconnection and lockout of electrical power when necessary. There are times when machines need to be lubricated or cleaned by personnel, and it is important that all power be turned off. This prevents personnel from being shocked if they come into contact with electrical terminals or if the cleaning fluids and equipment they are using come into contact with electrical terminals. All hydraulic or pneumatic power should also be disabled at this time by bleeding the pressure to zero. This prevents accidental activation of a hydraulic or pneumatic actuator, which could cause a cylinder to extend and crush someone.

Other dangerous conditions that could occur due to exposure to dangerous chemicals and fuels must also be controlled. This also includes exposure to fuel intakes and outlets and exposure to extreme heat, such as from steam pipes or other exposed heating elements.

13.1
A Typical Lockout, Tag-out Policy

Each factory must create a policy for lockout, tag-out and ensure that the policy is being enforced. The following illustration may be used as an example, but it does not necessarily include all facets of your shop.

The policy ensures that electrical disconnects for all equipment shall be locked out to the deenergized state where possible while maintenance work is being completed on said equipment. If it is not possible to lockout the electrical disconnect, equipment must be tagged out to protect against possible operation by unauthorized personnel during maintenance procedures. In all cases, all the power sources for the equipment, such as electrical, hydraulic, pneumatic, and gravity, must be disabled and made inoperable to prevent injury to personnel. Employees must never try to operate or reapply any of the power sources that have been locked out or tagged out by another person. The person that applies the lockout or tag-out equipment is the only person authorized to remove these safety devices unless the procedure continues past the shift of that person. If the work continues past the shift of the person applying the lockout device, the person should remove his or her lockout and allow the new shift workers to place their devices on the equipment. If the device must be removed by personnel on a different shift, they must follow the procedure for such work.

13.2
Determining the Lockout, Tag-out Policy and Procedure

The lockout and tag-out procedures should be determined and enforced by each individual company. There are many sources for examples of what should be in-cluded in these procedures. This chapter provides a basic version of a typical procedure. It is limited to a few of the things that you should do during the lockout, tag-out procedure; it is not intended to be a complete procedure. It is important to understand that each work site has specific machines and equipment that may require additional procedures.

When a company is designing its lockout and tag-out procedures, those in charge must keep in mind the different types of energy that the system uses and the way the machine is engineered to operate, such as if it uses gravity for part of the machine operation. While personnel are writing the procedure, they must also have a plan for evaluating how well the procedure is working to prevent accidents, and finally they must also design a means of enforcing their procedures to ensure they are being used at all times by all personnel. The evaluation and enforcement part of the procedure should also include continual education. This ensures that all personnel will remain aware of the potential dangers that are always present and will learn to perform the procedures automatically at all times.

13.3
Identifying Sources of Kinetic Energy and Potential Energy

Energy in and around a machine may take the form of potential or kinetic energy. *Kinetic energy* is energy that is in motion. This is the energy that is very apparent to see and understand when you are around a machine. For example, the energy that a plastic injection molding machine uses to move its mold (clamp) open and closed is easy to see when the clamp is moving. *Potential energy* is energy that is stored and is available for use after the power source is removed. The major problem with potential energy is that its dangers are not always so apparent. For example, a hydraulic system may be designed to store hydraulic fluid in a pressurized accumulator, where it is waiting to be released when the hydraulic valve is activated; the stored pressure then moves a hydraulic cylinder. What is dangerous about this form of energy is that it continues to be present even after the hydraulic pump is turned off and locked out.

13.3.1
Other Forms of Potential Energy

Other forms of potential energy include gravity, which may be used by a machine for part of its operation. For example, during each cycle, a metal stamping press may raise a die several feet above the parts that are being stamped from the die. If the machine stops with the die in the top dead-center position, it has the potential of falling due to gravity when power is turned off. In this type of application, any part of the machine that could move due to gravity must be supported in such a way that it cannot move when power is removed. Another problem arises when the machine or the die must be checked or tested while it is running.

Another example of potential energy is the energy that is stored in rotating parts of a machine, such as a rotary saw blade or rotary grinding wheel. In these applications, the machine uses an electric motor or similar power source to provide large amounts of energy to get the saw blade or grinding wheel up to top speed. When the blade or wheel is spinning at top speed, it has a large amount of energy stored in its mass, which causes it to continue to spin even after the power to the motor is turned off. In most conditions, the heavier the blade or wheel and the higher the speed, the more energy it will store and the longer it will continue to spin when power is turned off. In these applications, the company must evaluate the potential for danger and provide braking to stop the blades or wheels as quickly as possible. Another method of protection is to provide sufficient guards so that personnel cannot come into contact with the moving parts. In some cases the guards are interlocked with the moving parts to cause them to stop moving immediately any time the guard is moved or removed.

Another type of potential energy is the energy that is stored in a spring under tension. In some machines, springs are used to provide an additional source of energy. Anytime a spring is under compression or extension, it has the potential to cause physical harm. This type of energy is not always apparent, because some springs are inside spring-return-type pneumatic cylinders. When the ends of the cylinder are removed during maintenance, the spring may release its energy and cause harm.

13.4
Designing Safety into the Machine

It is important that the personnel developing lockout, tag-out procedures be aware of all these types of energy sources and the potential problems they present. It is also important that they evaluate the machine, provide guards and stops where necessary, and build in other safety equipment to minimize the dangers to personnel.

13.4.1
Documentation and Training

Documentation and training are essential parts of the lockout, tag-out program. Documentation ensures that each procedure is explained in detail so that personnel are continually reminded about using lockout, tag-out procedures at the appropriate times when they are on the job. The training is also essential to ensure that the employees understand the proper procedure and perform it correctly. Training will also help new employees to understand the dangers of working around equipment. It also reminds each employee that he or she has a personal responsibility in ensuring that lockout, tag-out procedures are used at all times.

13.5
OSHA Inspections

The Occupational Safety and Health Administration (OSHA) may become involved if accidents occur that are caused by not following procedures or because incorrect lockout, tag-out procedures have been used. The problem becomes extreme if OSHA finds that the employer does not have lockout, tag-out procedures or if they are not being enforced.

13.6
What Is Lockout?

Lockout procedures include manually disconnecting the electrical and hydraulic power and placing a locking device (usually a padlock) on the disconnect so that it cannot be reconnected without removing the padlock. If more than one employee is working on the equipment, each should place his or her padlock on the disconnect. A special device that allows multiple padlocks to be connected on the same disconnect is shown in Figure 13–1. From this diagram you can see that when the clasp is closed, up to six padlocks can be installed at one time. The clasp is secured around the disconnect, and it cannot be opened until the last padlock is removed.

The lockout, tag-out procedure also requires that each padlock and lockout device be identified with a picture and the name of the employee who placed the lockout on the disconnect. It is important to understand that the lockout devices must be supplied by the employer and are to be maintained by the employee.

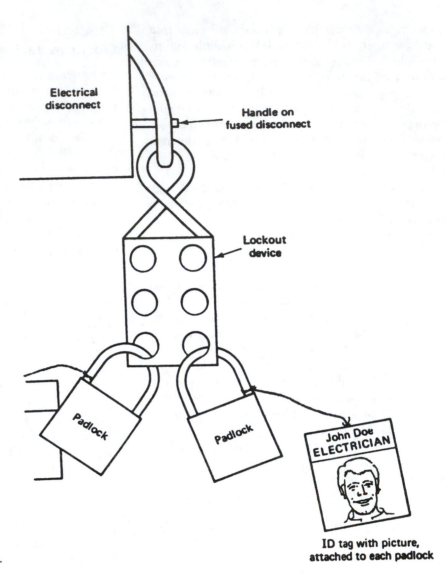

Figure 13–1 A typical lockout device that can accept multiple padlocks.

In some cases, the work on a machine involves multiple personnel, and multiple padlocks are applied to the disconnect. When all work is completed and it is time to return power to the machine, all the locks must be removed, and all personnel must be accounted for so that they are not in danger.

13.7
What Is Tag-out?

Figure 13–2 shows a typical tag used for tag-out procedures. From this figure you can see that the tag has a written warning that explains why the machine has been tagged. This warning also explains that the power should not be applied to the machine until the person working on the machine has completed his or her work and is ready for power to be reapplied.

It is important that all tags used in the lockout, tag-out procedure be uniform and standardized so that they are easily recognized by all employees.

13.8
Designing a Lockout, Tag-out Procedure for Your Machines

The first step in a typical lockout, tag-out procedure is to evaluate each machine and determine all dangerous conditions that are present when the machine is running and when it is locked out for maintenance. This evaluation includes identifying all possible sources of power to the machine and listing them by type, such as electrical, hydraulic, and pneumatic. It also includes identifying any hazardous material, chemicals, sources of radiation, or exposure to sources of thermal (heat) energy.

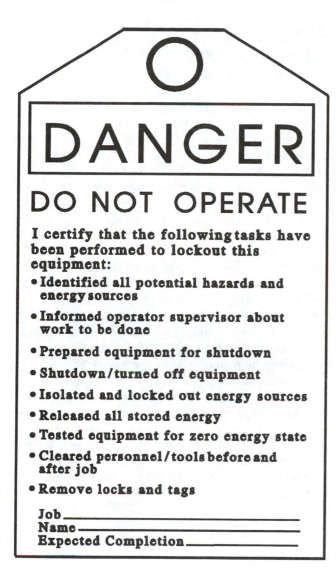

DANGER

DO NOT OPERATE

I certify that the following tasks have been performed to lockout this equipment:

- **Identified all potential hazards and energy sources**
- **Informed operator supervisor about work to be done**
- **Prepared equipment for shutdown**
- **Shutdown/turned off equipment**
- **Isolated and locked out energy sources**
- **Released all stored energy**
- **Tested equipment for zero energy state**
- **Cleared personnel/tools before and after job**
- **Remove locks and tags**

Job _____

Name _____

Expected Completion _____

Figure 13–2   Typical tag-out tag.

The next step includes identifying all employees who may work on or around the machine. These include operators, inspectors, supervisors, and others who may work with the machine. All employees on this list should always be notified when the machine is going to be turned off, locked out, and tagged out. It is also important that there is a way to notify affected personnel who work on other shifts. The notification should include the name of the person who shut the machine down and what they are working on. Other information could include the expected time when the machine will be in operation again.

The third step includes the actual shutting down of the equipment. This process includes pressing the stop button, moving any other switches to the off position, and making sure all motion on the machine is stopped. This step also includes securing all sources of energy in the off condition and ensuring that all forms of potential

energy are neutralized so that they do not present a hazard. After you have identified all sources of energy and you have ensured they are in the off position, you can apply lockout, tag-out hardware to ensure that they remain in the off position until you are ready to turn them back on.

It is important at this time to identify several methods of disrupting power that are unacceptable because they can easily be bypassed and pose a danger to workers. For example, if you simply remove the fuses to a machine and do not lockout the fuse box, someone could replace the fuses and apply power to the machine while you are still working on it. If you turn off a pneumatic hand valve but do not lock valve in the off position by a chain or other blocking device, it could be turned on again and cause serious injury.

It is also important to identify any interlocking power sources or auxiliary power sources for each machine. For example, if a machine uses a robot to load and unload it, it is possible that some electrical signals from the robot cabinet are sent to the electrical cabinet of the machine. If the machine is locked out but the robot is not, you could have problems with the robot moving into the machine space, or you could have problems with electrical wires from the robot that terminate in the machine cabinet still being powered. Because many of these signals are 110 VAC, they can pose an electrical shock hazard.

After all power is shut down and locked in the off position, you should make a number of tests to ensure that all power is, in fact, turned off. You should also test all blocks that are in place to ensure that they are secure and holding the machine in a safe condition. After all tests are made, you can begin your work.

13.9
Removing the Lockout, Tag-out Devices and Returning Power to the Machine

After all work has been completed on the machine, you are ready to remove the lockout, tag-out devices and return power to the machine. It is important to understand that you must make several checks prior to removing the lockout, tag-out devices. The first step in this process is to ensure that all work is complete and that all tools and any test equipment has been removed. Return all guards and safety doors to the normal condition and test them to ensure that they operate correctly. Be sure that all parts of the machine that have been disassembled are reassembled properly.

When you have completed these tasks, you must physically account for all personnel who are in the area to ensure that they know and understand that you are ready to turn on power to the machine again. This is probably the most important part of the procedure, especially if more than one person is working on the machine or if several employees routinely work on the machine when it is in normal operation.

When you are ready to apply power again, be sure you follow the operational check list so that power is returned to the parts of the machine in the proper order. For example, some machines must have circulating pumps running before steam valves are returned to their open positions.

13.9.1
Removing a Coworker's Lockout and Tag-out Equipment

The general rule is that the person who placed the lockout on the machine must be the one who removes it. But there will be times when work is started on the day shift and completed 14 hours later on another shift. In this case the person who placed the lockout, tag-out equipment on the machine is not present to remove it. When this occurs, it is permissible for a supervisor or person of responsibility to remove the lock. The personnel on each shift who may remove a lock should be identified when the lockout, tag-out procedure is written. The supervisor may have a master key; it may also be permissible to cut the lock off with bolt cutters. The person who removes the lock should always file a report to indicate why the lock was removed and who performed the action.

13.10
Review of Lockout, Tag-out Procedures

The lockout, tag-out procedure is designed to ensure all sources of energy are locked out and tagged out whenever maintenance procedures require that power must be removed from the machine. The lock ensures that the disconnect switch will remain in the off position, and the tag provides a means of identifying who placed the lock on the disconnect. When the procedure is created, a comprehensive review of all safety aspects of a machine should be accounted for. The lockout, tag-out practices should include training and education to ensure that all employees understand the safety hazards present with each machine and that they continually use correct safety procedures when working on and around the machine. The procedures should identify all sources of kinetic and potential power as well as sources of hazardous material, sources of thermal energy (steam), and sources of radiation.

You should also remember that lockout and tag-out procedures are mandated and enforced by OSHA, and all procedures, equipment, and training are to be provided by the employer. Each employee is responsible for utilizing proper lockout and tag-out procedures and for maintaining the equipment that has been issued to them. Be sure to use the correct lockout and tag-out procedure any time you are working around equipment.

Questions for This Chapter

1. Explain what a lockout, tag-out procedure is.
2. For what do the initials OSHA stand?
3. What is kinetic energy?
4. What is potential energy?
5. Who is allowed to remove a padlock from a lockout?

4. The lockout device should have a picture of the owner on it in case the lockout is lost.
5. It is permissible to have electrical power applied to a machine if you need to make voltage readings as part of your troubleshooting procedure.

True or False

1. A vertical press that is in the open position so that its platen could fall is an example of potential energy.
2. A fluid accumulator that is under pressure for a hydraulic press is an example of kinetic energy.
3. Lockout devices are available that allow more than one padlock to be connected to the same lockout.

Multiple Choice

1. Who is responsible for creating the lockout, tag-out procedure in a factory?
   a. OSHA.
   b. Each worker should create his or her own procedure.
   c. Each factory should create their own procedure.
2. The forms of energy that you should be aware of during a lockout procedure include _____.

a. hydraulic and electrical power.

b. hydraulic, electrical, and potential energy sources such as gravity.

c. only electrical power sources.

3. If an employee on the day shift places a padlock on a lockout device and a second shift employee completes the repair, _____.

  a. the second-shift supervisor can remove the first-shift padlock.

  b. the second-shift employee can remove the first-shift padlock.

  c. no one on the second shift can remove the padlock, and they will have to wait until the employee that placed the padlock on the lockout comes back to work.

4. Prior to removing a padlock from a lockout device and restarting a machine, you should _____.

  a. replace all guards and turn on the power.

  b. alert all employees in the area who work on this machine that you are turning on the power and starting the machine.

  c. call the operator to start the machine.

5. The major problem with potential energy sources during a lockout procedure is _____.

  a. the hazard they present may not be readily apparent; for instance, a machine part that is supported by hydraulic pressure may drop when the pressure is released.

  b. potential energy is much more dangerous than electricity.

  c. potential energy is not a real threat, as is kinetic energy, so it can be ignored.

Problems

1. You are asked to create a lockout, tag-out procedure for a plastic injection molding machine that has both electrical power and hydraulic power. List all of the things of which you should be aware for the lockout procedure.

2. List four sources of potential energy of which you should be aware when you devise a lockout procedure.

3. List the things you should do prior to restarting a machine when you are ready to remove the padlock from the lockout device.

4. You are asked to help implement a lockout, tag-out procedure for your factory. Explain what you would consider as part of the education and training of all employees.

5. You are asked to help develop an inspection procedure for the lockout, tag-out procedures in your factory. Explain what you would include in creating the inspection part of the lockout, tag-out program.

◀ Chapter 14 ▶

Motor-Control Devices and Circuits

Objectives

After reading this chapter you will be able to:

1. Identify pilot devices in a control circuit and their symbols in electrical diagrams.
2. Identify the load circuit and control circuit.
3. Explain the operation of a push-button switch, limit switch, flow switch, level switch, and pressure switch as they are used as pilot devices.
4. Identify a two-wire control circuit.
5. Identify a three-wire control circuit.
6. Explain the operation of a drum switch as it is used to reverse the operation of an AC or DC motor.
7. Explain the operation of a single-element and a dual-element fuse.
8. Identify the proper enclosure for specific industrial applications.

14.0
Introduction

In this chapter we explain basic motor-control circuits commonly in use on the factory floor. It is very important to understand that many of the electronic circuits are used to supplement or replace standard motor-control circuits that have been in operation for the last 30 years. These motor-control circuits have been named and are easily identified so that when you hear about them on the job or when you see them on the job you will easily recognize them. You must also understand that in the past few years, companies have changed their concept of an electronic technician, and they now

expect to hire each person to troubleshoot and repair the motor-control systems as well as the electronic systems. Traditionally the motor-control systems have been tested by electricians and the electronics technician. In the past 10 years, companies have streamlined their personnel so that they expect the maintenance technician to work on motor-control circuits. For this reason it is vitally important that you learn as much as you can about standard motor-control circuits so that you can compete for the best-paying jobs.

The circuits in this chapter are identified so that you can come back and study them when you are on the job if you need to review them. This will be extremely useful when you find machinery that incorporates four or five of these basic circuits with additional complex circuits. In these circuits you will be able to gain knowledge that you can transfer to more difficult circuits for troubleshooting and repair. These circuits will include pilot devices, motor starters, and different types of motors. If you do not understand the function or operation of relays and motor starters, you may need to review the chapters where they were introduced.

14.1
Pilot Devices

Pilot devices are a group of components that include push-button switches, limit switches, and other switches that are commonly found in motor-control circuits. These switches are called pilot devices because they are rated for control-circuit voltage and current. That is, these switches normally have 115 VAC and less than 1 A current flowing through their contacts. The control circuits use pilot duty switches to energize or deenergize the coil of a relay, motor starter, solenoid, or indicator lamp. All the loads in these circuits use low current. Again, you must understand that the L1–N voltage in your part of the country may be any value between 110 VAC and 125 VAC. In this chapter all these voltages are referred to as 115 VAC.

Figure 14–1 shows a set of multiple push-button switches and selector switches that are pilot devices used to control a large industrial machine. These push buttons are the most commonly used switches to start and stop a machine, and selector switches are used to select the operation of the machine for such functions as manual or automatic operation. Figure 14–2 shows

an exploded view of a typical push-button switch installation. From this diagram you can see that the switch is actually an assembly of a number of parts, including the push-button actuator on which the operator presses to activate the switch, a number of seals to keep moisture and other contaminates out of the switch assembly, and the contact block and contacts where the wires are actually connected. When the push-button activator is depressed, the contacts in the bottom of the switch transition. If the switch has normally open contacts, they will go closed, and if the switch has normally closed contacts, they will go open. Some switches will have one set of contacts, whereas other switches have multiple sets of contacts. The important point to recognize is that all wires are connected to the contact block, and the other parts of the switch can be changed without removing any wiring. This allows switches to be changed in the field with a minimum of downtime for rewiring.

14.1.1
Limit Switches and Other Pilot Switches

A variety of pilot switches are available for use in control circuits. One of the most popular is called a *limit switch*. Figure 14–3 shows examples of several types of limit switches. These switches are physically mounted in the machine so that the motion of the machine or the parts around the switch will cause the switch to activate. Figure 14–3a shows a yoke-type limit switch. The yoke has a roller on each section of its activator arm. This arrangement is used where machine motion is in two directions. When the machine moves in one direction, it strikes the right-hand roller and causes it to switch to the right. This action causes the right roller to be pressed downward, and the left roller snaps to the up position, where it is in position to detect machine motion when the machine moves back to the left. The yoke

Figure 14–1 A typical set of push-button switches and selector switches used to operate a machine.
(Courtesy of Rockwell Automation's Allen Bradley Business)

Figure 14–2 An exploded view of a typical switch. Notice that the switch activation pushbutton is shown above the panel where it is accessible to the operator, and the remainder of the switch, including rubber seals, and the switch contacts are shown below the panel.
(Courtesy of Rockwell Automation's Allen Bradley Business)

arm on the switch will continue to detect motion to the right and then to the left. This type of switch is very useful for surface grinders, where the part being finished is moved back and forth under the grinding wheel. Each time the part moves to the end of its stroke, the switch is activated in the other direction.

Figure 14–3b shows a limit switch with a single roller arm. This type of switch detects motion in only one direction. When the machine moves past the roller arm, it moves the arm to the right so that the switch is activated. Figure 14–3c shows a top-roller-type limit switch. This type of switch has a small roller mounted on a plunger-type arm. When the machine moves past the roller, it causes the plunger to depress to activate the switch. When the machine moves past the switch, a spring causes the plunger to move back upward into place to be activated again. This type of switch can detect motion in either direction. Another variation of this type of switch uses a blunt end on the plunger, which means the switch can detect only motion that is directed down on the plunger. This type of switch is generally mounted at the very end of machine travel, and

when the machine reaches the end of its stroke, it depresses the plunger.

Figure 14–3d shows a wobble-lever-actuated limit switch that is sometimes called a *cat whisker*–type limit switch. This type of limit switch can detect motion in any direction. This means that any movement of the cat whisker can cause the switch to activate. This type of switch is also much more sensitive than other types of limit switches. Figure 14–3e shows a side-roller-type limit switch. This type of limit switch is mounted in such a way as to detect end of travel. When the machine reaches the end of its travel, it depresses the plunger and causes the switch to activate.

14.1.2
Symbols for Limit Switches and Other Motor-Control Devices

The limit switch can be connected as a normally open or a normally closed switch. It can also be used on a machine in such a way that the machine travel will hold the switch in the normally open or normally closed position before the machine starts its travel. Figure 14–4 shows the four electrical symbols for a limit switch. The top left switch symbol is for a *normally open limit switch*. In this symbol notice that the switch's contact arm is shown below the terminal on the right side of the switch and that the switch is shown as a normally open switch. The *normally open, held-closed limit switch* is shown in the bottom left diagram. You should notice this is similar to the normally open switch in that the switch contact arm is shown below the switch's output terminal on the right side. The major difference with this switch is that it is shown with normally closed contacts. The reason it is shown with normally closed contacts is because the machine motion will keep this switch in its normally closed position.

Figure 14–3 (a) Fork-lever-roller yoke-type limit switch; (b) roller-arm-type limit switch; (c) top-roller-type limit switch; (d) wobble-lever-actuated cat whisker limit switch; (e) side-roller-type limit switch.

(Courtesy of Honeywell's Micro Switch Division)

LIMIT SWITCHES	
Normally Open	Normally Closed
(symbol)	(symbol)
Held Closed	Held Open

Figure 14–4 The electrical symbol for normally open limit switch, normally open, held-closed limit switch, normally closed limit switch, and the normally closed, held-open limit switch.

NOTICE!

It is important to remember that when you are purchasing limit switches that the switch is available only as a normally open or as a normally closed switch. The held-open and held-closed conditions occur when the switch is mounted in place on machinery and the location of the switch causes the machine to hold the switch in the activated position.

The switch in the top right section of this figure is a *normally closed limit switch*. This symbol is different in that the switch contact is shown on top of the switch's output terminal. When the machine motion activates the switch, it moves the contact arm upward to its open position. When this switch is wired in the field, you should select the normally closed contacts.

The symbol for the *normally closed, held-open limit switch* is shown in the bottom left corner of this figure. You can see this symbol shows the switch's contact arm is shown in the open position, and it is shown above the output terminal. When you wire this switch in the field, you use the normally closed set of contacts, and the position of the machine when it is at rest keeps the switch held in the normally open position.

14.1.3
Electrical Symbols for Other Pilot Devices

Other pilot devices, such as pressure switches, flow switches, float switches, temperature-activated switches, time-delay relays, and lamps, are also used in motor-control circuits. The symbols for these devices are shown in Fig. 14–5. These symbols are a standard for pilot devices used in motor-control circuits. When you encounter a typical motor-control diagram on a blueprint, you can use the symbols in this figure to identify each type of switch. You should notice that the symbol for the part that causes the switch to activate is shown at the bottom of each symbol. For example the symbol for a liquid-level switch is a circle, which represents the ball-type float typically found on this type of switch, and the symbol for the pressure-type switch looks like a pressure dome. The symbol for each switch should look very similar to the actuator for that switch. You should also notice that the contact arm for each switch is shown in the normally open or normally closed position.

When you examine each symbol carefully, observe whether its contacts are normally open or normally closed. You should also understand that the switch arm can be drawn on top of the terminals or below the terminals. Each of these conditions allows four combinations of switches to be identified. It is important to understand these subtle differences. For example, Fig. 14–6 shows a pressure switch drawn with its contact arm in each of the four ways described previously, so each switch is different from the other. Figure 14–6a shows the switch arm normally closed and above the contact terminal; the switch is activated by pressure. This indicates that the switch is a normally closed switch, and an increase in pressure causes the switch to open. It may help to remember the switch symbol by the type of application for which the switch would be used. For example, this type of switch is used as a high-pressure safety switch. When the pressure gets too large in a system, the switch opens and shuts off a compressor that is adding pressure to the system. The pressure switch in Fig. 14–6b shows a normally open switch with the switch arm above the contact terminal. This type of switch activates a circuit when pressure falls. This type of switch can be used to turn on a compressor when air pressure drops below a minimum so the system will sustain the pressure. Because the switch arm is drawn above both these switches, they may also be called *high-pressure switches* when equipment manufacturers refer to them, because they typically sense high pressure.

Figure 14–6c shows a normally closed pressure switch with the switch arm below the contact terminal. This symbol indicates that when pressure decreases, the switch contacts will fall to the open position. This type of switch can be used as a low-pressure safety switch. For example, if a system is to maintain a minimum oil pressure, the switch will open if the oil pressure drops below the setpoint. Figure 14–6d shows a normally open pressure switch with the switch arm below the contact terminal. When pressure increases, it causes the switch arm to go up and close. This type of switch can be used to start a motor when the pressure gets too high. Because the switch arm is shown below the contact on both these switches, they may be referred to as *low-pressure switches*.

It is important to remember that the pressure switches are sold only with normally open or normally closed contacts or the combination of one set of normally open and one set of normally closed contacts. Again, as with the limit switches, the conditions in which the switches are used when they are mounted on machinery may be referred to in the name the machine manufacturer gives the switch. For example, the switch may be called the compressor high-pressure cutout safety, or it may be called the low-oil-pressure protection switch. In both cases, the switch is still a pressure switch, and a set of normally open or normally closed contacts are used to control circuit current when pressure increases or decreases.

You should now understand that whichever way the contact arm is drawn, a decrease in the physical para-

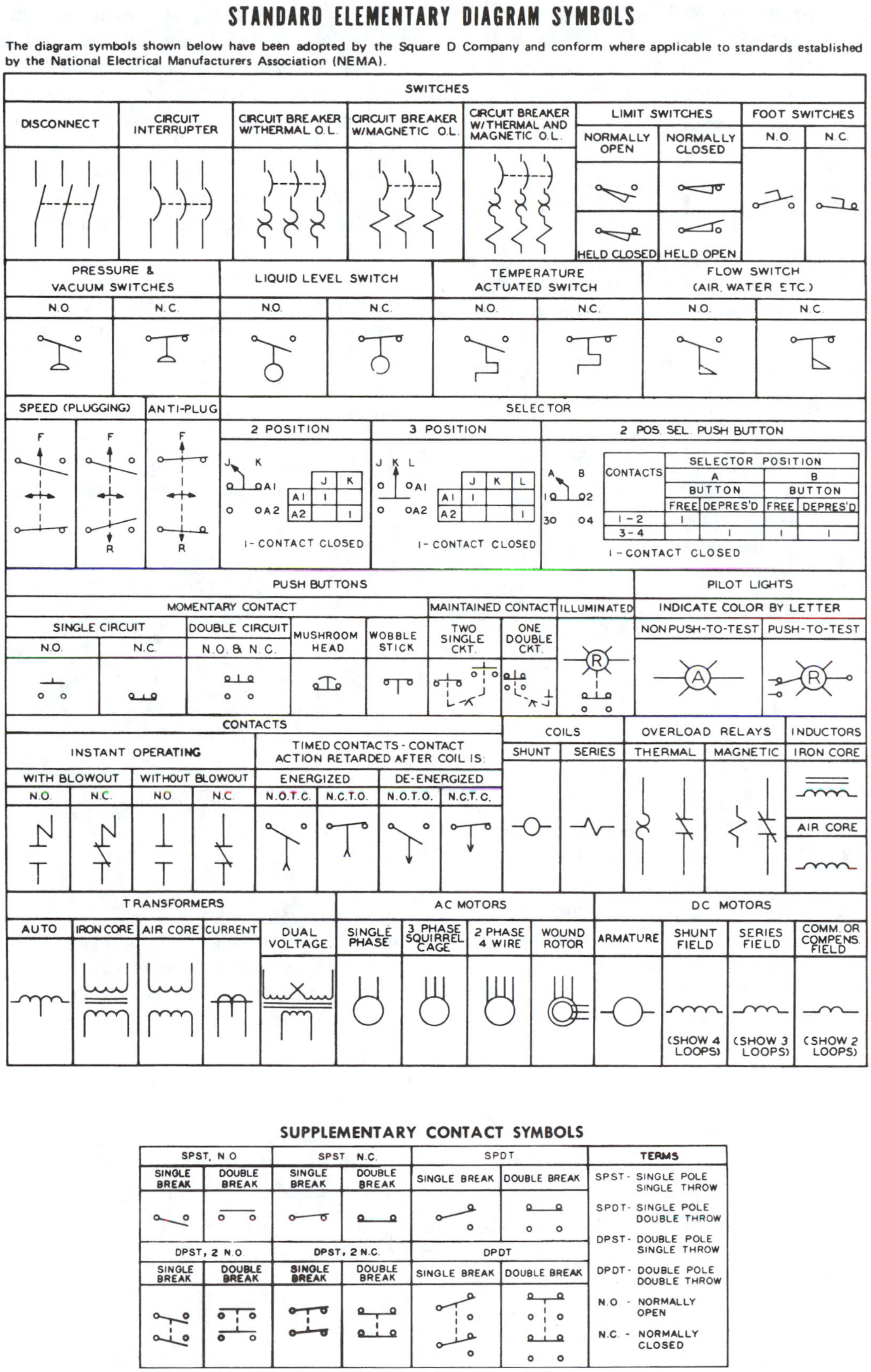

STANDARD ELEMENTARY DIAGRAM SYMBOLS

The diagram symbols shown below have been adopted by the Square D Company and conform where applicable to standards established by the National Electrical Manufacturers Association (NEMA).

Figure 14–5 Electrical symbols for pilot devices and other motor-control devices.

(Courtesy of Square D Company)

meter the switch is measuring will cause the switch arm to drop, and an increase will cause the arm to raise. The next section shows how all the pilot devices are used together to make common motor-control circuits.

14.2
Control Circuits and Load Circuits

In motor-control circuits, the components are broken into two groups: the *control components*, which are the switches and coils that control the load, and the *load components*, such as motors that cause the system to perform some type of action. The part of the circuit that has the controls is called the *control circuit*, and the part of the circuit that has the motor in it is called the *load circuit*. Typically, the switches and wires in the control circuit use lower voltages than the load and smaller currents than the load. In fact, in most circuits, the load is a single-phase or three-phase motor.

Figure 14–7 shows a ladder diagram of a control circuit and a load circuit. You should notice in this figure that the control circuit is shown on the left, and it has a float switch connected to the relay coil. A normally closed set of overload (OL) contacts are shown connected in series with the motor-starter coil. The load circuit is shown on the right side of the figure, and you can see that it consists of a three-phase motor starter and a three-phase motor. (*It is important to note that the relay coil is technically a load, but it is considered part of the control circuit because of the function it performs.*)

The same control and load circuits are shown as a wiring diagram in Fig. 14–8. When this circuit is shown as a wiring diagram, you will notice that some of the

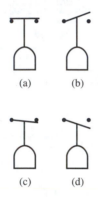

Figure 14–6   (a) Normally closed pressure switch. (This switch may be referred to as a normally closed, high-pressure switch.) (b) Normally open pressure switch. (This switch may also be referred to as normally open, high-pressure switch.) (c) Normally closed pressure switch. (This switch may be referred to as a normally closed, low-pressure switch.) (d) Normally open pressure switch. (This switch may be referred to as a normally open, low-pressure switch.)

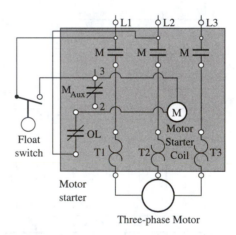

Figure 14–8   An example of a wiring diagram that shows a control circuit and a three-phase motor in the load circuit. Notice the motor starter is shaded in the wiring diagram so that you can locate it and all its terminals.

Figure 14–7   A ladder diagram of a control circuit and a load circuit.

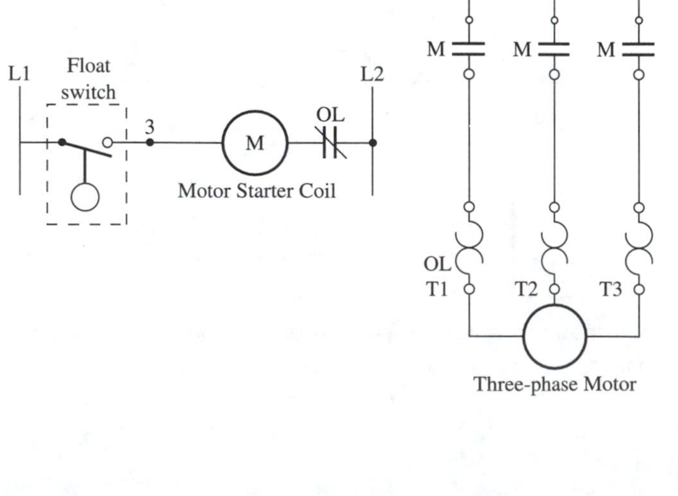

control-circuit components, such as the motor-starter coil, are drawn where they are physically located, on the motor starter. Because the motor-starter contacts are part of the load circuit, it is hard to determine in a wiring diagram what components are part of the load circuit and what components are part of the control circuit. In some diagrams the load circuit is usually drawn with darker and wider lines than the wires of the control circuit to indicate the wires are larger. The remaining part of this chapter explains the different types of control-circuit and load-circuit configurations.

14.2.1
Two-Wire Control Circuits

The two-wire control circuit is commonly used in applications where the operation of a system is automatic and basically two wires are used to provide voltage to the load. This may include such applications as sump pumps, tank pumps, electric heating, and air compressors. In these systems you typically close a disconnect switch or circuit breaker to energize the circuit, and the actual energizing of the motor in the system is controlled by the operation of the pilot device.

The circuit is called a two-wire control because only two wires are needed to energize the motor-starter coil. The circuit controls that are located prior to the coil will provide the operational and safety features, and the overload contacts that are located after the coil are used to protect the circuit against over-currents.

Figure 14-7 shows a ladder diagram of the two-wire control circuit for a pumping station and Fig. 14-8 shows the control circuit as a wiring diagram. You will notice that it is difficult to see the sequence of operation in the wiring diagram, so you generally use the ladder diagram to determine the sequence of operation. The wiring diagram shows the location of all wires and components.

Voltage to the pumping station is provided through a fused disconnect (not shown). After the disconnect is closed, the float switch is in complete control of the motor starter. Notice that the float control is identified as a float switch. When the level of the sump rises past the set point on the control, its contacts energize, and the motor starter coil becomes energized. The coil of the motor starter energizes and pulls in the motor-starter contacts, which energize the motor. As the motor operates, it pumps the liquid, which reduces the level in the sump. When the level of the sump is lowered, the level switch senses the change in level and switches its contacts off, which deenergizes the motor-starter coil. When the motor-starter coil is deenergized, the motor-starter contacts are opened, and the motor is turned off.

This on-off sequence continues automatically until the disconnect is switched off or unless the motor starter overloads are tripped. If the motor draws too much current when it is pumping, the heaters trip the overload contacts and open the control circuit, which interrupts current flow to the motor-starter coil. Because the overloads must be reset manually, the circuit does not become energized automatically when the overload is cleared. In this case, someone must physically come to the motor starter and press the reset button. At this time, the system should be thoroughly inspected for the cause of the overload. It is also important to use a clamp-on ammeter to determine the full-load amperage (FLA) of the motor and cross reference this to the size of the heaters in the motor starter to see that they match. The FLA should be checked against the motor's data plate. You should also take into account the service factor listed on the data plate when you are trying to determine if the pump motor is operating correctly.

Figure 14-9 shows a more complex two-wire control circuit that is used to control an air compressor. Notice that the wiring diagram for the control and load circuit is shown in Fig. 14-9a, and the ladder diagram of the

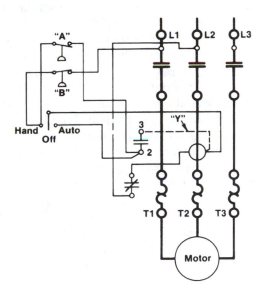

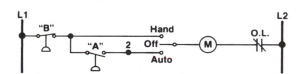

Figure 14-9   (a) A wiring diagram of a two-wire control circuit for a compressor motor. The compressor motor is controlled by a motor starter that is energized by one of two pressure switches. (b) A ladder diagram of the same circuit shown in the wiring diagram. In this circuit you can see a selector switch provides hand or auto control. Pressure switch A is in the circuit when the selector switch is in auto, and pressure switch B is in the circuit when the selector switch is in manual.

(Courtesy of Rockwell Automation's Allen-Bradley Business)

control circuit is shown in Fig. 14–9b. This circuit is still a two-wire control circuit, and it is used to turn the air compressor on automatically when the pressure drops below 30 psi and to turn off the compressor when the pressure reaches 90 psi. Pressure switch A in this circuit controls the operation of the control at these pressures. Its high and low pressures are adjustable so that the system can be energized and deenergized at other pressures if the need arises. A hand switch is provided in this circuit that allows the system to be pumped to a predetermined pressure. The hand switch is intended to be used when the automatic switch is not functioning properly or when you need to test the system.

Pressure switch B in this diagram acts as a safety for this circuit. This pressure switch is set at 120 psi and is not adjustable. It is in the circuit to prevent the tank pressure from rising too high. This can occur if the operational pressure switch becomes faulty and will not open when the pressure reaches 90 psi. It can also be used to protect the system against too much pressure when the switch is in the hand position. Generally, the safety switch is meant to protect the system against component failure or control failure.

If a pressure control were to fail, the compressor would continue pumping air into the tank, which would allow its pressure to rise to an unsafe level. Because the operational switch would still be closed, the air pressure in the tank could be increased to a level where the pump would cause the motor to stall if the safety switch were not in the circuit. This could build up pressures to several hundred pounds, which could cause lines and fittings in the air system to explode.

The nonadjustable pressure switch acts as a backup to the operational switch and trips off any time the pressure reaches 120 psi. This switch is also different in that it is interlocked when it trips, so that it must be reset manually by having someone pressing the reset button. If you find the reset button activated on the safety switch, you must test the system thoroughly to determine why the operational switch did not control the circuit.

In normal operation, the motor starter cycles the air compressor on and off to keep the pressure in the tank between the high and low set points on the operational control. The motor-starter overloads could also trip this circuit and require manual reset. This could occur if the motor were incurring over-current problems. The over-current could occur due to bad bearings or because someone has physically depressed the switch. When this push button is released, the normally open contacts return to their normally open state.

The normally open auxiliary contacts from the motor starter close when the motor-starter coil is energized. When they close they provide an alternative path around the push-button contacts for current to get to

the coil. These contacts are called *seal* or *seal-in* contacts when they are used in this manner. Sometimes the seal contacts are said to have memory, because they maintain the last state of the push buttons in the circuit.

14.2.2
The Three-Wire Control Circuit

The *three-wire control circuit* is the most widely used motor-control circuit. This circuit is similar to the two-wire circuit except it has an extra set of contacts that are connected in parallel around one of the original pilot switches to seal it in. The extra set of parallel contacts provides the third wire, which also gives this circuit its name. You should fully understand this circuit and learn to recognize it when it is shown as a ladder diagram and as a wiring diagram. In this way you will be able to understand the operation of the circuit wherever it appears.

One variation of the three-wire circuit is shown in Fig. 14–10. In this figure, a stop and a start push button

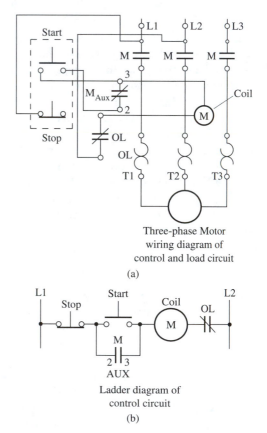

Three-phase Motor wiring diagram of control and load circuit

(a)

Ladder diagram of control circuit

(b)

Figure 14–10 A typical three-wire control system. This control circuit gets its name because of the auxiliary contacts that are connected in parallel with the start button. The auxiliary contacts seal in the circuit to keep the coil energized after the start push button is released.

(Courtesy of Rockwell Automation's Allen-Bradley Business)

are used as the pilot devices for control. Figure 14–10a shows the wiring diagram for the three-wire circuit, and Fig. 14–10b shows the ladder diagram for the three-wire control circuit. In the ladder diagram you can see that the stop push button is wired normally closed. When the start push button is depressed, voltage flows through its contacts to energize the motor-starter coil. When the coil becomes energized, it pulls its main contacts closed to cause the motor to become energized, and it also pulls its normally open auxiliary contacts that are connected across the start push button to their closed positions. When the auxiliary contacts close, they provide an alternative route for voltage to travel around the start push-button contacts, which will return to their normally open position when the switch is no longer depressed. The auxiliary contacts are called the *seal-in* circuit or, in some cases, the *memory circuit*. If the auxiliary contacts are not used, you need to keep your finger on the start button to keep the circuit energized.

You should notice that in the three-wire circuit, the stop button is used to unseal the circuit and deenergize the motor-starter coil. Any time the stop button is depressed, the coil becomes deenergized and the seal-in circuit drops out. This means that the start push button must be depressed again to energize the circuit.

It is also important to understand in the three-wire circuit that the stop push button traditionally is always the first switch in the circuit. Most people assume this has something to do with safety. In reality, it has to do with economics. When you look at a motor starter, you will notice that the auxiliary contacts are physically located near the coil. If the start button is connected in the circuit next to the coil, the wire that is connected to the auxiliary contact on the right side (terminal 3) can actually be a short jumper wire that is connected directly to the left side of the coil. If the start button is not located next to the coil, the wires that connect the auxiliary contacts in parallel with the start push button need to be as long as the distance between the start push button and the motor starter. This may be a distance of several hundred feet.

14.2.3
A Three-Wire Start-Stop Circuit with Multiple Start-Stop Push Buttons

Additional start and stop push buttons may be needed on a system that is very large. For example, if the system is installed over a large area, such as for a long conveyor system that may be more than several hundred feet long, it would be inconvenient and unsafe to have one start and stop button at only one end of the system. To make the system more safe and to make it more convenient to start and stop the system, push-button

switches can be installed every 40 ft. Each additional start button should be connected in parallel with the first start button, and all additional stop buttons should be connected in series with the original stop button. Figure 14–11 shows examples of the additional start and stop push buttons in a three-wire circuit.

14.2.4
Three-Wire Control Circuit with an Indicating Lamp

An indicator lamp can be added to a three-wire circuit to show when the coil in the circuit is energized or deenergized. The indicator lamp can be green to show when the system is energized and red to show when the circuit is deenergized. The lamp is usually mounted where personnel can easily see it at a distance. Sometimes the indicators are used where one operator must watch four or five large machines. After a machine has been set up, the operator will move on to the next machine.

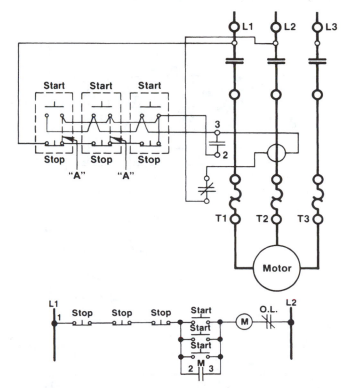

Figure 14–11 The wiring diagram for a three-wire control circuit with additional start and stop push buttons added to the circuit is shown at the top of this figure, and the ladder diagram of just the control circuit is shown at the bottom.

(Courtesy of Rockwell Automation's Allen-Bradley Business)

Because the installation is very large and spread out over a distance, the operator can watch for the indicator lamps to see if the machine is still in operation.

The indicator lamps can also be used by maintenance personnel when a machine has more than one motor starter. In this type of application, each motor starter has an indicator lamp to indicate its coil is energized. This helps the maintenance personnel begin testing for faults in the correct part of the circuit when the system has stopped or is not operating correctly. This is especially useful if the motor starter is mounted in a NEMA (National Electrical Manufacturers Association) enclosure, where the technician cannot see through the enclosure door to verify if the starter is energized or deenergized.

Figure 14–12 shows a wiring diagram and a ladder diagram of a circuit with an indicator lamp connected in the control circuit. The lamp is called a *pilot light*, and it is connected in parallel with the seal-in contacts on the motor starter. When the motor starter closes, the lamp is energized to indicate that the motor-starter contacts are closed. Anytime the lamp is deenergized, the operator and maintenance personnel know that the motor starter is not energized. A press-to-test lamp can also be used in this circuit. The press-to-test lamp allows the operator and maintenance personnel to put their fingers on the lamp and depress the lens at any time to test it to see if it is operational. When the lamp lens is pressed, it causes a special set of contacts in the base of the lamp holder to provide voltage instantly to the lamp and illuminate it. If the bulb is burned out, the lamp does not illuminate, and the maintenance personnel or the operator can change the bulb. If the indicator is energized most of the time, such as in a continuous operation, the press-to-test lamp may not be necessary, because the indicator is energized most of the time.

Indicator lamps are available for 120 V, 240 V, 480 V, and 600 VAC, which means they are available for any control-circuit voltage. Various colored lenses are also available to indicate other conditions with the machine. These indicators can be connected across different individual motor starters in the machine to provide other information, such as hydraulic pumps running, heaters energized, conveyor in operation, and other conditions that are vital to the machine. Figure 14–13 shows an

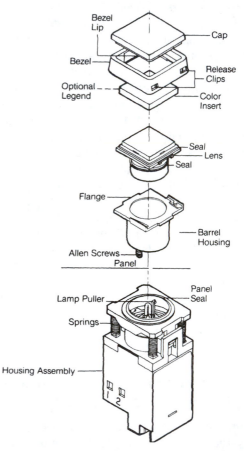

Figure 14–13   An exploded view of a typical indicator lamp. Notice that the lens is replaceable, so different-color lenses can be used. The field wiring is connected to a terminal section that allows the lamp to be changed when it is damaged without removing and replacing wiring.

(Courtesy of Rockwell Automation's Allen-Bradley Business)

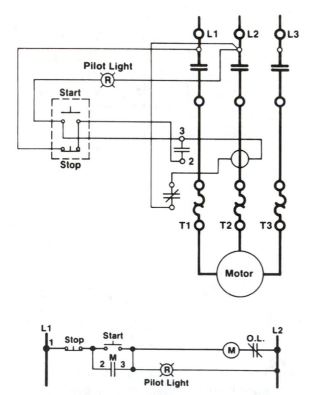

Figure 14–12   A wiring diagram and a ladder diagram of a three-wire control circuit with an indicator lamp added to show when the motor-starter coil is energized.

(Courtesy of Rockwell Automation's Allen-Bradley Business)

exploded-view diagram of a typical indicator lamp. In this diagram you can see that the lens is replaceable, and the part of the lamp that has wires connected to it can easily be removed from the socket part of the lamp in case the socket needs to be replaced.

14.2.5
Reversing Motor Starters

In previous chapters you have found that DC and AC motors can be reversed. These chapters provided the terminal connections for each type of DC motor, AC single-phase motors, and AC three-phase motors. The circuit shown in Fig. 14–14 shows a forward and reversing motor starter for an AC three-phase motor. When you must install or troubleshoot this circuit, you must be sure to provide interlocks so that the motor

cannot be energized in both the forward and reverse directions at the same time.

From this diagram you can see that the control circuit has forward and reverse push buttons. The forward and reverse push buttons each have a normally open and a normally closed set of contacts. In the ladder diagram each button switch has a dashed line, which indicates that both sets of contacts are activated by the same button. The forward and reverse push buttons are better defined in the wiring diagram, which shows that each switch has an open set and a closed set of contacts. The stop button is wired in series with the open contacts of both of these switches, so the motor can be stopped when it is operating in either the forward or the reverse direction.

The operation of the push buttons is best understood through the use of a ladder diagram. The ladder diagram

Figure 14–14 A wiring diagram and ladder diagram of a forward and reversing motor starter. Notice that you can clearly see the interlock system in the ladder diagram.

(Courtesy of Rockwell Automation's Allen-Bradley Business)

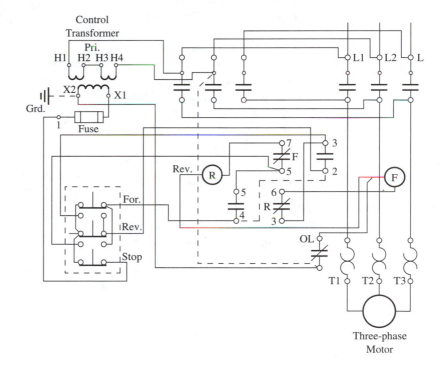

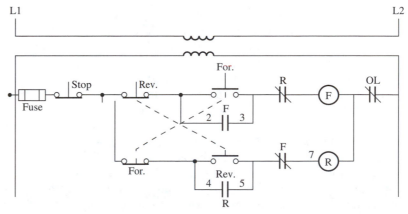

is used to show the sequence of the control circuit, whereas the wiring diagram shows the operation of the load circuit, which includes the motor and heaters for the overloads. The overload contacts are connected in the control circuit, where they will deenergize both the forward and reverse circuit if the motor is pulling too much current.

From the wiring diagram you can also see that two separate motor starters are used in the circuit. The forward motor starter is shown on the right side of the reverse motor starter. Each starter has its own coil and auxiliary contacts, which are used as interlocks. The location of the auxiliary contacts is shown in the wiring diagram, but their operation is difficult to determine there. When you see the auxiliary contacts in the ladder diagram, their function can be more clearly understood.

This comparison of the ladder diagram and the wiring diagram should help you understand that you need both diagrams to work on the equipment. The ladder diagram will be useful in determining what should be tested, and the wiring diagram is useful in showing where the contacts you want to test are located.

You should also notice that the control circuit is powered from a control transformer that is connected across L1 and L2. The secondary side of the transformer is fused to protect the transformer from a short circuit that may occur in either coil.

14.2.6
Reversing Motors with a Drum Switch

The drum switch is a manual switch that allows you to manually reverse the direction in which a motor is turning. The switch contacts are open and closed manually by moving the drum switch from the *off* position to the *forward* or *reverse* position. Figure 14–15 shows a picture of the drum switch, and Fig. 14–16 shows a diagram of the drum-switch contacts. Figure 14–16a shows the drum-switch contacts when the switch is in the reverse position, Fig. 14–16b shows the contacts when the switch is in the off position, and Fig. 14–16c shows the contacts when the switch is in the forward position. When the switch is in the *reverse position*, you should notice that terminal 1 is connected to terminal 2, terminal 3 is connected to terminal 4, and terminal 5 is connected to terminal 6. In the *off* position, all contacts are isolated from all other contacts. In the *forward position*, terminal 1 is connected to terminal 3, terminal 2 is connected to terminal 4, and terminal 5 is connected to terminal 6.

After you understand the operation of the drum switch in its three positions and methods of reversing each type of motor, these concepts can be combined to develop manual reversing circuits for any motor in the factory as long as its full-load and locked-rotor amper-

Figure 14–15 A drum switch. Notice the handle requires the operator to manually change the position of the switch from forward to reverse, or to the off position. (Courtesy of Eaton Corporation Cutler-Hammer Products)

Handle end		
Reverse	Off	Forward
1 o——o 2	1 o o 2	1 o——o 2
3 o——o 4	3 o o 4	3 o——o 4
5 o——o 6	5 o o 6	5 o——o 6
(a)	(b)	(c)

Figure 14–16 (a) Contacts of a drum switch when it is switched to the reverse position. (b) Contacts of a drum switch when it is switched to the off position. (c) Contacts of a drum switch when it is switched to the forward position.

age (FLA and LRA) do not exceed the rating of the drum switch. Figure 14–17a shows an AC single-phase motor connected to a drum switch. Notice that the start winding must be reversed for the motor to run in the reverse direction, so the start winding is connected to terminals 3 and 2.

Figure 14–17b shows a three-phase motor connected to a drum switch. You should remember that with this switch, any two lines to the motor can be swapped so the motor will change direction of rotation. Figure 14–17c shows a DC motor connected to a drum switch. You should notice with the DC motor that the direction of current flow through the field is reversed to

Figure 14–17 (a) A single-phase AC motor connected to a drum switch; (b) a three-phase AC motor connected to a drum switch; (c) a DC motor connected to a drum switch.
(Courtesy of Eaton Corporation Cutler-Hammer Products)

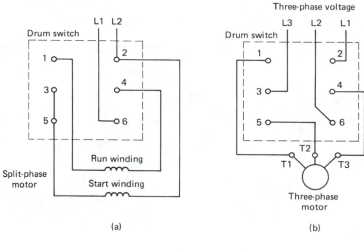

(a)

(b)

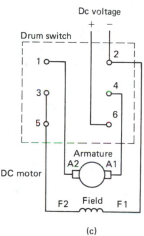

(c)

make the motor run in the opposite direction. You may need to study the connections inside the drum switch when it is in the forward or reverse position to fully understand how the drum switch is used with each of these motors to reverse the direction of their rotation.

These diagrams are especially useful for installation and troubleshooting of these circuits. The drum switch can be tested by itself or as part of the reversing circuit. The motors can also be disconnected from the drum switch and operated in the forward and reverse directions for testing or troubleshooting if you suspect the switch or motor is malfunctioning.

14.2.7
Industrial Timers

Timers that are used in industrial applications are designed to be easy to operate and provide a wide range of functions that fall under the categories of *time delay on* and *time delay off*. The time delay for these timers may be provided by air escaping a pneumatic bellows or by a motor-driven timer. More advanced types of solid-

state timers use some type of electronic time-delay circuit, such as the 555 chip, to generate their time base. The main feature of the industrial timer is that it must be designed to be easily adjusted by a machine operator who does not understand anything about electronics. The second feature of industrial timers is that they should be able to handle the high temperatures and strong vibrations that are present on the factory floor. Industrial timer can be interfaced to relay or solid-state logic circuits, and they use one of four basic types of operators that provide the amount of time delay. The time-delay operators can be classified as *pneumatic* (air chamber), *motor and cam, solid state* (time constant), or *programmable* (microprocessor). The operation of each type of timer is explained in this section, and examples of common applications are given. Figure 14–18 shows examples of industrial-type timers.

The electrical symbol for time-delay devices uses the *head* or *tail* of an arrow. The tail of the arrow is used to indicate a time-delay-on function, and the head of the arrow pointing down is used to indicate a time-delay-off function. Figure 14–19 shows examples of the timer symbols.

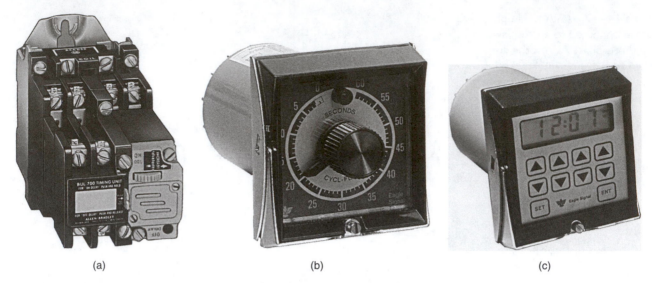

(a)   (b)   (c)

Figure 14–18   (a) A pneumatic-type industrial timer; (b) a motor-driven-type industrial timer; (c) a solid-state-type industrial timer.

((a) Courtesy of Rockwell Automation's Allen-Bradley Business; (b) and (c) Courtesy of Danaher Controls)

TIME DELAY-ON   TIME DELAY-OFF

Figure 14–19   Examples of time-delay-on and time-delay-off symbols.

14.2.8
On-Delay Timers

Industrial timers have contacts and an operator that causes the time delay. On-delay timers change the position of contacts after power is energized to the operator of the timer. One of the largest uses of timers in industrial applications is to control the starting time of larger motors so that they do not all try to start at the same time. When a motor starts, it draws a larger current than when it is running at full speed. This starting current is called *locked-rotor amperage* (LRA) or *inrush current*, and it is caused by the armature not turning when power is first applied to the motor. As the motor shaft begins to rotate, the current drops to normal levels as the motor comes up to full speed. If a machine has several large hydraulic pump motors, they will all try to start at the same time when power is turned on. The starting cycle of the motors can be staggered by using on-delay timers so that the effects of inrush current are minimized.

The on-delay timer may also be called a *delay-on* timer. When power is applied to the motors, the first mo-

tor is allowed to start normally, but power to the second motor starter is controlled through the on-delay timer. If the timer is set for 10 s, the timer contacts will close after a 10-s delay. The time-delay period starts when power is applied to the time-delay operator. Figure 14–20 shows an example of this time-delay-on circuit.

14.2.9
Off-Delay Timers

In some applications it is necessary to ensure that a motor continue to operate for several minutes after power is turned off to the main system. For example, in large industrial heating systems, the fan may need to run for up to 2 min after the heating element or gas valve has been deenergized. The additional time the fan is allowed to run after the heat source is turned off allows the system to capture all the heat that has built up in the heating chamber and use it. This provides a degree of efficiency, because this heat would be lost if the fan were turned off at the same time as the heating element. The time delay also prevents the heating system from *overshooting* the temperature set point on the controller. If the temperature set point is set for 150°F and the controller deenergizes the heating element when the temperature reaches 150°, the heat that remains in the heating chamber would cause the system to overheat. The additional heat that remains in the chamber acts like a flywheel and continues to add heat to the application, which causes the over-temperature condition.

The off-delay timer allows the heating control to deenergize the heating element several degrees prior to

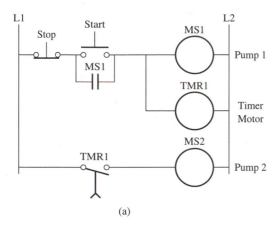

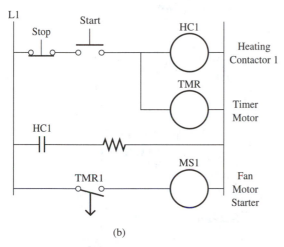

Figure 14–20 (a) Time-delay-on-type timer used to delay the time two pump motors start to limit the effects of inrush current. (b) Time-delay-off-type timer used to delay the time a furnace fan stays on after the heating element is turned off. The additional time the fan runs allows extra heat to be removed from the heating chamber.

the set point and allow the fan to continue to operate for several additional minutes to dissipate the remaining heat. This type of timer control may also be called a *time-delay-off timer*. Figure 14–21 shows an example of the time-delay-off timer used to control a furnace fan.

14.2.10
Normally Open and Normally Closed Time-Delay Contacts

Time-delay-on and time-delay-off timers can have normally open, normally closed, or both types of contacts. Figure 14–21 shows the symbol for each of the four possible timers. This figure also shows a timing diagram for each type of timer contact.

An on-delay-type timer can have normally open and normally closed contacts. The symbol for an on-delay

timer with normally open contacts is called *normally open, timed close* (NOTC). The contacts for this type of timer start out open and, once the timing element is energized, the contacts close after the delay time has elapsed. The symbol for the delay-on-type timer with normally closed contacts is called *normally closed, timed open* (NCTO). In Fig. 14–21 you should notice that the NOTC-type timer shows the contacts opening downward, and the tail of the arrow is used to indicate it is an on-delay-type timer. The NCTO-type timer symbol shows the contact closed.

Timing diagrams for the four types of timers are also shown in Fig. 14–21. The timing diagram shows the condition of the contacts as either open or closed during the four periods of the timer operation. The first timer period shows the contacts prior to timing. This represents the time *before* the timer operator is energized. The second period represents the time *during timing*, and the third period represents the time *after timing* but when power is still applied to the timer operator. The fourth period represents the time after timing when power is turned off to the operator. This condition is called *reset*, and it actually occurs twice during the cycle, once at the very beginning of the cycle during the first timing period before timing and again at the very end of the cycle. This means the very first part of the timing cycle may be called the time before timing, or reset. You should also notice that the timing diagrams show a diagram for the timing operator as well as a diagram for each type of contact used. The diagram that shows power applied to the operator is provided as a reference to indicate when the timing cycle actually begins. The time-delay period for each timer should be counted from the point where power is applied to the time-delay operator.

An off-delay-type timer can also have normally open and normally closed contacts. If an off-delay timer uses normally open contacts, it is called *normally open, timed open* (NOTO). The first question that comes to mind about this type of timer is how it can be normally open and then also time open. This situation occurs because one of the states the off-delay timer goes through is omitted. The actual timing cycle for the off-delay timer includes three distinct steps—normally open, *energized closed*, and then timed open. The timing diagram for this timer shows the contacts are normally open during the time prior to timing (reset). When power is initially applied to the circuit, the time-delay operator is powered and the normally open contacts move to the closed position. When power to the time-delay operator is deenergized, it begins the time-delay period, and the contacts open at the end of the timing period. Thus the name normally open, timed open fits.

If an off-delay timer uses a closed set of contacts, it is called *normally closed, timed closed* (NCTC). Again,

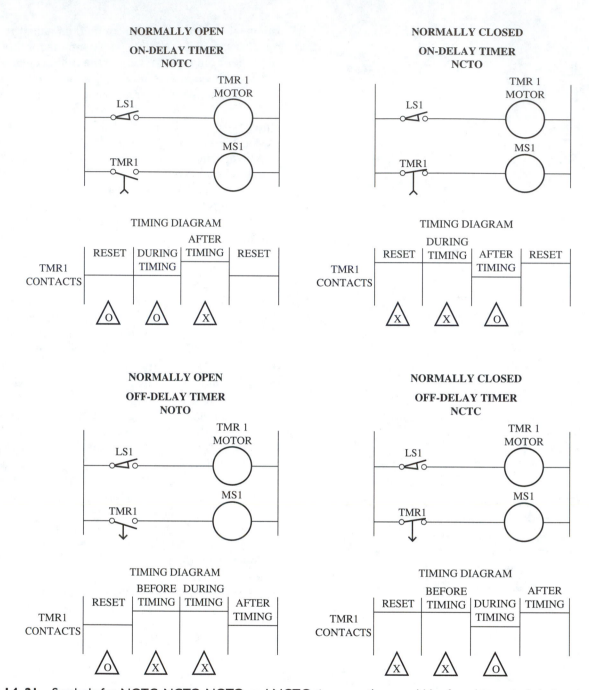

Figure 14–21 Symbols for NOTC, NCTO, NOTO, and NCTC timers as they would be found in typical timing circuits. Timing diagrams are provided for each type of timer.

the name seems to be contradictory because the contacts start out closed and end the cycle being timed closed. The part that is missing is the middle part of the cycle, where the contacts are opened when power is applied to the time-delay element. On-delay and off-delay timers can have a single set of normally open or normally closed contacts, or they can have multiple sets of both normally open and normally closed contacts.

14.2.11
Pneumatic-type Timer Operators

Several different types of operators are used in industrial-type timers. The simplest type of operator is called a *pneumatic* operator. This type of time-delay operator is also called a time-delay element. The pneumatic operator or element uses a bellows that fills with air. Prior to the start of the timing cycle the bellows is completely filled with air; when the time delay begins, the air is released through a *needle valve*. The needle valve can be adjusted to regulate the time it takes for the air to move out of bellows. A spring is mounted against the bellows to apply pressure to it to help force the air out. The amount of time it takes for the air to empty out of the bellows determines the amount of time delay. When sufficient air has emptied out of the bellows, the spring pressure is strong enough to trip the mechanism that ends the timing cycle. The trip mechanism also actuates the timed contacts.

The pneumatic-type time-delay element is an *add-on device* that can be added to a traditional relay. Figure 14–22 shows a picture of this type of add-on timing device. Because the pneumatic time-delay element is added to a normal control relay, it is possible to convert any control relay that is used in a circuit into a time-delay relay. Because the pneumatic element is added to the relay, it is also possible to take advantage of the original relay contacts that still operate with the relay coil. These contacts open or close instantly when the coil is activated, so they are called *instantaneous contacts*.

The pneumatic time-delay element can be converted from time delay on to time delay off right in the field. The element is designed so that when it is mounted on the relay in the upright position, it will provide time-delay-on functions. If the element is turned upside down when it is mounted on the relay, it will provide time-delay-off functions. You can determine if the element is mounted for delay-on or delay-off functions by looking closely at the lettering on the element. If the element is mounted for delay-on functions, the words *on-delay* will be right side up and the words *off-delay* will be upside down. When the element is mounted for off-delay functions, the words *off-delay* will be right side up.

The time-delay element generally has two sets of contacts mounted on it. One set is usually normally open, and one set is normally closed. These contacts can be changed from normally open to normally closed, or vice versa, by loosening a mounting screw and turning the contact set upside down. Thus the contacts can be changed in the field. You can determine if the set of contacts is normally open or normally closed by checking them with an ohmmeter.

14.2.12
Motor-Driven-type Timers

Motor-driven-type timers are widely used in industrial applications. These timers are also called *synchronous motor-type timers* or *electromechanical*-type timers. The timer uses a motor to turn a shaft on which cams are mounted. The shaft has a clutch that engages it to or

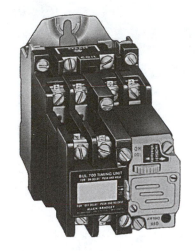

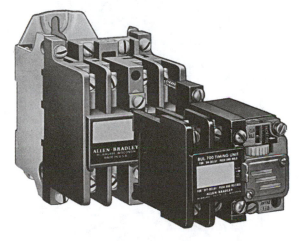

Figure 14–22 Pneumatic-type time-delay element. This element is added to a traditional relay to create a time-delay relay. Notice that the element can be mounted right side up for time-delay-on operation or it can be mounted upside down for time-delay-off operation. The delay contacts can also be mounted upside down to change from normally open to normally closed.
(Courtesy of Rockwell Automation's Allen-Bradley Business)

disengages it from the motor. When the shaft is engaged (energized), the motor turns the shaft and the cams actuate several set of contacts. When the time-delay period has elapsed, the time-delay contacts change from open to closed or from closed to open, and the timer motor resets for the next cycle. Figure 14–23 shows a picture and diagram of the motor-driven-type timer.

From the diagram in Fig. 14–23 you can see that the timer has a clutch coil and a motor that are connected in parallel with a common point identified as terminal 2. The timer has four sets of contacts. In the diagram contacts 9–10–C and 6–7–8 are located directly under the clutch coil to indicate they are energized instantaneously by the clutch when power is applied to the timer. Sets 11–12–A and 3–4–5 are located in the dia-

Schematic diagram

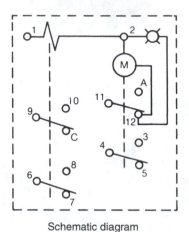

Terminals and wiring diagram
on rear of timer case

Figure 14–23 Synchronous motor-driven-type timer with a diagram of the internal circuit of the electric contacts.

gram directly under the timer motor to indicate they are controlled by the timer motor. These contacts are called the *delay contacts*.

One typical application uses a limit switch to control a timer. When the limit switch is closed, power is applied through it to the clutch coil. This same power energizes the motor, but it must go through normally closed time-delay contacts 11–12. When power reaches the clutch

coil, it immediately activates the instantaneous contacts and engages the clutch. Power also reaches the timer motor, and it begins to rotate its shaft. Because the clutch is engaged, the motor shaft turns the time-delay shaft, and when the correct amount of time delay has passed, the cams located on the shaft move with the shaft and activate both sets of time-delay contacts. Because the timer motor is connected through the normally closed set of time-delay contacts 11–12, the timer motor becomes deenergized when the contacts open and the timer is reset and becomes ready for the next cycle. The instantaneous sets of contacts controlled by the clutch coil remain energized until the limit switch is deenergized.

14.3
Jogging Control Circuits

In some applications, such as motion control, machine tooling, and material handling, you must be able to turn the motor on for a few seconds to move the load slightly in a forward or reverse direction. This type of motor control is called *jogging*. The jogging circuit utilizes a reversing motor starter to allow the motor to be moved slightly when the forward or reverse push button is depressed.

Another requirement of the jogging circuit is that the motor starters do not seal in when the push buttons are depressed to energize the motor when it is in the jog mode, yet operate as a normal motor starter when the motor controls are switched to the run mode. A diagram of a jogging circuit is provided in Fig. 14–24. The load circuit and control circuit are shown as electrical wiring diagrams, which give the location of each component, and the control circuit is shown again as a ladder diagram so you can see the sequence of operation for the forward and reversing motor starter with the jog function.

The wiring diagram in this figure gives you a very good idea of the way the jog-run switch operates. This switch is shown to the left of the motor starter in the diagram. You can see that it is part of the start-stop station. The jog-run button is a selector switch that is mounted above the forward-reverse-stop buttons.

When the switch is in the jog mode, the selector switch is in the open position. From the ladder diagram, you can see that the jog switch is in series with both the seal-in circuits, which prevents them from sealing the forward or reverse push buttons when they are depressed. This means that the motor operates in the forward direction for as long as the forward push button is depressed. As soon as the push button is released, the motor starter is deenergized. This jog switch also allows the motor to be jogged from one direction directly to the other direction without having to use the stop button.

Figure 14–24 A ladder diagram of a forward and reversing jogging circuit. Notice the interlock between the forward push buttons and the reverse push buttons so that you cannot energize the forward and reverse motor starters at the same time.
(Courtesy of Rockwell Automation's Allen-Bradley Business)

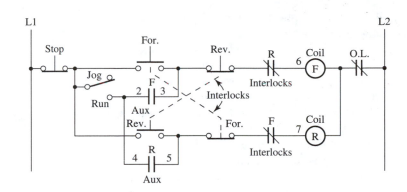

Figure 14–25 Motor starters are sequenced so that motor starter 1 must be on before motor starter 2 is started.
(Courtesy of Rockwell Automation's Allen-Bradley Business)

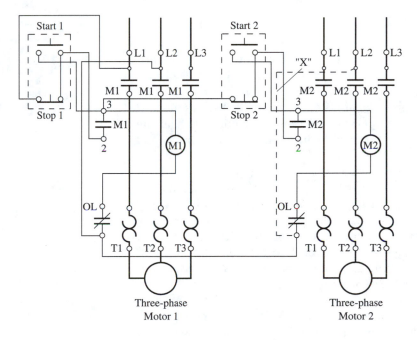

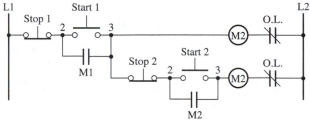

The motor is protected by the overloads that are connected in series with the forward and reverse motor-starter coils. If the overload trips, the overload contacts in the control circuit open, and neither coil can be energized until it is reset.

These two diagrams allow you to understand the operation of the jog circuit. You can make a forward and reversing motor-starter circuit into a jogging circuit by adding a jog switch, but you must be sure that the motor and the motor starters are rated for jogging duty. *Remember:* some motor starters and motors cannot take the heat that builds up when a motor is started

and stopped continually during a jogging operation. The motor and the motor starters will be rated for jogging or plugging if they can withstand the extra current and heat.

14.3.1
Sequence Controls For Motor Starters

Sequence control allows a motor starter to be utilized as part of a complex motor-control circuit that use one set of conditions to determine the operation of another circuit.

Figure 14–25 shows an example of this type of circuit. The circuits in this figure are presented in wiring-diagram and ladder-diagram form. You can really begin to see the importance of the ladder diagram, because it shows the sequence of operation, which would be very difficult to determine from the wiring diagram. The wiring diagram is still very useful, because it shows the field wiring connections and the locations of all terminals that will need to be used during troubleshooting tests.

The circuit in Fig. 14–25 shows two conveyors that are controlled by two separate motor starters. Conveyor 1 must be operating prior to conveyor 2 being started. This is required because conveyor 2 feeds material onto conveyor 1, and material would back up on conveyor 2 if conveyor 1 were not operating and carrying it away. The ladder diagram shows a typical start-stop circuit with an auxiliary contact being used as a seal-in around the start button. When the first start button is depressed, M1 is energized, which starts the first conveyor in operation. M1 auxiliary contacts seal the start button and provide the circuit power to the second start-stop circuit.

Because the second circuit has power at all times after M1 is energized, its start and stop buttons can be operated at any time to turn the second conveyor on and off as often as required without bothering the first conveyor's motor starter. Remember, this circuit requires conveyor 1 to be operating prior to conveyor 2, because conveyor 2 feeds material onto conveyor 1.

The circuit also protects the sequence if conveyor 1 is stopped for any reason. When it is stopped, the M1 motor starter becomes deenergized, and the M1 auxiliary contacts return to their open condition, which also deenergizes power to the second conveyor's start-stop circuit.

You should also notice that the power for this control circuit comes from L1 and L2 of the first motor starter. This means that if supply voltage for the first conveyor motor is lost for any reason, such as a blown fuse or opened disconnect, the power to the control circuit is also lost, and both motor starters are deenergized, which stops both conveyors. If the first motor draws too much current and trips its overloads, it causes an open in the motor starter's coil circuit, which causes the auxiliary contacts of the first motor starter to open and deenergizes both motor starters.

If you need additional confirmation that the belt on the first conveyor is actually moving, a motion switch can be installed on the conveyor; its contacts would be connected in series between the first start button and M1 coil. This would cause the first motor-starter coil to become deenergized any time the conveyor belt was broken or slipping too much.

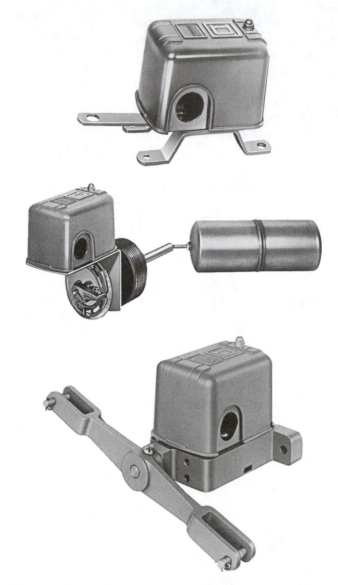

Figure 14–26 Three types of level switches. Applications and electrical diagrams of these switches are shown in Figure 14–27.
(Courtesy of Square D Company)

14.4
Other Types of Pilot Devices

A wide variety of other pilot devices are commonly used in motor-control circuits that are interfaced to industrial electronic circuits. These switches include pressure, level, and temperature switches, as well as other types of switches. Figure 14–26 shows pictures of three types of level switches, and Fig. 14–27 shows a diagram of each switch. Level switches are sometimes called *float switches*.

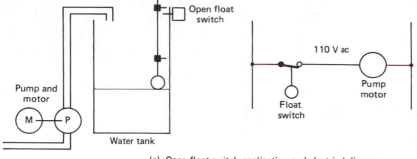

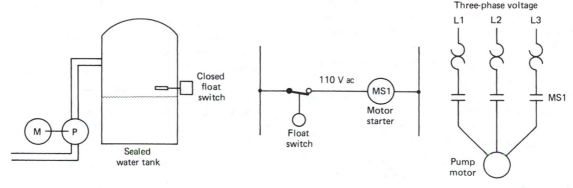

(a) Open float switch application and electrical diagram

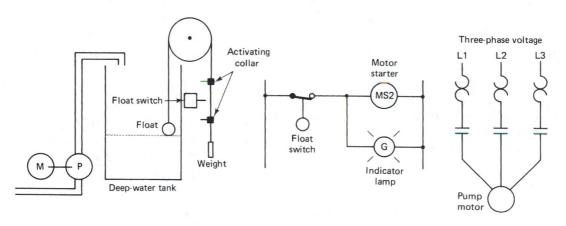

(b) Closed float switch application and electrical diagram

(c) Rod and chain float switch application and electrical diagram

Figure 14–27 (a) A float and rod are used with the level switch to turn it on and off. (b) A closed-type flow switch is mounted at the specific level that is to be controlled. (c) A float is attached to one end of a cable and a weight is attached to the other end. When the float moves up or down, the switch is activated.
(Courtesy of Square D Company)

The first switch, in Fig. 14–26a, is used with a ball-type float that is attached to a long rod. The rod has adjustable cams attached at the bottom and the top of the rod. When the level in the sump raises, the float lifts the rod, and the cam that is attached to the lower part of the rod moves high enough to trip the switch handle to the up position, which closes the level-switch contacts. When the pump runs, it pumps the sump level down and allows the float to drop. When the float drops low enough, the cam that is attached to the top of the rod causes the switch handle to move to the down position, which turns the pump off.

The switch in Fig. 14–26b uses a sealed-type float to activate the level switch. In this application, the switch is mounted in the tank at the precise level desired. When the level of the liquid in the tank is above this level, the float lifts and causes the switch contacts to close and turn on a pump. When the pump runs, the level of the tank is decreased, which allows the float to drop and turn the switch off. It is important to understand that the length of the switch arm determines the amount of travel the float must make to turn the switch on and off. This distance is called the *control band*, or *dead band*.

The float switch in Fig. 14–26c is a pivot-arm-type level switch. An arm or a wheel can be attached to the pivot to cause it to activate. An arm is shown in the picture of this switch, and a wheel is shown in the diagram. A cable is threaded around the wheel or arm. A float is attached to one end of the cable and a weight is attached to the other. When the level of liquid in the tank changes, the cable will move up or down and cause the switch to turn on or off.

It is important to understand that because these switches are pilot switches, they can safely switch only 10 to 15 A. This means that they can be used to control small motors or they can control larger motors by switching power on and off to a motor-starter coil.

Another type of pilot switch is called a *pressure switch*. This type of switch is used for sensing air pressure or water pressure and turning a set of contacts on or off. Figure 14–28 shows a typical pressure switch. The pressure switch may have a single point where it is activated on or off, or it may provide a span between the point where it turns on and off. If the switch has a span adjustment and a low-pressure adjustment, you should set the low-pressure adjustment where you want the switch to turn on and adjust the span to determine where you want the switch to turn off. For example, if you want the switch to control the pressure in a reservoir of an air compressor, you would set the low pressure to turn on when the pressure reaches 20 psi and then set the span to 40 psi. The span of 40 psi would cause the switch to turn off when the air pressure reached 40 psi more than the low-level setting. This means that switch would turn on the compressor when

the air pressure dropped to 20 psi, and it would turn the compressor off when the pressure reached 60 psi. Even though the switch has two points that turn it on and off, it still has only one set of single-pole or double-pole contacts that are activated.

14.5
Enclosures

Enclosures are required to house disconnects, motor starters, and other motor controls. The enclosures are available in a variety of designs that will protect the devices inside from all types of environmental conditions. Figure 14–29 shows a list of typical enclosure types as classified by the National Electrical Manufacturers As-

NEMA Enclosures for Starters	
Type	*Enclosure*
1	General purpose—indoor
2	Drip-proof—indoor
3	Dust-tight, raintight, sleet-tight—outdoor
3R	rainproof, sleet-resistant—outdoor
3S	Dust-tight, raintight, sleetproof—outdoor
4	Watertight, dust-tight, sleet resistant—indoor
4X	Watertight, dust-tight, corrosion-resistant—indoor/outdoor
5	Dust-tight—indoor
6	Submersible, watertight, dust-tight, sleet-resistant—indoor/outdoor
7	Class 1, group A, B, C, or D hazardous locations, air-break—indoor
8	Class 1, group A, B, C, or D hazardous locations, oil-immersed—indoor
9	Class II, group E, F, or G hazardous locations, air-break—indoor
10	Bureau of Mines
11	Corrosion-resistant and drip-proof, oil-immersed—indoor
12	Industrial use, dust-tight, and driptight—indoor
13	Oiltight and dust-tight—indoor

Figure 14–28 Example of a typical pressure switch that is used as a pilot device.
(Courtesy of Square D Company)

Figure 14–29 List of enclosure types provided by the National Electric Manufacturer's Association (NEMA).
(Courtesy of Rockwell Automation's Allen-Bradley Business and NEMA [National Electric Manufacturer's Association])

sociation (NEMA), and Fig. 14–30 shows examples of each type of enclosure.

Type 1 enclosures are designed for general-purpose applications in indoor locations. Thus the enclosure should not be exposed to extreme conditions, such as excessive moisture. The type 2 enclosure is rated for drip-proof conditions that exist indoors. This means that some moisture may come in contact with the enclosure, but it is not approved where equipment must be washed down or steam-cleaned daily.

Type 3 enclosures are designed for dust-tight, rain-tight, sleet-tight conditions that exist outdoors; type 3R enclosures are rainproof and sleet-resistant, and type 3S enclosures are designed for dust-tight, rain-tight, and sleetproof conditions. These enclosures are intended for use where protection is needed against falling rain, sleet, or dust. They are not intended to prevent condensation from forming on the inside of the enclosure or on internal components in the enclosure.

If the application is located where windblown water or sleet may be encountered, a type 4 enclosure should be used. The type 4 and 4× enclosures are designed to protect against windblown dust, rain, or water from direct-hose-down conditions. This enclosure is specified by NEMA for watertight, dust-tight, sleet-resistant indoor and outdoor applications. The 4× enclosure also provides protection in environments where corrosion resistance is required. The covers of these enclosures have protective gaskets that provide a barrier against these conditions. All hardware for the conduit must match the requirements to maintain the integrity of the design throughout the installation.

NEMA type 5 enclosures are designed for dust-tight applications located indoors. This type of enclosure is designed to protect the switch and contacts from a buildup of dust that may prevent proper operation of the switches and starters that are enclosed.

Type 6 enclosures are designed to provide submersible, watertight, dust-tight, sleet-resistant indoor and outdoor applications. The submersible watertight specification allows the switch to be fully submersed in water to a limited depth and maintain watertight integrity. This type of enclosure also depends on gaskets to maintain the watertight condition. As with other enclosures, conduits and connectors must meet the same NEMA standard so that the entire installation maintains the stated protection.

Type 7 enclosures provide class I protection for group A, B, C, or D hazardous locations, with air-break protection for indoor locations. These enclosures are designed to prevent dangerous gases from penetrating the enclosure and coming into contact with the open parts of switches or other controls. This provides protection in explosive atmospheres. Class I locations are hazardous because flammable gases or vapors may be present in sufficient quantities to cause explosions. Type 8 enclosures provide class 1, group A, B, C, or D hazardous locations, oil-immersed for indoor applications. This type of enclosure is very similar to the type 7 enclosure except that it provides protection against oil immersion instead of air-break. This allows the enclosure to be used where oil and machine coolants are used extensively.

Type 9 enclosures provide protection against class II, group E, F, or G hazardous locations and air-break for

Figure 14–30 Examples of NEMA-rated enclosures.

(Courtesy of Rockwell Automation's Allen-Bradley Business)

Size	Temperature Rating of Conductor. See Table 310-13.						Size
	60°C (140°F)	75°C (167°F)	90°C (194°F)	60°C (140°F)	75°C (167°F)	90°C (194°F)	
AWG kcmil	Types TW†, UF†	Types FEPW†, RH†, RHW†, THHW†, THW†, THWN†, XHHW† ZW†	Types TBS, SA SIS, FEP†, FEPB†, MI, RHH†, RHW-2, THHN†, THHW†, THW-2†, THWN-2†, USE-2, XHH, XHHW†, XHHW-2, ZW-2	Types TW†, UF†	Types RH†, RHW†, THHW†, THW†, THWN†, XHHW†	Types TBS, SA, SIS, THHN†, THHW†, THW-2, THWN-2, RHH†, RHW-2, USE-2, XHH, XHHW†, XHHW-2, ZW-2	AWG kcmil
	Copper			Aluminum OR Copper-clad Aluminum			
18	....	....	18	....	....	....	....
16	....	....	24	....	....	....	....
14	25†	30†	35†	....	....	....	....
12	30†	35†	40†	25†	30†	35†	12
10	40†	50†	55†	35†	40†	40†	10
8	60	70	80	45	55	60	8
6	80	95	105	60	75	80	6
4	105	125	140	80	100	110	4
3	120	145	165	95	115	130	3
2	140	170	190	110	135	150	2
1	165	195	220	130	155	175	1
1/0	195	230	260	150	180	205	1/0
2/0	225	265	300	175	210	235	2/0
3/0	260	310	350	200	240	275	3/0
4/0	300	360	405	235	280	315	4/0
250	340	405	455	265	315	355	250
300	375	445	505	290	350	395	300
350	420	505	570	330	395	445	350
400	455	545	615	355	425	480	400
500	515	620	700	405	485	545	500
600	575	690	780	455	540	615	600
700	630	755	855	500	595	675	700
750	655	785	885	515	620	700	750
800	680	815	920	535	645	725	800
900	730	870	985	580	700	785	900
1000	780	935	1055	625	750	845	1000
1250	890	1065	1200	710	855	960	1250
1500	980	1175	1325	795	950	1075	1500
1750	1070	1280	1445	875	1050	1185	1750
2000	1155	1385	1560	960	1150	1335	2000

Figure 14–31 American Wire Gauge (AWG) wires sizing table provided by the National Electrical Code (NEC). (National Electrical Code®, and NEC®, are registered trademarks of the National Fire Protection Association, Quincy, MA 02269.)

Ambient Temp. °C	Correction Factors						Ambient Temp. °F
	For ambient temperatures other than 30°C (86°F), multiply the allowable ampacities shown above by the appropriate factor shown below.						
21-25	1.08	1.05	1.04	1.08	1.05	1.04	70-77
26-30	1.00	1.00	1.00	1.00	1.00	1.00	78-86
31-35	.91	.94	.96	.91	.94	.96	87-95
36-40	.82	.88	.91	.82	.88	.91	96-104
41-45	.71	.82	.87	.71	.82	.87	105-113
46-50	.58	.75	.82	.58	.75	.82	114-122
51-55	.41	.67	.76	.41	.67	.76	123-131
56-60	....	.58	.71	....	.58	.71	132-140
61-70	....	.33	.58	....	.33	.58	141-158
71-80	....	....	.41	....	....	.41	159-176

†Unless otherwise specifically permitted elsewhere in this *Code*, the overcurrent protection for conductor types marked with an obelisk (†) shall not exceed 15 amperes for No. 14, 20 amperes for No. 12, and 30 amperes for No. 10 copper; or 15 amperes for No. 12 and 25 amperes for No. 10 aluminum and copper-aluminum.

Figure 14–31 *(continued)* American Wire Gauge (AWG) wires sizing table provided by the National Electrical Code (NEC). (National Electrical Code®, and NEC®, are registered trademarks of the National Fire Protection Association, Quincy, MA 02269.)

indoor applications. Class II locations are hazardous because of the presence of combustible dust in quantities sufficient to explode or ignite.

Type 10 enclosures are rated for all applications within mines. This type of enclosure must be able to protect switch gear against explosive conditions. Type 11 enclosures provide corrosion resistance and dripproof, oil-immersion protection for indoor applications. This type of enclosure is used where the vapors and fumes may be corrosive to switch gear or other motor controls that are mounted inside. The exterior of the enclosure is also resistant to corrosion from these fumes. This type of enclosure is also used for applications where machining operations are performed.

Type 12 enclosures are intended for indoor applications and provide protection against dust, falling dirt, and dripping noncorrosive materials. They are not intended for use against direct spraying of these materials. The last classification of enclosures is type 13.

14.6
Conductors

One part of the power-distribution system that is involved at all points in the system is the conductor. Conductors can be made of aluminum or copper wire. Aluminum is generally used for long-distance distribution because of its lighter weight. Once the power is inside the plant, copper conductors are generally used for distribution. Copper conductors are solid in bus bar and busway applications, and they are stranded conductors for all other applications.

The conductors are sized by the amount of amperage they can carry. A table of typical conductor sizes and their ampacities is provided in Fig. 14–31. The standard used to determine the size of each conductor is called American wire gage (AWG). These tables have been established by the National Electrical Code (NEC), and they are used to select the proper size of conductor to carry the load.

Conductors are also classified by the type of insulation used as a cover. The type of cover also determines the voltage rating of the wire. Typical voltage ratings for conductors used in motor-control applications include 300, 600, and 1,000 V. The voltage and current ratings of the conductor should not be exceeded under any circumstances.

Types of wire covering and abbreviations for each of these coverings are listed in the NEC tables and they help determine the type of wire that should be selected for each application. The outer covering of the wire serves several purposes, including protecting the conductors from coming in contact with metal in the cabinet or other conductors. The cover also provides a location to stamp all specification data regarding the wire, including the voltage rating, AWG size, temperature specification, and type of covering.

Questions for This Chapter

1. Identify five types of limit switches and explain where each would be used.

2. Use the diagrams in Fig. 14–17 and explain how the drum switch can reverse the direction of rotation for a single-phase motor, a three-phase motor and a DC motor.

3. Use the diagram in Fig. 14–24 to explain how the jog circuit must modify the traditional three-wire control circuit so it will not latch in.

4. Explain the operation of each of the three-level pilot devices shown in Fig. 14–27.

5. Use Fig. 14–29 to select the correct enclosure for the following applications: dust-tight indoors, rainproof and sleet-resistant outdoors, general-purpose indoors, oil-tight and dust-tight indoors.

True or False

1. A jogging circuit is used to interlock a motor starter during a sequence move.

2. A three-wire control circuit uses normally open auxillary contacts of the motor starter to seal in the start push button.

3. In the diagram in Fig. 14–8, the float switch will close when the level that is being sensed increases.

4. A drum switch can be used to reverse the direction of a single-phase AC motor, a three-phase AC motor, or a DC motor.

5. The auxillary contacts found between terminal 2 and terminal 3 on a motor starter are used to seal in the start push button in a three-wire control circuit.

Multiple Choice

1. In the diagram in Fig. 14–25, motor starter M1 will always be running before motor starter M2 can become energized because _____ .
   a. the M1 coil requires a larger voltage than the M2 coil.
   b. the M1 coil is in series with the M2 coil.
   c. the M1 contacts are in series with the M2 coil.

2. When two extra stop push buttons are added to a start-stop circuit, the stop buttons must be wired _____ .
   a. in series with the other stop push button.
   b. in parallel with the other stop push button.
   c. in parallel with the start push button.

3. In Fig. 14–9, which pressure switch will protect the circuit when it is in the auto position?
   a. only pressure switch B
   b. only pressure switch A
   c. both pressure switches A and B

4. When two extra start push buttons are added to a start-stop circuit, the start push buttons must be wired _____ .
   a. in series with the other stop push button.
   b. in parallel with the other stop push button.
   c. in parallel with the start push button.

5. A 12-gauge wire with THWN insulation (according to the table in Figure 14–31) can safely carry _____ A.
   a. 30
   b. 25
   c. 20

Problems

1. Draw the symbols for a normally open limit switch, a normally open, held-closed limit switch, a normally closed limit switch, and a normally closed, held-open limit switch.

2. Draw the symbols for normally open and normally closed pressure switches that are used for a high-pressure application and normally open and normally closed pressure switches that are used for a low-pressure application.

3. Draw a two-wire control circuit of a limit switch and a motor as a ladder diagram.

4. Use the table in Fig. 14–26 to select a wire that will carry 30 A and has THHN insulation that can withstand 90°C.

5. Draw a two-wire control circuit of a limit switch and a motor as a wiring diagram.

◀ **Chapter 15** ▶

Programmable Controllers

Objectives

After reading this chapter you will be able to:

1. Describe the four basic parts of any programmable controller.
2. Explain the four things that occur when the PLC processor scans its program.
3. Explain the function of an input module and describe the circuitry used to complete this function.
4. Explain the function of an output module and describe the circuitry used to complete this function.
5. Explain the function of an internal control relay.
6. Discuss the classifications of PLCs.
7. Describe what happens when a PLC is in run mode.
8. Identify the input instruction and output instruction for a PLC.
9. Explain the operation of a PLC on-delay timer.
10. Explain the operation of a PLC up counter and down counter.

15.1
Overview of Programmable Controllers

The programmable controller has become the most powerful change to occur in factory automation. A programmable controller (P/C) is also called a programmable logic controller (PLC). Because a personal computer is also called a PC, a programmable controller is referred to as a PLC to prevent confusion. As a mainte-

nance technician, you will run into PLCs in a number of places, such as on the factory floor. In this chapter you study PLCs from two different perspectives. First, you will see how the PLC is programmed to perform logic functions that control machines much the same way that motor controls do. Second, you will see how simple it is to troubleshoot large-machine controls and automation with a programmable controller. You will see how to use the status indicators on PLC modules to help you troubleshoot systems.

Today in industry you may find PLCs as stand-alone controls or as part of a complex computer-integrated manufacturing (CIM) system. In these large, integrated manufacturing systems the PLC controls individual machines or groups of machines. They may also provide the interface between machines and robots or machines and color graphics systems, which are called human-machine interfaces (HMI).

In the 1970s and 1980s companies were able to hire both an electrician and an electronics technician to install, interface, program, and repair PLCs. Since the late 1980s the number of PLCs has grown so large that many companies have found that it is better to hire one individual as a maintenance technician to make minor programming changes and troubleshoot. If you have never heard of a PLC or if you have had only a brief introduction to them, this chapter provides you with all the information necessary to work successfully with them. A typical PLC is shown in Fig. 15–1. This chapter uses generic PLC addresses wherever possible. When specific applications are provided, the Allen-Bradley MicroLogix, PLC5, and SLC500 are used as example systems. The early

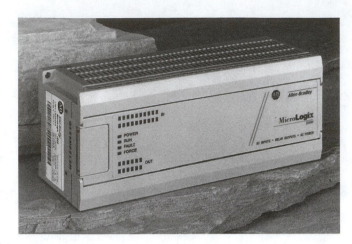

Figure 15–1 Typical programmable controller.
(Courtesy of Rockwell Automation's Allen-Bradley Business)

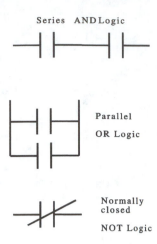

Figure 15–2 Logic for AND, OR, and NOT functions.

examples in this chapter are not specific to any brand of PLC to help you understand the functions that are generic to all controllers.

15.1.1
The Generic Programmable
Logic Controller

A programmable logic controller (PLC) is a computer that is designed to use logic (AND, OR, NOT) that specifically controls industrial devices, such as motors and switches, and that allows other control devices of varied voltages to be easily interfaced to provide simple or complex machine control. Figure 15–2 shows examples of AND, OR, and NOT logic. Switches that are connected in series are called *AND logic* in a PLC, and switches that are connected in parallel are called *OR logic*. If closed contacts are used, they are called *NOT logic*. At this point do not let the term *logic* confuse you. Logic is a word used to describe the actions that switches and contacts perform when they are wired in series or parallel.

A PLC is designed to have industrial-type switches, such as 110-V push-button and 110-V limit switches connected to it through an optically isolated interface called an *input module*. The PLC can also control 480-V, three-phase motors through 120-V motor-starter coils with a similar interface, called an *output module*. The computer part of the PLC, which is called a *central processing unit* (CPU), allows a program to be entered into its memory that represents the logic functions. The program in the PLC is not a normal computer programming language, such as BASIC. Instead the PLC program uses contact and coil symbols to indicate which switches should control which outputs. These symbols look very similar to the typical relay ladder diagram shown in Chapter 14. You can view the program through a hand-

held terminal, or the program can be displayed on a computer screen, where the action of switches and outputs are animated as they turn motor-starter coils and other outputs on or off. The animation includes *highlighting* the input and output symbols in the program when they are energized.

In short, the PLC executes logic programs just like the hard-wired circuits you used in Chapter 14. The PLC was designed to combine the technology of logic control with the simplicity of an electrical diagram so that its inputs and output are easy to troubleshoot. Because the logic is programmable, it is easily changed. As you know, it may take several hours to make a change such as adding a limit switch in a hard-wired circuit. In a PLC, you will be able to make changes to the operation of a machine in a matter of minutes. The logic can also be easily stored on a disk so that it can be loaded into a PLC on the factory floor any time the program logic for a controller needs to be changed, as when different parts are made or when the machine function needs to be changed.

In most factories today, it is important for an expensive machine to make more than one part. Sometimes it is just a different size or style of the same part, which requires the machine to have a different set of switches. The PLC is designed to make these changes as easy as possible.

When PLCs were first introduced, they allowed the factory electrician to use them to help troubleshoot large, automated systems. As PLCs evolved, they began to provide many of the more complex logic functions, such as timing, up-down counting, shift registers, and first-in, first-out (FIFO) functions.

Figure 15–3 shows an example of a PLC program. Notice that the program looks exactly like an *electrical relay diagram*. As the program (diagram) gets larger,

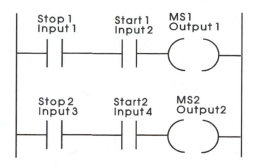

Figure 15–3 Example of a typical programmable controller program that is called ladder logic. This program uses the generic addresses of input 1, input 2, output 1, output 2, etc.

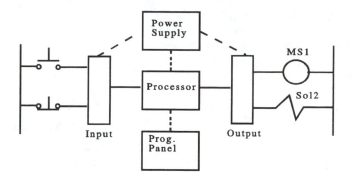

Figure 15–4 Block diagram showing the four major parts of a programmable controller. The programming panel is the fifth part of the system, but it is not considered a basic part of a PLC, because it can be disconnected when it is not needed.

additional lines are added. This gives the program the appearance of a wooden ladder that has multiple rungs. For this reason the program is called a *ladder diagram* or *ladder logic*.

The original function of the PLC was to provide a substitute for the large number of electromechanical relays that were used in industrial control circuits in the 1960s. Early automation used large numbers of relays, which were difficult to troubleshoot. The operation of the relay was slow, and they were expensive to rewire when changes were needed in the control circuit. In 1969 the automotive industry designed a specification for a reprogrammable electronic controller that could replace the relays. This original PLC was actually a sequencer-type device that executed each line of the ladder diagram in a precise sequence. In the 1970s the PLC evolved into the microprocessor-type controller used today. Several major companies, such as Allen-Bradley, Texas Instruments, and MODICON, produced the earliest versions. These early PLCs allowed a program to be written and stored in memory, and when changes were required to the control circuit, changes were made to the program rather than to the electrical wiring. It also became feasible for the first time to design one machine to do multiple tasks by simply changing the program in its controller. Today the major controllers come from Allen-Bradley, Modicon, GE, Siemens (formerly Texas Instruments), and Square D.

15.1.2
Basic Parts of a Simple Programmable Controller

All PLCs, regardless of brand name, have four basic major parts: the power supply, the processor, input modules, and output modules. A fifth part, a programming device, is not considered a basic part of a PLC, because the program can be written on a laptop computer and

downloaded to the PLC or the program can be written to an EPROM (erasable, programmable read-only memory) chip. Figure 15–4 shows a block diagram of a typical PLC. From this diagram you can see that input devices, such as push-button switches, are wired directly to the input module and the coils of the motor starter or solenoids are connected directly to the output module. Before the PLC was invented, each switch was directly connected (hard-wired) to the coil of the motor starter or solenoid it was controlling. In the PLC the switches and output devices are connected to the modules, and the program in the PLC determines which switch controls which output. In this way, the physical electrical wiring needs to be connected only one time during the installation process, and the control circuitry can be changed an unlimited number of times through simple changes to the ladder logic program.

15.1.3
The Programming Panel

At the bottom of the diagram in Fig. 15–4 you can also see a programming panel. A programming device is necessary to program the PLC, but it is not considered one of the parts of a PLC, because the programming panel or device can be disconnected after the program is loaded and the PLC will run by itself. The programming device is used so humans can make changes to the program, troubleshoot the inputs and outputs by viewing the status of contacts and coils to see if they are energized or deenergized, save programs to a disk, or load programs from a disk.

The programming panel can be a dedicated device or it can be a personal or portable computer with PLC programming software loaded on it. The ladder logic program can be displayed on the programming device, where it can become animated. This feature is unique to

Figure 15–5 Diagram of the overhead view of a conveyor sorting system that is controlled by a generic programmable controller.

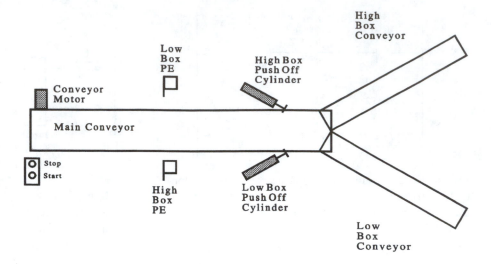

a PLC, and it helps a technician troubleshoot very large logic circuits that control complex equipment. When a switch that is connected to the PLC input module is turned on, the program display can show this on the screen by highlighting the switch symbol everywhere it shows up in the program. The display can also highlight each output when it becomes energized by the ladder logic program. When the on-off state of a switch on the display is compared with the electrical state of the switch in the real world, a technician can quickly decide where the problem exists.

Another feature that makes the PLC so desirable is the fact that it has a wide variety of logic functions. This means that functions such as timing and counting can be executed by the PLC rather than using electromechanical or electronic timers and counters. The advantage of using timer and counter functions in the PLC is that they do not involve any electronic parts that could fail, and their preset times and counts are easily altered through changes in the program. Additional PLC instructions provide a complete set of mathematical functions as well as manipulation of variables in memory files. Modern PLCs, such as the Allen-Bradley PLC-5, have the computing power of many small mainframe computers as well as the ability to control several hundred inputs and outputs.

15.2
An Example Programmable Controller Application

It may be easier to understand the operation of a PLC when you see it used in a simple application: sorting boxes according to height as they move along a con-

veyor. It is also important to understand that the address-numbering schemes for inputs and outputs of some PLCs may be difficult to understand when you are first exposed to them. For this reason the first examples of the PLC presented in this section use generic numbering that does not represent any particular brand name. This allows the basic functions of the PLC that are used by all brand names of PLCs to be examined in a simplified manner. Later sections of this chapter go into specific numbering systems for several major brands of PLCs.

Figure 15–5 shows a pictorial diagram of the top view of the conveyor sorting system. The location of the photoelectric switches that detect the boxes and the pneumatic cylinders used to push the boxes off the conveyor are also shown in this diagram. Each air cylinder is controlled by a separate solenoid. The conveyor is powered by a three-phase electrical motor that is connected to a motor starter. Start and stop push buttons are used with photoelectric switches to energize the coil of the motor starter, which turns the conveyor motor on and off. Figure 15–6 shows all the switches connected to an input module and the solenoids and motor starter connected to an output module. The inputs are generically numbered input 1, input 2, etc. The outputs are numbered output 10, output 11, etc. The ladder logic program that determines the operation of the system is shown in Fig. 15–7. The diagram of the switches and outputs connected to the input and output modules uses standard electrical symbols for each switch and output device.

Because each switch in the program is represented by a normally open contact symbol, —] [—, or a normally closed contact symbol, —] \ [—, each symbol must have a number so you can tell the switches apart. Each output is identified by the symbol –( )—; outputs must also be numbered so you can tell them apart.

Figure 15–6 Input and output diagram that shows all switches that are connected to the PLC input module and the coil of motor starters and solenoids that are connected to the PLC output modules. The input and output numbering is generic and does not represent any brand name of programmable controller. Note that the electrical devices are represented by standard electrical symbols in this type of diagram.

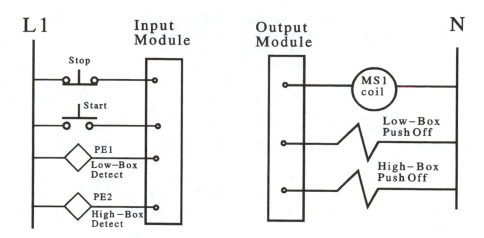

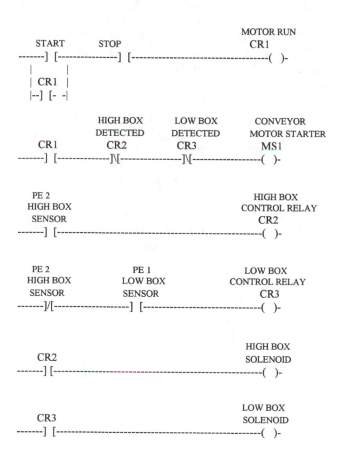

Figure 15–7 Ladder logic diagram showing the logic control for the conveyor sorting system. The diagram is generic and does not represent any particular PLC brand. Notice that all switches and inputs are represented by the contact symbol, and all outputs are identified by - ( ) -.

15.2.1
Operation of the Conveyor Sorting System

When the conveyor system is ready for operation, the start push button must be depressed. Because the stop push button is wired as a normally closed switch, power will pass through its contacts in the program to the start push-button contacts. When the start push button is depressed, the *motor-run* condition will be satisfied in the first rung of the program and the motor-run control relay in the program will be energized. (It should be noted at this time that the PLC can have hundreds of control relays in its program like the motor-run relay. The advantage of having control relays in a PLC program is that it provides the logic functions of a control relay, which can have multiple sets of normally open and normally closed contacts, yet they exist only inside the PLC, so they can never fail or break.) A set of normally open contacts from CR1 is connected in parallel with the start push button to seal it in, because the start push button is only momentarily closed. The photoelectric switches are used to energize and deenergize the conveyor motor starter in the second line of the logic. Line 2 of the diagram is designed so the start-stop circuit can turn the motor starter on or off without affecting the start-stop circuit. It should be noted at this time that the photoelectric switches send out a beam of light that is focused on a reflector mounted on the opposite side of the conveyor. As long as a box does not interrupt the beam of light, the photoelectric switches are energized.

When the motor-run control relay coil in the first rung of the program is energized, its normally open contacts in the second rung of the program become "logically" closed. Because no boxes are being sensed by the photoelectric switches, they are energized and their contacts in the program, "Low-Box Detected PE" and "High-Box Detected PE," pass power to the output in the second rung, called "Conveyor Motor Starter," which directly controls the coil of motor starter 1. When motor starter 1 is energized, the conveyor motor is energized and the conveyor belt begins to move.

As boxes are placed on the conveyor, they travel past the photoelectric switches. If the box is a low box, photoelectric switch PE1, which is mounted to detect a low box, becomes energized. This causes three things to happen. First, the normally open contacts of "PE1 Low-Box Sensor" in rung 4 close. Because the box is not a high box, the normally closed contacts of "PE2 High-Box Sensor" in rung 4 remain closed to pass power to "PE1 Low-Box Sensor" contacts (that are now closed) and on to the coil of CR3 ("Low-Box Control Relay"). Second, the coil of CR3 is energized and all the contacts in the program that are identified as CR3 change state. This means the normally closed CR3 contacts in rung 2 open and deenergize MS1, the conveyor motor starter. When MS1 is deenergized, the motor starter becomes deenergized, and the conveyor motor stops. The third event to happen in this program when the coil of CR3 is energized by "PE1 Low-Box Sensor" is that the output for the low-box solenoid is energized and the rod of the low-box push-off cylinder is extended; the low box is then pushed off the main conveyor onto the low-box conveyor.

After the low box is pushed off the main conveyor, it no longer activates PE1, and the coil of CR3 becomes deenergized in rung 3. When the coil of CR3 is deenergized, the normally closed CR3 contacts in rung 2 return to their closed state, which energize the output for MS1 and starts the conveyor motor again.

When a high box moves into position on the conveyor, it blocks both the high-box photoelectric switch, PE2, and the low-box photoelectric switch, PE1. This could create a problem of having both the low-box control relay, CR3, and the high-box control relay, CR2, energized at the same time. The logic in rung 4 is specifically designed to prevent this. Because the contacts of the high-box sensor, PE2, are programmed normally closed in rung 4, they open when the high box is detected and do not allow power to pass to the coil of CR3. This type of logic is called an *exclusion*, because it prevents both control relay coils from energizing at the same time, even though both photoelectric switches are activated by the high box.

When the high box is detected by PE2, the normally closed PE2 contacts in rung 3 close and energize the coil of the high-box control relay, CR2. When the coil of CR2 is energized, all the contacts in the program that are identified as CR2 change state. The normally closed CR2 contacts in rung 2 open and deenergize the coil of MS1, the conveyor motor starter. The normally open CR2 contacts in rung 5 close at this time and energize the high-box solenoid, and air is directed to the high-box pneumatic cylinder; the high box is then pushed off the conveyor.

After the box has been pushed off the conveyor, the photoelectric switch that detected the high box returns to its normal state, high-box contacts in rung 2 return to their normally closed logic state, and the motor starter becomes energized again and starts the conveyor motor. The high-box contacts in rung 3 return to their normally open logic state and the push-off cylinder is retracted. This allows the conveyor to return to its normal operating condition.

15.3
How the PLC Actually Works

When the PLC is in the *run mode*, its processor examines its program line by line, which is the way it solves its logic. The processor in the PLC actually performs several additional functions when it is in the run mode. These functions include reading the status of all inputs, solving logic, and writing the results of the logic to the outputs. When the processor is performing all these functions, it is said to be *scanning* its program. Figure 15–8 shows an example of the program scan.

When the PLC checks its inputs, it needs to determine only if they are energized or deenergized. When the contacts of a switch are closed, they allow voltage to be sent to the input-module circuit. The electronic circuit in the input module uses a light-emitting diode and a photo transistor to take the 110-VAC signal and reduce it to the small voltage signal used by the processor.

The contacts in the program that represent the switch change state in the program when the module receives power. If you use the programming panel to examine the contacts in the PLC program, they highlight when the switch they represent is energized. The PLC

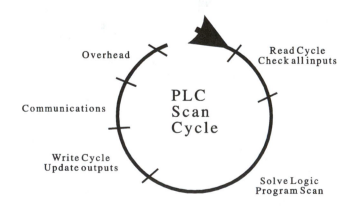

Figure 15–8 Example of a typical programmable controller scan. The processor continually reads the status of all inputs, solves logic, writes the results of the logic to all outputs, and provides communications and overhead. Overhead includes the functions of interfacing with the programming panel, printer, color screens, etc.

transitions the state of the programmed contacts at this time. For example, if the contacts are programmed normally open, when the real switch closes and the module is energized, the PLC causes the programmed contact to transition from open to closed. Thus the programmed contacts will appear to *pass power* through them to the next set of contacts. Because the contacts in line 1 of the program are in series, the output will become energized when both sets of contacts are closed and passing power.

Another important point to understand is that the term *set* is used to describe contacts or functions in the PLC that are energized, and term *reset* is used to describe contacts or functions that are off.

15.3.1
Image Registers

One of the least understood parts of the PLC is the *image register*. The image register resides in the memory section that is specifically designed to keep track of the status (on or off) of each input. In some controllers this image register is also called an *image table* or an *input data file*. Figure 15–9 shows an example of the image register. From this diagram you can see that all the inputs and outputs for the PLC each have one memory location where the processor places a 1 or 0 to indicate whether the input or output connected to that address is energized or deenergized. The processor determines the status of all inputs during each scan cycle by check-

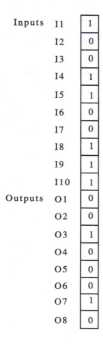

Figure 15–9   Example of an image register for a PLC. The image register is 1 bit wide and has a memory location for each input and output in the PLC. A *bit* is a binary digit.

ing each one to determine if the switch connected to it is energized. If the switch is energized and the input module sends a signal to the processor, the processor places a *binary 1* in the memory location in the image register that represents that input. If the switch is off, the processor places a *binary 0* in the memory location for the image register. The word *binary* is derived from the word meaning two, because each input switch can have only one of two states, on or off. The PLC represents a number of things in its memory with 1s or 0s.

Another important word is also derived from the word binary. This word is *bit*, and it comes from the combination of the word *binary* and the word *digit*. A bit is any value in the computer that is restricted to a value of 1 or 0.

Now that you have a better idea of what a *bit* is, you can understand how easy it is for the PLC to use the image register to keep an exact copy of the status of all its inputs during the *read* part of the scan. The image register is important to you as a troubleshooter, because it will tell you if the PLC is receiving voltage from an input switch or if the PLC is sending power to an output device, such as a motor starter coil. In a hardwired control system, you would have to use a voltmeter to test all switches to tell which ones are on or off. In the PLC, you can view the image register and see quickly the status of all switches at one time to determine which ones have a 1 in the image register and are on and which ones have a 0 and are off.

In the second part of the PLC scan, the processor uses the values in the image register to solve the logic for that particular rung of logic. For example, if input 1 is ANDed with input 2 to turn on output 12, the processor looks at the image register address for input 1 and input 2 and solves the AND logic. If both switches are high and a 1 is stored in each image register address, the processor will solve the logic and determine that output 12 should be energized.

After the processor solves the logic and determines the output should be energized, the processor moves to the third part of the scan cycle and writes a 1 in the image register address for output 12. This is called the *write cycle* of the processor scan. When the image register address for output 12 receives the 1, it immediately sends a signal to the output module address. This energizes the electronic circuit in the output module, which sends power to the device—such as the motor starter—that is connected to it.

It is important to understand that the PLC processor updates all the addresses in the image register during the read cycle even if only one or two switch addresses are used in the program. When the processor is in the *solve logic* part of the scan, it solves the logic of each rung of the program from left to right. This means that if a line of logic has several contacts that are ANDed with several more contacts that are connected in an OR condition, each of these logic functions is solved as the processor looks at the logic in the line from left to right. The processor also solves each line of logic in the program from top to bottom. This means that the results of logic in the first rung are determined before the processor moves down to solve the logic on the second rung. Because the processor continues the logic scan on a continual basis approximately every 10 to 20 ms, each line of logic will appear to be solved at the same time. The only time the order of solving the logic becomes important is when the contacts from a control relay in line 1 are used to energize an output in line 2. In this case the processor actually takes two complete scan cycles to read the switch in line 1 and then write the output in line 2. In the first scan, the processor reads the image register and determines if the start push button has closed. When the logic for rung 1 is solved, the control relay is energized. The processor does not detect the change in the control relay contacts in the second line until the second scan, when it reads the image register again. The output in the second line is finally energized during the write cycle of the second scan, and power is sent to the motor-starter coil that is connected to the output address. It is also important to understand that the processor writes the condition of every output in the processor during the *write* part of the scan even if only one or two outputs are used in the program.

The image register is also useful to the troubleshooting technician, because the contents of each address can be displayed on the hand-held programmer or on a CRT, and the technician can compare the image register with the status lights on the input and output modules and the actual switches and solenoids that are hardwired to the modules. If an input has voltage from a switch, its status light will be illuminated, and a 1 should show up in the image register location for the input. This gives the technician two places to tell if a switch is on or off without using a voltmeter.

A separate image register or data file is also maintained in the processor for the *control relays*. In some systems these relays are called *memory coils* or *internal coils*. A control relay has one coil and one or more sets of contacts that can be programmed normally open or normally closed. When the coil of a control relay is energized, all the contacts for the relay change state. Because control relays reside only in the program of the PLC and do not use any hardware, they cannot have problems such as dirty contacts or an open circuit in their coil. Thus if the contacts of a control relay will not change from open to closed, you should not suspect the coil of having a malfunction; instead you should look at the contacts that control power to the control relay coil in the program. One or more of them will be open, which will cause the coil to remain deenergized.

When the processor is in the RUN mode, it updates its input, output, and control-relay image registers during the write-I/O part of every scan cycle. Because the image registers are continually updated approximately every 20 ms to 40 ms, it is possible to send copies of them to printers or color graphic terminals at any time to indicate the status (on or off) of critical switches for the machine process. This allows the machine operator to do some minor troubleshooting when a problem occurs before a technician is called. For example, if an operator is running a press or machine that must have a door or gate closed before it will begin its cycle, a fault monitor on a color graphic display can show the status of the door switch. If the door is ajar and the cycle will not start, the operator can look at the fault monitor; the copy of the input image register from the PLC would indicate that the door switch was open, and the operator could take the appropriate action and close the door instead of calling a technician. In this manner, the PLC not only controls the machine operation, it also helps in troubleshooting problems.

15.3.2
The RUN Mode and the Program Mode

When the PLC is in the *RUN mode*, it is continually executing its scan cycle. Thus it monitors its inputs, solves its logic, and updates its outputs. When the PLC is in the *program mode*, it does not execute its scan cycle. Perhaps a better name for the program mode is the *NOT RUN* mode, or the *NO-SCAN* mode. The name *program mode* was originally given to this mode because the programming of older PLCs could not be changed while they were executing their scan cycle. Programs of modern PLCs can be edited and changed while the processor is in the RUN mode. At first glance changing the PLC program while it is in the RUN mode looks dangerous, especially when the machine the PLC is controlling is in automatic operation. But you will find in larger control applications, such as pouring glass continually or continuous steel-rolling mills, it is not practical to stop the machine process to make minor program changes such as the changes to a timer. In these types of applications the changes are made while the PLC is in the RUN mode. As a rule, however, you should switch the PLC to the program mode if possible when any program changes are being made.

15.3.3
On-Line and Off-Line Programming

The programming software for a PLC may allow you to write the ladder logic program in a personal computer and later download the program from the personal computer to the PLC. If you are connected directly to a PLC and you are writing the ladder logic program in the PLC memory, it is called *on-line programming*. If you are writing the ladder logic program in a personal computer or programming panel that is not connected to a PLC, it is called *off-line programming*. Most modern PLC programming software allows the program to be written on a personal computer or laptop computer without being connected to the PLC. This allows new programs or program changes to be written at a location away from the PLC. For example, you may write a program in Detroit at an office, save the changes on a floppy disk, and mail the disk to a company in Chicago, where the PLC is connected to automated machinery.

If you are using a *hand-held programmer*, it must be connected directly to a PLC so that you are writing the program in the PLC memory. It is also important to understand that some small PLCs, such as the Allen-Bradley SLC100, do not have a microprocessor chip, so you cannot write the PLC program on-line with a laptop computer. Because the SLC100 uses a sequencer chip instead of a microprocessor chip, all programming must be completed off line in the laptop, and then you must download the program changes to the PLC memory. An alternative method is to use a hand-held programmer to program the PLC on line.

15.4
Features of a Programmable Controller

The examples in Fig. 15–5 through 15–7 show several unique features of the PLC. First, you should notice that the ladder diagram allows you to analyze the program in the sequence that it will occur. Secondly, the PLC allows you to program more than one set of contacts in the program for each switch. For example, the photoelectric switch may be in the program both normally open and normally closed and placed in the program as many times as the logic requires. The PLC gives you the capability of solving AND, OR, and NOT logic by connecting contacts in series or parallel, and using normally open and normally closed contacts as needed. Figure 15–10 shows examples of the five basic logic circuits as you would see them in ladder logic. As complex as some ladder logic programs look, you must remember that the logic program can consist of combinations of only these five basic functions.

Another feature of the PLC is the *control relay, internal coils*, or *memory relays*. As stated previously in the discussion of image registers, the control relay acts like the control relays that were used in relay panels years ago. The relay has one coil, and it can have any

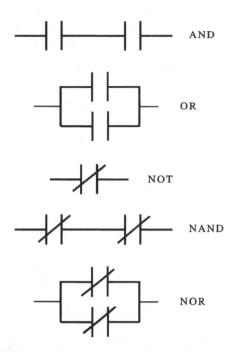

Figure 15–10 Five basic ways to program contacts in a PLC: in series (AND), in parallel (OR), normally closed contacts (NOT), series normally closed contacts (NAND), and parallel normally closed contacts (NOR).

number of normally open or normally closed contacts. When the coil is activated, all the contacts in the program that have the same address number as the coil will change state. The control relay exists only inside the PLC, which means it can never break or malfunction physically. Also, because control relays exist only in the PLC memory, you can have as many of them as the memory size of the PLC can accommodate. This means you can add control relays to a PLC program to accommodate changes to the machine-control logic as long as the PLC has space remaining in its memory.

It may help to see a case where a control relay is used in an actual circuit to better understand its function. For example, in Fig. 15–7, you can see the low-box-detected control relay (CR3) is used to indicate the logic has determined that a low box has been detected. The control relay in this rung of the program does nothing more than represent the conditions that indicate a low box has been detected. This condition exists when "Low-Box PE1" is on and "High-Box PE2" is off. Any time CR3 is energized, a low box has been detected, and it will change the state of all the contacts throughout the remainder of the program that have its number to represent that a low box has been detected.

In many cases each control relay in a program represents a *condition* for the machine operation, such as ready to go to auto, all safety doors closed, or ready to jog. If you think of the condition the control relay represents each time you see the control-relay contacts, you will find it easier to read and interpret the program.

Another feature of the PLC that makes complex factory automation easier to troubleshoot is the way that each input switch or each output load is electrically connected to its interface module. In Fig. 15–6 you can see that when a switch is connected to an input module, only two wires are involved. The outputs also use only two wires to connect the load device, such as the coil of a motor starter to the module. This means that when there is a malfunction in the electrical part of the circuit, the troubleshooter has only the switch and two wires to check for any individual input and the coil and two wires to check for any output. When you compare this type of wiring to a complex series-parallel circuit in an older-style relay panel, you can see it is much easier to troubleshoot two wires and a switch or two wires and a solenoid. In the older-style relay panels, a typical signal might pass through 10 to 15 relay contacts and switches and all the wiring that connects them together before it gets to a solenoid. Thus an open could occur in this complex circuit if any set of contacts opens, if any switch opens, or if a wire has a fault where it becomes open.

To make troubleshooting the switch and two wires that connect it to an input module even easier, a small indicator light called a *status indicator* is connected to each input circuit; it is mounted on the face of the module so it is easy to see. The troubleshooter can look at the status indicator and quickly determine if power is passing through the contacts or not. If the switch contacts are closed but the status indicator is not illuminated, the troubleshooter needs to check only the two wires and the switch for that circuit to see where the problem is. This feature is particularly valuable when the automated system the PLC is controlling has several hundred switches that could cause a problem. A status indicator is also connected to each output circuit and mounted on the face of the output module. Anytime the PLC energizes an output, the status indicator for that circuit is illuminated.

15.4.1
Classifications of Programmable Controllers

PLCs are generally classified by size. Small-size systems cost around $500 to $1,000, and they have a limited number of inputs and outputs. As a general rule the small PLCs have less than 100 inputs and outputs with approximately 12 inputs and 8 outputs mounted locally with the processor. In some systems additional inputs and outputs can be added through remote I/O racks to accommodate the remaining inputs and outputs. In other systems, the number of inputs and outputs are limited, but several small PLCs can be networked together to provide a larger number of I/Os that are controlled together. The smaller PLCs generally have 2K to 10K of memory that can be used to store the user's logic program.

Medium-size PLCs cost between $1,000 and $3,000 and have extended instruction sets that include math functions, file functions, and PID process control. These PLCs are capable of having up to 4,000 to 8,000 inputs and outputs. They also support a wide variety of specialty modules, such as ASCII communication modules, BASIC programming modules, 16-bit multiplexing modules, analog input and output modules that allow interfaces to both analog voltages and currents, and communication modules or ports that allow the PLC to be connected to a local-area network (called a LAN).

Large-size PLCs were very popular in the early 1980s before networks were perfected. The concept for the large-size PLC was to provide enough user memory and input and output modules to control a complete factory. Problems occurred when minor failures in the system brought the complete factory to a halt.

The advent of local-area networks brought about the concept known as *distributive control*, where small- and medium-size PLCs are connected together through a network. In this way, the entire factory is brought under the control of a number of PLCs, but a failure in one system does not disturb any other system. The majority of PLCs that you will encounter today are small- and medium-size systems that are used as stand-alone control or as part of a network.

15.5
Operation of Programmable Controllers

Now that you have a better understanding of how the basic PLC operates, you are ready to understand some of the more complex operations. You may need to review several of the previous diagrams to help you get a better understanding of these functions. In Fig. 15–5 you can see that the PLC has a processor, input modules, output modules, and power supply. The processor has several sections, including resident memory that contains the instruction set, user memory reserved for programs, and addressable data memory that has room to store the timer, counter, and other variables. The variables may be stored as binary data, as integers (whole numbers), or as floating-point numbers (numbers that have decimal points and can be expressed with exponents).

When the processor is in the RUN mode, it continually scans the inputs to see if they are on or off and updates the input image register. Next it sequentially solves the ladder logic that is stored in its user program area. It completes this part of its cycle by reading the ladder diagram. The processor goes through a *fetch cycle* to get each address found in the ladder diagram and puts it into the logic solving area. Each address is compared to the value found in its image register address, and this value is used to solve the logic. During the next part of the cycle, the processor writes the results of the logic to the output part of the image register. If the output in the ladder logic has contacts assigned to it, like the RUN CR1 in the second rung of the diagram in Fig. 15–7, they will reflect the status of the output image register address when the processor solves that line of logic. If the output address is connected to the output module, the signal is sent to the module to ensure that the device connected to it is energized.

The final section of each processor-scan cycle is called *overhead*. During the overhead part of the scan, the processor does all of its work outside of solving logic, such as updating the computer screen (monitor) if one is connected, sending messages to its printer, interacting with the network, and other similar functions. The main purpose of the PLC is to read inputs, solve logic, and write to its outputs, so less than 10% of the scan cycle is devoted to the overhead. The amount of time for the typical scan cycle for a generic PLC is approximately 20–60 ms (thousandths of a second). Newer enhancements to the PLCs try to lower the scan cycle to less than 10 ms.

15.6
Addresses for Inputs and Outputs for the MicroLogix PLC

Up to this point in our discussion we have identified inputs and outputs in generic terms by name only. In reality the processor must keep track of each input and output by an address or number. Each brand of PLC has devised its own numbering and addressing scheme. These addressing schemes generally fall into one of two categories: sequentially numbering each input and output or numbering according to location of input and output modules in the rack. The sequential numbering system is used in most small-size PLCs and about half of the medium-size systems. Allen-Bradley, which is one of the most widely used brands, has two different addressing systems. In its medium-size PLCs (PLC2, PLC5, and SLC500) the address for an input or output uses numbers that indicate the location of the input or output module in the rack. In the smaller PLCs, such as the

MicroLogix, a sequential numbering system for addresses is used.

From this point on, our applications use the address and numbering scheme of the Allen-Bradley MicroLogix. If you are using a different brand name of PLC you will need to convert the addresses to your system if you wish to try the program examples. Figure 15–11 shows an example of the numbering system used in the Allen-Bradley MicroLogix small-size PLC. From this example you can see that the input addresses start with address I0 (I stands for input and zero is the first address) and continue through address I19. Output addresses start with address O0 (O stands for output and zero is the first address) and continue through address O11. The first group of I/Os are mounted locally on the processor. It is very important to understand that each I/O address refers to a specific circuit in the hardware module and to a specific address in the processor's memory. The addresses in memory are assigned to *memory blocks*. The diagram that displays all the addresses in a memory block is called a *memory map*.

The Micrologix PLC is available in three sizes: units that have 10 inputs and 6 outputs, 12 inputs and 8 outputs, and 20 inputs and 12 outputs. At this time

Inputs	Outputs
I0	O0
I1	O1
I2	O2
I3	O3
I4	O4
I5	O5
I6	O6
I7	O7
I8	O8
I9	O9
I10	O10
I11	O11
I12	
I13	
I14	
I15	
I16	
I17	
I18	
I19	

Figure 15–11   Input and output addresses for the Allen-Bradley MicroLogix.

the MicroLogix does not support expansion modules; instead it provides a means of networking multiple PLCs so that they can provide common control to a larger group of inputs and outputs.

The MicroLogix also provides up to 512 internal relays. As you know, the internal relays act like a control relay in that they perform logic but do not directly energize an output.

NOTICE!

You can view the entire User Manual for the Allen-Bradley MicroLogix PLC by visiting Allen-Bradley's site on the Internet, http://www.ab.com/. You can visit their sections "Publications On Line" and "Manuals On Line."

15.7
Example Input and Output Instructions for the MicroLogix

Figure 15–12 shows an example of all the input and output instructions the MicroLogix uses. From this figure you can see that normally open and normally closed contacts are provided for input instructions under a number of different names to indicate if they are connected in series or parallel in the circuit. The first column in this figure shows a symbol for each instruction. The symbol shows the way the instruction looks when it is displayed in a program, as you would see it with a hand-held terminal display or when the program is printed out. The second column shows the mnemonic (pronounced *new-mon-ic*) for each instruction. A mnemonic is the code that you will enter in the hand-held programmer, and it is similar to an abbreviation for each instruction. The function code is the code you can use with the mnemonic. The fourth column is the name of the instruction. You should notice the name of the instruction is very similar to the mnemonic. The last column in this table gives the purpose of the instruction, which explains the function of each instruction.

From this figure you can see that there are several different instructions for normally open and normally closed contacts. If the normally open contact is the first contact in the line of logic, it is identified by the term *LOAD* and has the mnemonic LD. The LD mnemonic is used to indicate the contact is the start of a new rung. The AND instruction is used for all other contacts in the line of logic that are connected in series. If a contact is connected in parallel in the line of logic, it is identified by the OR instruction and mnemonic.

If the contacts are normally closed, the term *inverted* is added to the AND instruction so that normally

closed contacts that are connected in series will have the mnemonic ANI and normally closed contacts that are connected in parallel will have the mnemonic ORI.

Sometimes Allen-Bradley uses the name *examine on* for the normally open contacts, —] [—, and *examine off* for the normally closed contacts, —] \ [—. The term *examine* was chosen as part of this name to remind you that the processor looks at (examines) the electronic circuit in the input module to see if it is energized (on) or deenergized. The processor keeps track of the status of the circuit by placing a 1 in the image register address if it is energized and a 0 if it is deenergized. The word *examine* is also used to remind you to look at the status light on the front of the module so that you can also determine if the input circuit is energized or deenergized. The branch instructions are used to allow contacts to be connected in parallel to each other in the ladder diagram.

The output instruction in this figure includes the traditional *coil*, which uses the symbol -( )-. This instruction is named *Output* or *Output Energize*. One specialized output instruction is called the *Output latch*, —(L)—, and it has a mnemonic called *Set*. Its companion is called the *Output unlatch*, —(U)—, which has the mnemonic *Reset*. The latch output maintains its energized state after it is energized by an input even if the input condition becomes deenergized. The unlatch output must be energized to toggle the output back to its reset (off) state. The latch coil and unlatch coil must have the same address to operate as a pair.

Latch coils are used in industrial applications where a condition should be maintained even after the input that energized the coil returns to its deenergized state. It is important to understand that the latch coil bit will remain HI even if power to the system goes off and comes back on. This feature is important in several applications, such as fault detection. Many times a system has a fault such as an over-temperature condition or over-current condition. After the fault is detected, the machine is automatically shut down by the logic. By the time the technician or operator gets to the machine, the condition may have returned to near-normal conditions, and it is difficult to determine why the system shut down. If a latch coil is used in the detection logic, the fault bit will remain HI until the fault-reset switch energizes the unlatch instruction. Another application for the latch coil is for feed-water pumps or fans that are located in remote locations in the building, such as in the ceiling. A latch coil is used in these circuits because the motors must return to their energized condition automatically after a power loss. In some parts of the country, it is common for power to be lost for several seconds during severe lightning storms. If a regular type of coil is used in the program, a technician has to go to each fan or pump and depress the start push button to start the motor again. If a latch coil is used, the coil remains in the state it was in when the

Basic Instructions

HHP Display	Mnemonic	Code	Name	Purpose
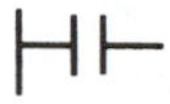 ⊣ ⊢	LD	20	Load	Examines a bit for an ON condition. It is the first normally open instruction in a rung or block
 ⊣⁄⊢	LDI	21	Load Inverted	Examines a bit for an Off condition. It is the first normally closed instruction in a rung or block.
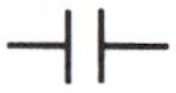 ⊣ ⊢	AND	22	And	Examines a bit for an On condition. It is a normally open instruction placed in series with any previous input instruction in the current rung or block..
 ⫫	ANI	23	And Inverted	Examines a bit for an Off condition. It is a normally closed instruction places in series with any previous input instruction in the current rung or block.
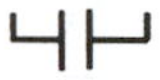 ⊔ ⊔	OR	24	Or	Examines a bit for an On condition. It is normally open instruction placed in parallel with any previous input instruction in the current rung or block.
 ⊔⁄⊔	ORI	25	Or Inverted	Examines a bit for an Off condition. It is a normally closed instruction placed in parallel with any previous input instruction in the current rung or block.
 ⊣LDT⊢	LDT	26	Load True	Represents a short as the first instruction in a rung or block.
 ⊣ORT⊢	ORT	27	Or True	Represents a short in parallel with the previous instruction in the current rung or block.
 ⊣OSR⊢	LD OSR	28	One-Shot Risisng	Triggers a one time event
 ⊣OSR⊢	AND OSR	29	One-Shot Rising	Triggers a one time event
 —( )—	OUT	40	Output (Output Energize)	Represents an output driven by some combination of input logic. Energized (1) when conditions preceding it pemit power continuity in the rung, and de-energized after the rung is false.
 —(L)—	SET	41	Set (Output Latch)	Turns a bit on when the rung is executed, and this bit retains its state when the rung is not executed or a power cycle occurs.
 —(U)—	RST	42	Reset (Output Unlatch)	Turns a bit off when the rung is executed, and this bit retains its state when the rung is not executed or when power cycle occurs.
	TON	0	Timer-On Delay	Counts timebase intervals when the instruction is true.
	TOF	1	Timer-Off Delay	Counts timebase intervals when the instruction is false.
	RTO	2	Retentive Timer	Counts timebase intervals when the instruction is true and retains the accumulated value when the instruction goes false or when power cycles occur.
	CTU	5	Count Up	Increments the accumulated value at each false-to-true transition and retains the accumulated value when the instruction goes false or when power cycle occurs.
	CTD	6	Count Down	Decrements the accumulated value at each false-to-true transition and retains the accumulated value when the instruction goes false or when power cycle occurs.
	RES	7	Reset	Resets the accumulated value and status bits of a timer or counter. Do not use with the TOF timers.

Figure 15–12   Example input, output, timer, and counter instructions available in the Allen-Bradley MicroLogix.

```
          I:1           I:1           I:1              O:0
----------]   [----------]   [----------]   [--------------(   )-----
          1             3             6                3
```

Keystrokes to enter the program through the Hand Held Programmer.
LD I1
AND I3
AND I6
OUT O3

Figure 15–13   Example of a PLC program with contacts in series.

power went off. This means that if the latch coil is energized and the motor is running when power goes off, the latch coil remains energized when power returns, and the motor begins to run again automatically. The latch coil is very useful in these circumstances, especially when a factory has dozens of fans and pumps and it would take several hours to reset them all.

Other output instructions have a rather specific function that is beyond traditional relay logic. These instructions include the timer, counter, sequencer, and zone control. Each of these instructions is covered in detail.

15.7.1
More about Mnemonics: Abbreviations for Instructions

In Fig. 15–12 you can see that each instruction has a mnemonic, which is an abbreviation for the function of the instruction. The mnemonic for each instruction is selected because it sounds like the name of the function that the instruction provides. It also is limited to two or three letters so that it is easy to program. The mnemonic also comprises the letters the PLC's microprocessor recognizes as instructions. You will need to learn the mnemonic for each instruction so that you can read and write PLC programs. The mnemonics for different PLC brand names may be slightly different, but they will always describe the instruction they represent.

15.8
Example of a Program with Inputs Connected in Series

A note is needed about the way the inputs, outputs, and timers are numbered and displayed in a printed program. As you can see in the program in Fig. 15–13, the

File Type	Identifier	Number
Output	O	0
Input	I	1
Status	S	2
Bit	B	3
Timer	T	4
Counter	C	5
Control	R	6
Integer	N	7

Figure 15–14   File types, identifiers, and numbers used in the Allen-Bradley MicroLogix PLC.

first input is number 3, and it is identified as I:1/3. The I indicates the contact is controlled by an input. The :1 indicates the processor keeps the status of this contact (on or off) in file number 1, and the /3 indicates this is input number 3. The output is identified in the program as O:0/3. The O:0 indicates this is an output whose status is stored in file 0, the /3 indicates this is output number 3. The format for numbering files is used because it is the format that Allen-Bradley uses in its other PLCs, such as the PLC5.

When you enter inputs with a hand-held programmer, you simply enter I1 for the first input, I3 for the second input, and I6 for the third input. You use O3 for the output. The PLC automatically does the remainder of the formatting, such as adding the file number.

Figure 15–14 shows the format Allen-Bradley uses for file types, identifier, and numbers to store information about inputs, outputs, timers, and counters in the MicroLogix PLC.

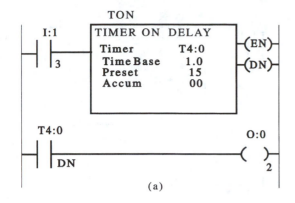

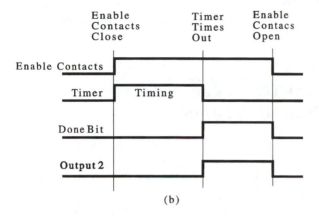

Figure 15–15 (a) Example of a TON timer. (b) The timing diagram for a TON timer and its done bit.

15.9
Using Timers in a PLC

Many industrial applications require time delay. The MicroLogix uses the mnemonic TON to represent the *timer-on-delay* instruction. The timer operates like the traditional hardware-type motor-driven timer discussed in Chapter 14. Figure 15–15 shows an example of a timer instruction used in a line of logic. The timers can use addresses T0–T39. Each timer has a *preset value* (PRE value) and *accumulative value* (ACC value). The preset value is the amount of time delay the timer provides, and the accumulative value is the actual time delay that has accumulated since the timer has been energized. The timer accumulative value *increments* (counts up) at a rate determined by the time base. The maximum preset value for timers in the MicroLogix is 32,767.

The time base for the MicroLogix timers can be 1.0 s or 0.01 s. When the timer accumulative value equals the preset value, the timer times out. When the

timer times out, it can cause a set of contacts to change state. These contacts need to have the same address as the timer, and they should be identified with the mnemonic DN, indicating they are controlled by the *done bit* for the timer.

The timer can control normally open or normally closed contacts with the done bit or two other status bits. Each of these bits is identified with a number as well as a mnemonic. The *done bit* is bit 13, and it is assigned the mnemonic DN. It changes state when the timer times out. The *timer timing bit* is bit 14, and it is assigned the mnemonic TT. It is set (ON) any time the timer is timing. The *enable bit* is bit 15, and it is assigned the mnemonic EN. It is set (ON) any time the timer is enabled. You will see that you can use either the bit number or its mnemonic when you are entering a program. In Figure 15–13 you can see the timer is timer T0, so it has the number T4:0 in the program. In the second line of the program you can see a set of contacts that are controlled by T4:0 DN, which is the done bit for this timer. When the timer times out, these contacts close and energize output O:0/2.

Another important part of each timer is the reset function. When a timer is reset, its accumulative value is reset to 0. Each timer must be reset at some point to make it ready to be used for the next timing cycle. The TON timer is reset any time power provided to it is interrupted. In the example in Figure 15–15, contacts I:1/3 control power to the timer, and any time these contacts are open, the timer resets its accumulative value. Any time these contacts are closed, the timer is enabled and its accumulative value increments (count up) at the rate determined by the time base, which in this example is 1 s. This may also be described by saying the timer's accumulative value increases 1 s with every tick of the timer time base. The contacts I:1/3 in this example are called enable/reset contacts, because they control the enable and reset functions of the timer.

15.10
Retentive Timer

Another type of timer provided in the PLC is called the *retentive timer*, and it has the mnemonic RTO. A retentive timer is different from the TON timer in that its accumulative value does not reset when the enable contacts are opened. In the retentive timer, the accumulative value is *retained*, and when the enable contacts are closed again, the accumulative value in the timer begins accumulating values again. This type of timer operates in a similar way to an hour meter on a piece of equipment. For example, this type of timer can be enabled by a set of contacts that control an air-compressor motor. If the contacts are

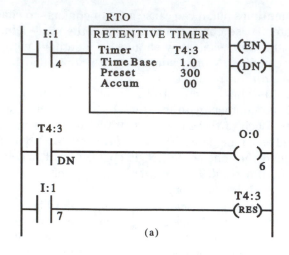

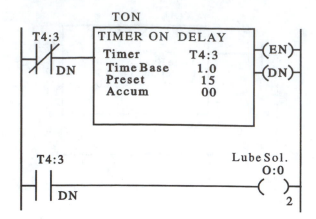

Figure 15–17   Example of a timer programmed as an automatic resetting timer that controls a lube solenoid.

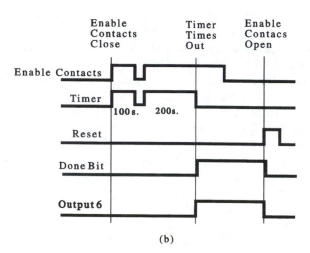

Figure 15–16   Example of an RTO timer and the RES instruction.

closed and the motor runs for 20 min, the timer keeps track of this time. When the contacts open and the motor stops, the timer also stops, but the accumulative value (20 min) does not reset. When the contacts close again and the motor runs an additional 30 min, the timer shows the total value of 50 min.

15.10.1
Resetting the RTO Timer

Because the enable contacts do not cause the RTO timer to reset, it needs a special reset instruction, whose mnemonic is RES. Figure 15–16 shows an example of the RTO timer and the RES instruction that will cause its accumulative value to reset. In this example, the timer adds time to its accumulative value any time the enable contacts, I:1/4, are closed. The timer resets its accumulative value any time contacts I:1/5 are closed and pass power to the RES instruction.

15.11
The TON as an Automatic Resetting Timer

A TON timer can be made to automatically reset itself. The automatic reset part of this circuit is accomplished by using a TON timer instruction and placing a set of the normally closed timer contacts in series with it and a set of normally open timer contacts in series with the timer reset instruction. Figure 15–17 shows an example of a timer programmed as an auto-resetting timer with its contacts controlling a lubrication solenoid. In this example the TON timer T4:3 instruction has a set of normally closed contacts controlled by the DN bit connected in series with it so that the timer begins to time any time the PLC is in the RUN mode. After 60 s it times out, and it closes the normally open T4:3/DN contacts in line 2 of the program to energize the lube solenoid. The normally closed T4:3/DN contacts that enable the timer open and cause the timer to reset and start its timing cycle over again automatically. It is important to understand that the lube system simply needs its solenoid pulsed to activate the lubrication system.

15.12
Off-Delay Timers

The off-delay timer is used in applications where the amount of time delay is timed from the time a condition turns off. For example, in a heating system, at the end of a cycle the fuel is turned off, and the fan remains running for several minutes to ensure all the residual heat in the heat exchanger is removed. The time-delay-off function of the heating system's fan can be controlled by an off-delay timer. For the off-delay timer to operate, its enable

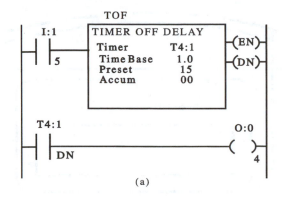

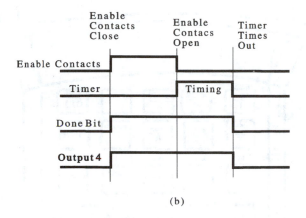

Figure 15–18   (a) A TOF instruction with its done bit controlling output O4; (b) a timing diagram for the TOF timer and its done bit.

contacts must be energized and then turned off. At the point where the enable contacts are turned off, the timer begins its delay cycle, which runs until the timer's accumulative value becomes equal to the preset value.

The off-delay timer in the PLC is called the timer-off-delay instruction, and it has the mnemonic TOF. The TOF timer and the TON timer are similar in that they both can use any of the 40 timers (T0–T39), and they both have preset values, accumulative values, time base values, and status bits. An example of a TOF instruction is shown in the top diagram of Figure 15–18. The important difference between the two is the timing cycle, which controls when the done bit is on or off. The bottom diagram in Figure 15–17 shows an example of the timing cycle of the done bit (DN) for the on-delay timer and the off-delay timer. In this figure you can see that at the start of a cycle, when the TON instruction is first energized by its enable contacts, the timer's done bit is off and remains off until the accumulative value becomes equal to the preset value and the timer times out. At that point, the done bit becomes energized and remains energized until the enable contacts to the timer are opened and the timer is reset.

The sequence for the TOF timer shows that its done bit (DN) contacts are off until the enable contacts for the timer are closed. At that point, the done bit is energized, and it remains energized until the timer cycle is complete. The TOF timer does not start its timing cycle until the enable contacts are opened again. After the enable contacts are opened and the timer runs its cycle and times out, the done bit is turned off.

15.13
Applications of Counters in Industry

Many industrial applications use counters. These applications can be separated into two basic operations. The first is called *totalizing*, and it occurs when a machine is making a large number of parts and the counter must keep track of the total. The second type of application is called *sequencing*. In this type of application the counter is used to keep track of a number of steps and then it resets itself to start the sequence all over again. An example of the sequence-type application is when a counter is used to keep track of the number of boxes being placed on each layer of a pallet and the number of layers the pallet has. For example, a pallet may require four boxes for each layer in a stack three layers high. One counter would use 1, 2, 3, 4 to keep track of the four boxes that are placed on each layer, and then it would reset to start the next layer. A second counter would count 1, 2, 3 and then reset to keep track of the layers on the pallet.

15.13.1
The Operation of a Counter in a PLC

A PLC has two types of counter instructions available for use. They include the up counter, whose mnemonic is **CTU,** which stands for *count up*, and the down counter, whose mnemonic is **CTD,** which stands for *count down*. The counters are similar to the timer instruction in that they have preset and accumulative values and they each can control a number of status bits. The up-counter instruction *increments* (add 1 to) its accumulative value when the enable contacts, which allow power to flow to it, transition from open to closed. Thus the enable contacts must return to their open state before they can transition for a second time to add another count to the counter's accumulative value. It is important to note that the counter does not care how long the contact stays closed, as does the timer instruction; rather, the counter looks only for the transition of the enable contacts. The down counter, CTD, decrements (subtracts 1) from its accumulative value each time power is transitioned to it.

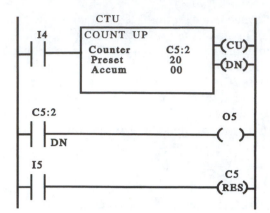

Figure 15–19 An example of an up counter with a reset instruction.

Both types of counters require a reset instruction (RES) to set the accumulative value back to zero. Figure 15–19 shows the instruction for an up counter that includes the reset operation. The enable contacts (I4) are not part of the counter, but they must be present for the counter to operate.

The preset value is the desired value for the counter and may represent how large a value the counter is expected to count. The maximum preset value for the up counter is +32,767, and the lowest value for the down counter is −32,768. Each counter also has several status bits under its control. Both types of counters use *bit 13* as a *done bit*. The done bit is energized (set) any time the accumulative value is equal to the preset value. The up counter uses bit 12 as an overflow bit. The overflow bit is set any time the accumulative value is larger than +32,767. The down counter uses bit 11 as an underflow bit. The underflow bit is set any time the accumulative value goes below −32,768. The up counter uses bit 15 as an enable bit and the down counter uses bit 14 as an enable bit. The bit number is used with the counter address (number) to create control logic. For example if you want counter number 6 (C5:6) to control a solenoid that places a piece of cardboard between layers on a pallet, you would use contacts that have the number C5:6/13 on them. The /13 indicates that the done bit for the timer controls the contacts, and they will become energized each time the counter accumulative value reaches the preset value.

15.13.2
Up-Down Counters

As you know, the PLC can use an up counter or a down counter. Some applications are easier to understand when up and down counters are used together, such as a good parts–bad parts counter. In this type of application the up-counter (CTU) and the down-counter (CTD) instructions are assigned to the same counter address. Each time the

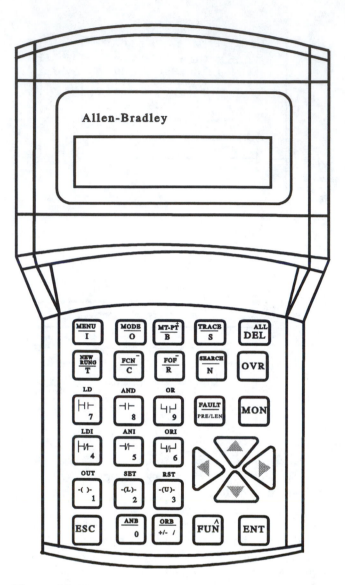

Figure 15–20 Hand-held programmer for the Allen-Bradley MicroLogix PLC.
(Courtesy of Allen-Bradley)

CTU instruction is transitioned, the counter will count up 1 value, and each time the CTD instruction is transitioned, the counter will count down 1 value. In this application both counters have the same accumulative value, because they use the same counter. At the end of the day, the number displayed in the accumulative value is the total difference between the good parts and the bad parts.

15.14
Entering Programs with a Hand-Held Programmer

After you understand how inputs, outputs, timers, and counters work, you are ready to enter a program into the PLC using the hand-held programmer. Figure 15–20

1. LANGUAGE
2. ACCEPT EDITS
3. PROG. CONFIG.
4. MEM. MODULE
5. CLEAR FORCES
6. CLEAR PROG.
7. COMMS.
8. CONTRAST

Figure 15–21   Choices listed under the Menu Function.

1. RPRG	Remote Program
2. RRUN	Remote Run
3. RCSN	Remote Test-Continuous Scan
4. RSSN	Remote Test-Single Scan
5. RSUS	Remote Suspend
6. FLT	Fault

Figure 15–22   Functions provided under the MODE function.

shows the hand-held programmer for the Allen-Bradley MicroLogix PLC. From the diagram you can see the hand-held programmer has an LCD display, a set of number keys (0–9), and a set of instruction keys, such as contacts and coils. The programmer also has a set of function keys that are used for diagnostics and troubleshooting. These keys are named Menu, Mode, Force On, Force Off, Trace, Search, and Fault. Other keys are named Enter, Escape, Monitor, Overwrite, and Delete; they are used for general editing of the PLC program.

The Menu function has eight choices, which are listed in the table in Fig. 15–21. Figure 15–22 shows the six choices listed under the MODE function.

15.15
Programming Basic Circuits in the PLC

At times you will be requested to program simple programs into the PLC through the hand-held programmer. These simple programs include contacts that are con-

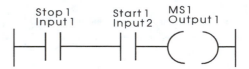

1. MON
2. ENT
3. NEW RUNG
4. ⊣⊢ I1 ENTER
5. ⊣⊢ I2 ENTER
6. ⊣( )⊢ O1 ENTER
7. Change MODE to RRUN and try switches and view input and output status lights.

Figure 15–23   Simple circuit with two normally open contacts.

1. MON
2. ENT
3. NEW RUNG
4. ⊣⊢ I4 ENTER
5. ⊣/⊢ I3 ENTER
6. ⊣( )⊢ O1 ENTER
7. Change MODE to RRUN and try switches and view input and output status lights.

Figure 15–24   Simple circuit with one normally open and one normally closed contact.

nected in series or parallel, timers, and counters. Figures 15–23 through 15–28 show six simple programs and all the keystrokes required to enter these programs. If you have a MicroLogix PLC available, you can use these examples to try these programs. If you have another brand name of PLC, your teacher can provide examples similar to these for your controller or you can E-mail the author (tkissell@mail.intelliworks.net) for examples for your brand of controller.

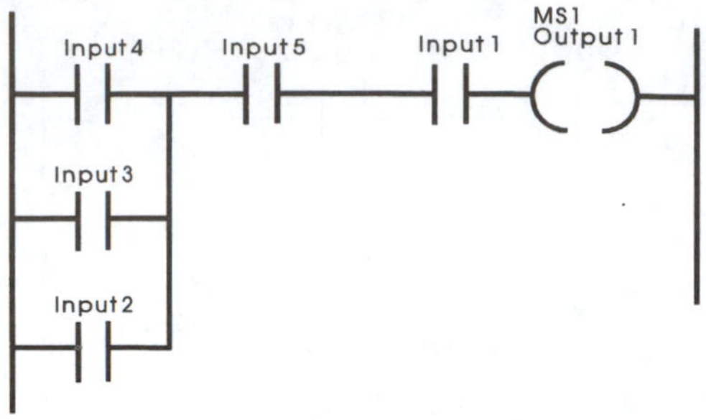

Input 4   Input 5   Input 1   MS1 Output 1

Input 3

Input 2

1. MON
2. ENT
3. NEW RUNG
4. ─┤├─ I4 ENTER
5. ─┘├┘─ I3 ENTER
6. ─┘├┘─ I2 ENTER
7. ─┤├─ I5 ENTER
8. ─┤├─ I1 ENTER
9. ─( )─ O1 ENTER
10. Change MODE to RRUN and try switches and view input and output status lights.

Figure 15–25 Simple parallel circuit.

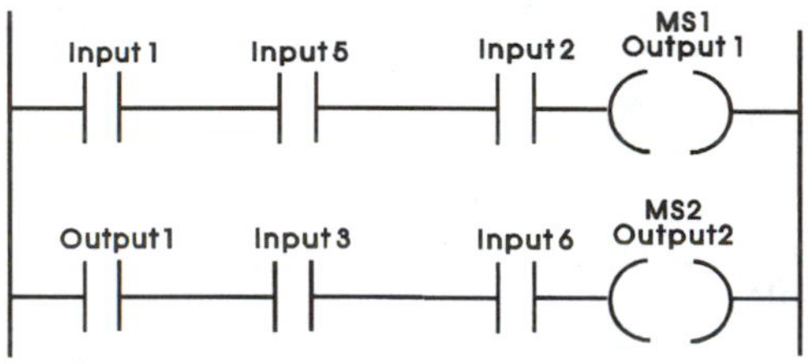

Input 1   Input 5   Input 2   MS1 Output 1

Output 1   Input 3   Input 6   MS2 Output 2

1. MON
2. ENT
3. NEW RUNG
4. ─┤├─ I1 ENTER
5. ─┤├─ I5 ENTER
6. ─┤├─ I2 ENTER
7. ─( )─ O1 ENTER
8. NEW RUNG
9. ─┤├─ O1 ENTER
10. ─┤├─ I3 ENTER
11. ─┤├─ I6 ENTER
12. ─( )─ O2 ENTER
13. Change MODE to RRUN and try switches and view input and output status lights.

Figure 15–26 Two series circuits.

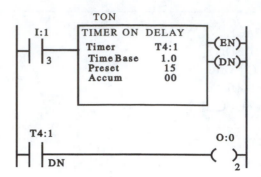

1. MON
2. ENT
3. NEW RUNG
4. ─┤├─ I3 ENTER
5. FUN
6. ANB 0 ENTER   (0=TON 1=TOF 2=RTO)
7. T1 ENTER
8. 15 (FOR PRESET VALUE)
   0 (FOR ACC VALUE)
9. △ ENTER TO GET 1.0 TIME BASE
10. NEW RUNG
11. ─┤├─ T1 ORB ENTER   THIS GIVES DN BIT
12. ─( )─ O2 ENTER
13. Change MODE to RRUN and try switches and view input and output status lights.

Figure 15–27 Simple timer program.

15.16
Wiring a Start-Stop Circuit to Input and Output Terminals on the PLC

All the inputs and outputs used in the PLC system must be connected to the input- or output-module hardware. The simplest circuit to understand is a simple start-stop circuit that controls a motor-starter coil. Figure 15–29 shows an example of this circuit, and you can see where a start push-button switch and a stop push-button switch are wired to two input addresses and a motor-starter coil connected to an output address. From this diagram you can also see the location of the screws on the MicroLogix PLC where the wires from the switch and coil for the motor starter are connected.

From the diagram you should notice that the start push button is wired as a normally open switch, and the stop push button is wired as a normally closed switch, just as in a hardwired control system. The major difference between the start-stop circuit in the PLC and a hardwired system is in the way the start and stop push

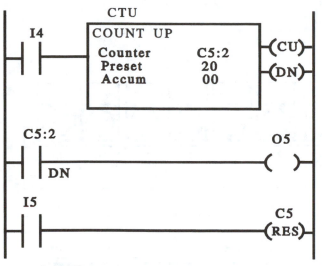

1. **MON**

2. **ENT**

3. **NEW RUNG**

4. ┤├ **I4 ENTER**

5. **FUN**

6. **ANB 5 ENTER   (5=CTU 6=CTD 7=RES)**

7. **C2 ENTER**

8. **20**   (FOR PRESET VALUE)
   0   (FOR ACC VALUE)

9. **NEW RUNG**

10. ┤├ **C2 ORB ENTER   THIS GIVES DN BIT**

11. ─( )─ **O5 ENTER**

12. **NEW RUNG**

13. ┤├ **I5 ENTER**

14. **FUN**

15. **ANB 7 ENTER (5=CTU 6=CTD 7=RES)**

16. **C2 ENTER**

17. **Change MODE to RRUN and try switches and view input and output status lights.**

Figure 15–28 Simple counter program.

buttons are programmed into the PLC. In the PLC program you should notice that the stop push button is programmed normally open. At first this seems to be incorrect, but if you look at the line of logic in the PLC, you will notice that because the stop push button is wired normally closed, it will pass power to the input module, which in turn sends a 1 to the image table. Because the stop contacts in the program are programmed normally open, they change state and pass power to the start contacts whenever the input module sends a 1 to the image table. If the stop contacts were programmed normally closed, they would open when the input module received power from the hardwired stop switch and the circuit would not operate.

> **IMPORTANT NOTICE!**
> The PLC program would work if you wired the stop push button normally open and programmed its contacts normally closed. This creates a severe problem: the stop push button would become a safety hazard if it were wired normally open and a wire came loose from its terminal or the stop button failed, because it could not send a signal to the PLC input module and stop the circuit. For this reason it is important that you always wire the stop push button as a normally closed switch so that if it fails, it will interrupt power to the input module and stop the circuit.

The hardwired circuit of a PLC must also be protected by a real relay called a *master control relay*. This relay is identified as a CRM (control relay master), and all the voltage to the input and output module is controlled by the contacts of the CMR. The coil of the CMR is connected to a hardwired start-stop push-button station that is not part of the PLC program. This allows the power to be shut off to the inputs and output by depressing the stop button. Power is connected directly to the processor so that it stays energized regardless of the state of the CRM.

15.17
Typical Problems a Technician Will Encounter with PLCs

As a technician you may encounter problems with PLCs at either the software or hardware level. At the software level, you may be required to lay out and enter the original ladder logic program. You may be also involved at the start-up of a machine when a program has been written for the PLC and a technician is required to test the program against the machine operation. In other cases, you may be involved in using a program to troubleshoot the system when a fault occurs. Sometimes you will be requested to make changes in the documentation that describes the operation of the system and identifies all the inputs and outputs. In each of these cases, you may need additional information to satisfactorily accomplish the job. This additional information may be provided in the manufacturer's manuals or in short 4- to 5-day training courses that are provided on specific PLCs.

You may also be involved with the hardware modules and I/O devices that are connected to PLCs. In some systems you will be requested to read the specification sheets for the signals that each module sends

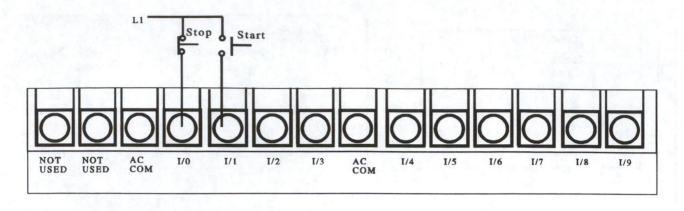

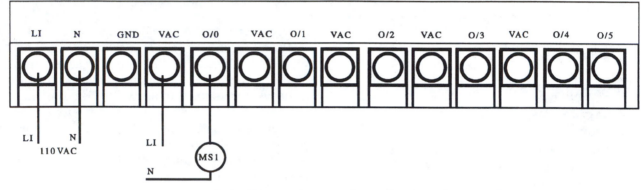

(a) Start/stop hardwired to PLC

(b) Start/stop program

Figure 15–29 (a) Wiring start and stop push-button switches to input terminals and wiring a motor-starter coil to the output terminals of a MicroLogix PLC; (b) the PLC program for the start-stop circuit.

or receives so that you can select the proper types of electronic devices to interface with the PLC. Sometimes you will be requested to analyze a signal to see that it meets the proper criteria for data transmission. If system component failure occurs, you will be required to use all the knowledge you have about electronics to test the hardware and make decisions. You now have a better idea of how electronics have been blended with the programmable controller to provide industrial automation.

15.18
Troubleshooting the PLC

When you are called to troubleshoot the PLC, you need to understand that each input module has a status indicator (light) for each input switch and an output status indicator for each output circuit. If the input switch is on and passing 110 VAC to the module, the status light will be on. The status light is provided on the PLC module so that you do not need a voltmeter to determine if

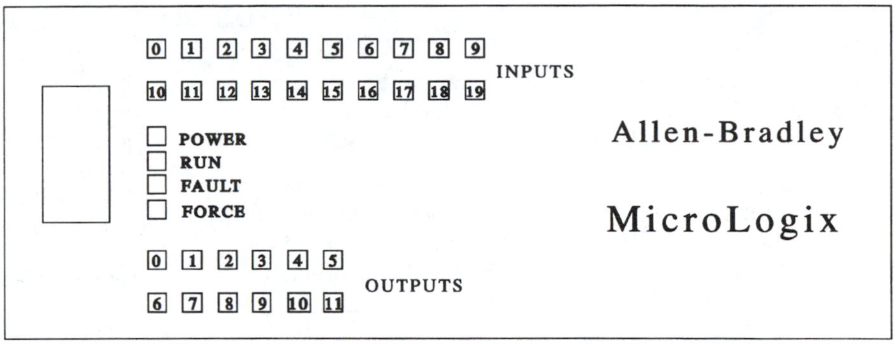

Figure 15–30 Example of status indicators on input and output modules for the MicroLogix PLC.

```
  0  1  2  3  4  5  6  7  8  9
 10 11 12 13 14 15 16 17 18 19    INPUTS

  □ POWER                          Allen-Bradley
  □ RUN
  □ FAULT
  □ FORCE                          MicroLogix

  0  1  2  3  4  5
  6  7  8  9 10 11    OUTPUTS
```

power is at the terminals. On the output module, the status light will be on whenever the output circuit is receiving power. Again, the status light can be used instead of a voltmeter to quickly tell if the circuit has power. Figure 15–30 shows an example of input and output status lights.

Questions for This Chapter

1. Describe the four basic parts of any programmable controller.

2. Explain the operation of a PLC up counter and a down counter.

3. Explain the four things that occur when the PLC processor scans its program.

4. Discuss the advantages a PLC-controlled system would have over a hardwired relay-type control system if you wanted to add one more photoelectric switch and output to detect a medium-size box in the box-sorting system described in Section 15.2.

5. The input module has a status indicator for each input circuit. Explain how the status indicator is used in troubleshooting.

True or False

1. One advantage of all PLCs is that you can use contacts with the same address more than once in the program.

2. When the PLC processor is in the RUN mode, it does not execute its scan cycle.

3. A mnemonic is a timer whose time delay is controlled by air pressure.

4. When a timer in a PLC program will not time, you should suspect the timer is broken, and you will need to program a replacement timer.

5. A control relay (memory coil) in a PLC program is different from an output instruction in that its signal does not control a circuit in an output module and it is used mainly to determine logic conditions.

Multiple Choice

1. When the input contact that enables a TON type timer is opened, the accumulated value in the timer _____ .
   a. resets (goes to zero).
   b. freezes (remains at its present value).
   c. goes to an undetermined value, so the timer must be reset manually.

2. An RTO-type timer has its accumulative value reset _____ .
   a. any time the enable contacts are opened.
   b. only when the reset for the timer is HI.
   c. any time the accumulated value in the timer exceeds 99.

3. When the PLC processor is in the PROGRAM mode, _____ .
   a. it executes its scan cycle.
   b. it does not execute its scan cycle.
   c. it is impossible to tell whether the processor is executing its scan cycle because you are not on line.

4. The stop push button should be wired _____ and programmed _____ when it is used as part of a start-stop circuit for a PLC system.
   a. normally closed; normally open
   b. normally open; normally closed
   c. It doesn't matter.

5. You can use the _____ on the face of a PLC to determine whether an input switch is on or off.
   a. power light
   b. input status indicators
   c. output status indicators

Problems

1. Enter the conveyor box-sorting program in Fig. 15–7 in a programmable controller and execute the inputs to watch the program operate. Describe the operation of each output.

2. Enter an on-delay timer and an off-delay timer in your PLC and describe the following: preset time, accumulative time, what must occur for the timer to run, what must occur for the timer to time out, and what contacts change when the timer times out.

3. Enter an up counter and a down counter in your PLC and describe the following: preset value, accumulative value, what must occur for each counter to count, what occurs when the counter's preset and accumulative values are equal, what happens when the counter's accumulative value exceeds the overflow value, what contacts change when the counter reaches its preset value, and what contacts change when the counter exceeds the overflow.

4. Program an auto-resetting timer and describe its operation.

5. Program an auto-resetting counter into your PLC and describe its operation.

◀ Chapter 16 ▶

Electronics for Maintenance Personnel

Objectives

After reading this chapter you will be able to:

1. Explain P-type and N-type material.
2. Identify the terminals of a diode and explain its operation.
3. Explain the operation of PNP and NPN transistors.
4. Identify the terminals of a silicon-controlled rectifier (SCR) and explain its operation.
5. Explain the operation of a triac.

16.0
Overview of Electronics Used in Industrial Circuits

Electronic devices such as diodes, transistors, and SCRs have become commonplace in industrial circuits because they provide better control than electro-mechanical devices and are less expensive to manu-facture. When you are troubleshooting you will find solid-state components in control boards and other controls. In this chapter you will gain an understanding of P-material and N-material, which are the building blocks of all electronic components. After you have a basic understanding of P- and N-material, you will be introduced to diodes, transistors, SCRs, and triacs, and you will see application circuits of each of these types of components. You will also learn the theory of opera-tion and troubleshooting techniques for each type of device.

16.1
Conductors, Insulators, and Semiconductors

In Chapter 2 you learned that atoms have protons and neutrons in their nuclei and electrons that move around the nucleus in orbits, which are also called shells. The number of electrons in the atom is different for each el-ement. For example, you learned that copper has 29 electrons, 3 of which are located in the outermost shell. The outermost shell is called the valence shell, and the electrons in that shell are called valence electrons. The atoms of the most stable material have eight valence electrons, and these eight valence electrons are found as four pairs. As an atom for materials that are used as conductors and insulators fills its valence shell with 8 electrons, it will open a new orbit with the additional electrons. This means that an atom may have five, six, or seven atoms and it will take less energy to add elec-trons to get a full shell (eight), or it may have one, two, or three electrons and it will take less energy to give up these electrons to get down to the previous full shell that has eight electrons.

A *conductor* is a material that allows electrons (electrical current) to flow easily through it, and an *in-sulator* does not allow current to flow through it. An ex-ample of a conductor is copper, which is used for elec-trical wiring. An example of an insulator is rubber or plastic. The atomic structure of a conductor makes it easier for electrons to flow through it and the atomic structure of an insulator makes it nearly impossible for any electrons to flow through it.

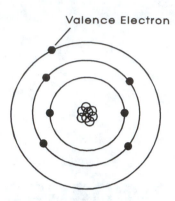

Figure 16–1   Atomic structure of a conductor.

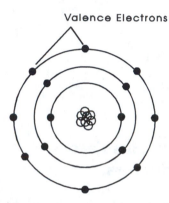

Figure 16–2   Atomic structure of an insulator.

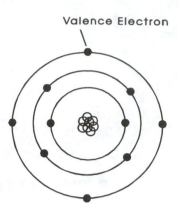

Figure 16–3   Atomic structure for semiconductor material.

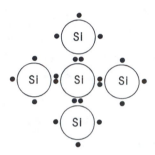

Figure 16–4   Atoms of silicon semiconductor material are combined to create a lattice structure.

Figure 16–1 shows the atomic structure of a conductor. Atoms of conductors can have one, two, or three valence electrons. The atom in this example has one valence electron. Because all atoms try to get eight electrons (four pairs) in their valence shell, it takes less energy for conductors to give up these electrons (one, two, or three), so the valence shell will become empty. At this point the atom becomes stable because the previous shell becomes the new valence shell, and it has eight electrons. The electrons that are given up are free to move as current flow.

Figure 16–2 shows the atomic structure of an insulator. Insulators have five, six, or seven valence electrons. In this example you can see that the atom has seven valence atoms. This structure makes it easy for insulators to take on extra electrons to get eight valence electrons. The electrons that are captured to fill the valence shell are electrons that would normally be free to flow as current.

Semiconductors are materials whose atoms have exactly four valence electrons. Because these atoms have exactly four valence electrons, they can take on four new valence electrons, like an insulator, to get

a full valence shell, or they can give up four valence electrons, like a conductor, to get an empty outer shell; then the previous shell, which has 8 electrons, becomes the valence shell. Figure 16–3 shows an example of the atomic structure for semiconductor material.

16.2
Combining Atoms

When solid-state materials or other materials are manufactured, large numbers of atoms are placed together. The structure that becomes most stable at this point is called a *lattice structure*. Figure 16–4 shows an example of the lattice structure that occurs when atoms are combined. In this diagram you can see that atoms of silicon (Si), which is a semiconductor material with four valence electrons, are combined so that one valence electron from each of the neighbor atoms is shared so that all atoms look and act as if they each have eight valence electrons.

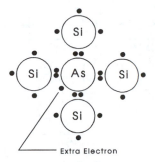

Figure 16–5   N-material formed by combining four silicon atoms with a single arsenic atom.

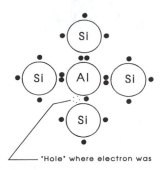

Figure 16–6   P-material formed by combining four silicon atoms with a single aluminum atom.

16.3
Combining Arsenic and Silicon to Make N-type Material

Other types of atoms can be combined with semiconductor atoms to create the special material that is used in solid-state transistors and diodes. Figure 16–5 shows a diagram of four silicon atoms that are combined with one atom of arsenic. From this figure you can see that arsenic has five valence electrons, and when the silicon atoms are combined with it they create a very strong lattice structure. You can see that each silicon atom donates one of its valence electrons to pair up with each of the valence electrons of arsenic atom. Because the arsenic atom has five valence electrons, one of the electrons is not paired up and becomes displaced from the atom. This electron is called a *free electron*, and it can go into conduction with very little energy. Because this new material has a free electron, it is called *N-type material.*

16.4
Combining Aluminum and Silicon to Make P-type Material

An atom of aluminum can also be combined with semiconductor atoms to create the special material that is called *P-type material.* Figure 16–6 shows a diagram of four silicon atoms that are combined with one atom of aluminum. From this figure you can see that aluminum has three valence electrons, and when the silicon atoms are combined with it they create a very strong lattice structure. You can see that each silicon atom donates one of its valence electrons to pair up with each of the valence electrons of aluminum atom. Because the aluminum atom has three valence electrons, one of the four aluminum electrons will not be paired up, and it will have a space where any free electron can move into to com-

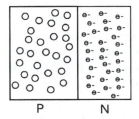

Figure 16–7   An example of a piece of P-material connected to a piece of N-material.

bine with the single electron. This free space is called a *hole*, and it is considered to have positive charge, because it is not occupied by a negatively charged electron. Because this new material has an excess hole that has a positive charge, it is called *P-type material.*

16.5
The PN Junction

One piece of P-type material can be combined with one piece of N-type material to make a *PN junction.* Figure 16–7 shows a typical PN junction. The PN junction forms to make an electronic component called a *diode.* The diode is the simplest electronic device. When DC voltage is applied to a PN junction with the proper polarity, it causes the junction to become a very good conductor; if the polarity of the voltage is reversed, the PN junction becomes a good insulator.

16.6
Forward Biasing the PN Junction

When DC battery voltage is applied to the PN junction so that positive voltage is connected to the P-material and negative voltage is connected to the N-material, the

Figure 16–8   (a) A forward-biased
PN junction; (b) a reversed-biased PN
junction.

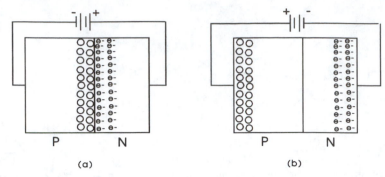

(a)   (b)

junction is forward biased. Figure 16–8a shows a battery connected to the PN junction so that it is forward biased. In this diagram you can see that the positive battery voltage causes the majority of holes in the P-type material to be repelled, so the free holes move toward the junction, where they come into contact with the N-type material. At the same time the negative battery voltage also repels the free electrons in the N-type material toward the junction. Because the holes and free electrons come into contact at the junction, the electrons recombine with the holes and cause a very low resistance junction, which allows current to flow freely through it. When a PN junction has low resistance, it allows current to pass just as if the junction were a closed switch.

It is important to understand that up to this point in this text, all current flow has been described in terms of *conventional current flow*, which is based on a theory that electrical current flows from a positive source to a negative return terminal. At this point you can see that this theory will not support current flow through electronic devices. For this reason *electron current flow theory* must be used when discussing electronic devices. In electron current flow theory, current is the flow of electrons, and it flows from the negative terminal to the positive terminal in any electronic circuit.

16.7
Reverse Biasing the PN Junction

Figure 16–8b shows a battery connected to the PN junction so that it is reversed biased. In this diagram you can see that the positive battery voltage is connected to the N-type material and the negative battery voltage is connected to the P-type material. The negative voltage on the P-type material attracts the majority of holes in the P-type material so that they move away from the junction and cannot come into contact with the N-type material. At the same time the positive battery voltage that is connected to the N-type material attracts the free electrons away from the junction. Because the holes

(a)   (b)

Figure 16–9   (a) PN junction for a diode. (b) Electronic symbol of a diode. The anode is the arrowhead part of the symbol, and the cathode is the other terminal.

and free electrons are both attracted away from the junction, no electrons can recombine with any holes, so a high-resistance junction is formed that will not allow any current flow. When the PN junction has high resistance, it will not allow any current to pass, just as if the junction were an open switch.

16.8
Using a Diode for Rectification

The electronic diode is a simple PN junction. Figure 16–9 shows the symbol for a diode. You can see that the symbol for the diode looks like an arrowhead that is up against a line. The part of the symbol that is the arrowhead is called the *anode*; it is also the P-type material of the PN junction. The other terminal of the diode is called the *cathode*, and it is the N-type material. Because the anode is made of positive P-type material, it is identified with a + sign. The cathode is identified with a − sign, because it is made of N-type material.

Figure 16–10 shows a diode connected in a circuit that has an AC power source. The AC power source produces a sine wave that has a positive half-cycle and then a negative half-cycle. The diode converts the AC sine wave to half-wave DC voltage by allowing current to pass when the AC voltage provides a forward bias to the PN junction, and it blocks current when the AC voltage provides a reverse bias to the PN junction. The forward-bias condition occurs when the AC voltage sine wave provides a positive voltage to the anode and a negative

Figure 16–10 A single diode used in a circuit to convert AC voltage to half-wave DC voltage.

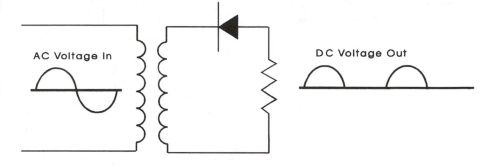

Figure 16–11 Four-diode, full-wave rectifier. This type of rectifier is often called a full-wave bridge rectifier, because the diodes are connected in a bridge circuit.

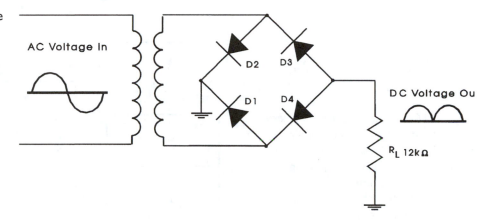

voltage to the cathode. During this part of the AC cycle, the diode is forward biased, and it has very low resistance so current can flow. When the other half of the AC sine wave occurs, the diode becomes reverse biased, with negative voltage applied to the anode and positive voltage applied to the cathode. During the time the diode is reverse biased, a high-resistance junction is created, and no current flows through it.

Rectification is the process of changing AC voltage to DC voltage. One of the main jobs of the diode is to convert AC voltage to DC voltage. Most electronic circuits used in industrial circuits need some DC voltage to operate. Because the equipment is connected to AC voltage, a power supply is required to provide regulated DC voltage for the solid-state circuits, and the diode is part of the power supply that rectifies the AC voltage to DC.

16.9
Half-Wave and Full-Wave Rectifiers

When one diode is used in a circuit to convert AC voltage to DC voltage, it is called a *half-wave rectifier*, because only the positive half of the AC voltage is allowed to pass through the diode, but the negative half is blocked. The rectifier shown in Fig. 16–10 is a half-wave

rectifier. The half-wave rectifier is not very efficient, because half of the AC sine wave is wasted.

If four diodes are used in the circuit, they can convert both the positive half-wave and the negative half-wave of the AC sine wave. Figure 16–11 shows a circuit with four diodes used to rectify AC voltage to DC voltage. Because the four diodes can convert both the positive half and the negative half of the AC sine wave, this type of rectifier is called a *full-wave rectifier*.

16.10
Three-Phase Rectifiers

Larger three-phase AC motors, pumps, and fans used in industrial systems can have their speed changed to run more efficiently by changing the frequency of the voltage supplied to them. In these applications, six diodes are used to convert three-phase AC voltage to DC voltage, and then a microprocessor-controlled circuit converts the DC voltage back to three-phase AC voltage; the frequency of this voltage can be adjusted to change the speed of the motors. Figure 16–12 shows a six-diode, three-phase rectifier. Notice that the supply voltage to the diodes is three-phase AC voltage, and the output voltage from the rectifier consists of six positive half-waves.

Figure 16–12   A six-diode bridge used to rectify three-phase AC voltage to DC voltage.

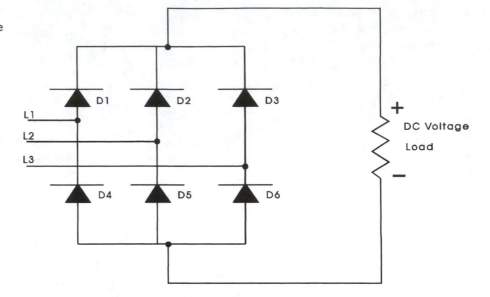

Figure 16–13   (a) Using the battery in an ohmmeter to forward bias a diode. The diode should have low resistance during this test. (b) Using the battery in the ohmmeter to reverse bias a diode. The diode should have high resistance during this test.

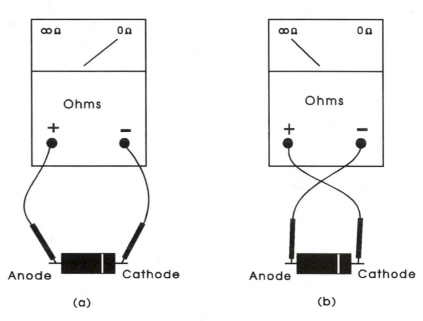

16.11
Testing Diodes

One of the tasks that you must perform as a technician is to test diodes to see if they are operating correctly. One way to do this task is to apply AC voltage to the input of the diode circuit and test for DC voltage at the output of the diode circuit. If the amount of DC voltage at the power supply is half of what it is rated for in a four-diode bridge rectifier circuit, you can suspect one of the diode pairs has one or both diodes opened. If this occurs, you can use an ohmmeter to test each diode to determine which one is faulty.

When you are testing the diodes with an ohmmeter, it is important that all power to the diode circuit be

turned off. You should remember from earlier chapters that the ohmmeter uses a battery as a DC voltage source. Because you know the diode can be tested for forward bias and reverse bias with a DC voltage source, you can use the ohmmeter as the voltage source and the meter to test for high resistance and low resistance through the diode junction. Figure 16–13a shows an example of putting the positive ohmmeter terminal on the anode of the diode and the negative ohmmeter terminal on the cathode of the diode to cause the diode to go into forward bias. During this test the diode is forward biased, and the ohmmeter should measure low resistance. When the ohmmeter leads are reversed, as in Fig. 16–13b, so that the negative meter lead is connected to the anode of the diode and the positive meter lead is connected to the

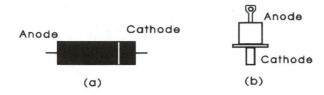

Figure 16–14 (a) Typical diode with anode and cathode identified; (b) power diode with anode and cathode identified.

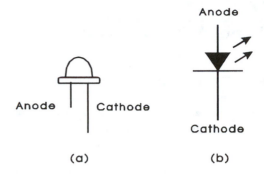

Figure 16–15 A typical light-emitting diode (LED) and the symbol for an LED. Notice the LED looks similar to a small indicator lamp.

cathode of the diode, the diode is reverse biased. When the diode is reverse biased, the ohmmeter should measure infinite (∞) resistance. If the diode indicates high and low resistance, it is good. If the diode indicates low resistance during the forward-bias test and the reverse-bias test, the diode is shorted. If the diode shows high resistance during both tests, it is opened.

16.12
Identifying Diode Terminals with an Ohmmeter

Because the ohmmeter can be used to determine if a diode is good or faulty, the same test can be used to determine which lead of a diode is the anode and which lead is the cathode. When you use the ohmmeter to test the diode for forward bias and reverse bias, you should notice that the ohmmeter will indicate high resistance when the diode is reverse biased and low resistance when the diode is forward biased. When the meter indicates low resistance, you know the diode is forward biased, so the positive lead is touching the anode and the negative lead is touching the cathode. This method works when you are testing any diode. If the diode has markings, you can identify the cathode end of the diode because it has a strip around it. Figure 16–14 shows two types of diodes; the anode is identified in each.

16.13
Light-Emitting Diodes

A light-emitting diode (LED) is a special diode that is used as an indicator because it gives off light when current flows through it. Figure 16–15 shows a typical LED and its symbol. From this figure you can see the LED looks like a small indicator lamp. You will probably encounter LEDs on various controls, such as thermostats. The major difference between an LED and an incandescent lamp is that the LED does not have a filament, so it can provide thousands of hours of operation without failure.

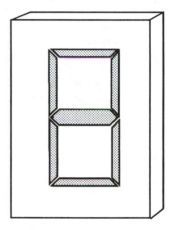

Figure 16–16 LEDs used in seven-segment displays. The seven-segment display can display the numbers 0–9.

An LED must be connected in a circuit in forward bias. Because the typical LED requires approximately 20 mA to illuminate, it is usually connected in series with a 600-Ω to 800-Ω resistor, which limits the current. Figure 16–16 shows a set of seven LEDs that are connected to provide a seven-segment display. The seven-segment display can display all numbers 0–9. Seven-segment displays are used to display numbers on thermostats and other electronic devices.

LEDs are also used in optoisolation circuits, where larger field voltages are isolated from smaller computer signals. The LED is encapsulated with a phototransistor. When the input signal is generated, it causes current to flow through the LED, and light from the LED shines on the phototransistor, which goes into conduction and passes the signal on to the computer.

16.14
PNP and NPN Transistors

Two pieces of N-type material can be joined with a single piece of P-type material to form an NPN transistor. A PNP transistor can be formed by joining two pieces of P-type material with a single piece of N-type material. Figure 16–17 shows the electronic symbol and the material for both a PNP and NPN transistor. The terminals of the transistor are identified as the emitter, collector, and base. The base is the middle terminal, and the emitter is the terminal identified by the arrowhead.

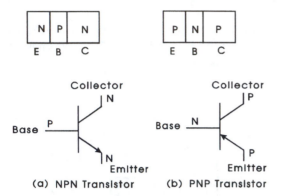

Figure 16–17 Electronic symbol and diagram of PNP and NPN material joined to form transistors.

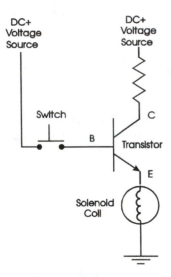

Figure 16–18 A transistor used as an electronic switch.

16.15
Operation of a Transistor

A transistor can be connected in a circuit to perform a wide variety of functions. The simplest function for a transistor to provide is the function of an electronic switch. Figure 16–18 shows a transistor as an electronic switch. In this type of application, the base terminal of the transistor provides a function like the coil of a relay, and the emitter-collector circuit provides a function like the contacts of a relay. When the proper amount and polarity of DC voltage is applied to the base of the transistor, the resistance between the collector and emitter is relatively low, which allows the maximum amount of circuit current to flow through the emitter-collector circuit. The transistor at this time acts like a relay that has its coil energized.

When the polarity of the voltage on the base of the transistor is reversed, the emitter-collector circuit is changed to a high-resistance circuit, which acts like the relay when the coil is deenergized. The major advantage of the transistor is that a very small amount of voltage or current on the base can switch the transistor from high resistance to low resistance. Because the base current is very small and the current flowing through the collector is very large, the transistor is called an *amplifier*. Transistors are used in a variety of applications, including thermostats, compressor-protection circuits, and variable-frequency motor drives.

Figure 16–19 shows a PNP transistor and an NPN transistor as two PN diode circuits. The equivalent diode circuits are shown with each transistor to give you the idea how the two junctions work together inside each transistor. Because each transistor is made from two PN junctions, each junction can be tested, just as with the single-junction diode, for forward bias (low resistance) and reverse bias (high resistance). If you must work on a number of systems that have electronic

Figure 16–19 PNP and NPN transistors shown as their equivalent PN junctions. Each PN junction can be tested for forward bias and reverse bias.

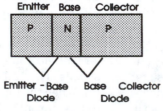

PNP Transistor

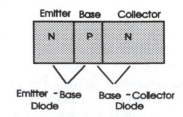

NPN Transistor

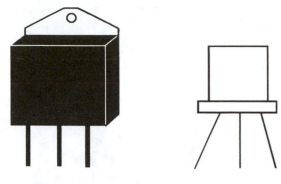

Figure 16–20 Typical transistors that are used for power control and switching.

circuits, you can purchase a commercial-type transistor tester that allows you to test the transistor while it is in the circuit or when it is out of the circuit.

16.16
Typical Transistors

You can identify transistors by their shapes. Small transistors are used for switching control circuits, and larger transistors are mounted to heat sinks so that they can easily transfer heat. Figure 16–20 shows examples of two types of transistors.

16.17
Troubleshooting Transistors

Transistors can be tested by checking each P and N junction for front-to-back resistance. Figure 16–21 shows these tests. You can see each time the battery in the ohmmeter forward biases a PN junction, the resistance is low, and when the battery reverse biases the junction, the meter indicates high resistance. You can test a transistor in this manner if it has been removed from the circuit. You can also test transistors while they are connected in circuit with a commercial-type transistor tester. The transistor tester performs a similar front-to-back resistance test across each junction.

16.18
Unijunction Transistors

There are approximately two dozen different types of electronic devices that are made by combining a number of different sections of P and N material together. Some of these devices are designed specifically for switching larger voltages and currents, and other de-

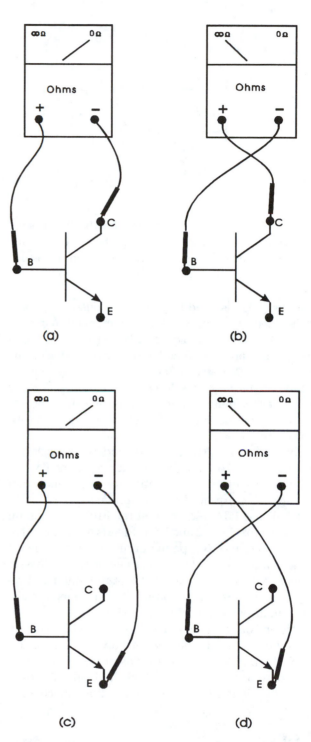

Figure 16–21 (a) Testing the base-collector junction of an NPN transistor for forward bias; (b) testing the base-collector junction of an NPN transistor for reverse bias; (c) testing the base-collector junction of a PNP transistor for forward bias; (d) testing the base-collector junction of a PNP transistor for reverse bias.

vices, such as the unijunction transistor, are designed to produce a small pulse of voltage that can be used to turn on or bias other switching devices. The unijunction

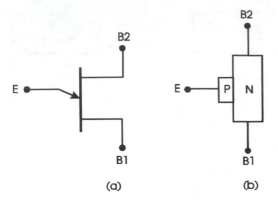

Figure 16–22 (a) Electronic symbol for the unijunction transistor. The terminals for the UJT are base 2, base 1, and emitter. (b) P- and N-material in the unijunction transistor.

transistor (UJT) is used to produce a pulse of voltage that is used to turn on silicon-controlled rectifiers (SCRs). SCRs are used to control large DC voltages and currents; they are introduced in the next section of this chapter. As a maintenance technician you may be more familiar with current-switching devices such as SCRs or triacs, but you must remember they will not operate without devices such as UJTs, which are used to turn them on and off.

The unijunction transistor is made of a single PN junction. Figure 16–22 shows a diagram of the PN material and the electronic symbol for the unijunction transistor. The terminals for the unijunction transistor are called base 1, base 2, and emitter. From the diagram you can see that base 2 and base 1 leads are mounted in the large section of N-type material. This means that if you measured the resistance between base 2 and base 1 terminals, you would measure the same amount of resistance regardless of the polarity of the meter leads. When you measured the resistance between base 2 and the emitter, you would find it is like a diode PN junction, and the polarity of the ohmmeter battery would cause the PN junction to be forward biased one way and reversed biased when you reversed the leads. The base 1–to–emitter junction also acts like a diode junction.

16.19
Operation of the UJT

The simplest way to explain the operation of a UJT is to show it in a typical circuit. Figure 16–23 shows the UJT in a circuit with a resistor and capacitor connected to the emitter terminal and a resistor connected to each of the base terminals. When voltage is applied to the resistor and capacitor that is connected to the emitter, the capacitor begins to charge. The size of the resistor con-

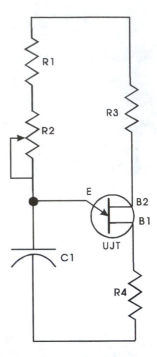

Figure 16–23 A UJT connected to an oscillator circuit to produce sharp output pulses. The size of the resistance and capacitance connected to the emitter controls the frequency of the pulses.

trols the time it takes for the capacitor to charge. This time is referred to as the *time constant* for the circuit.

The same circuit voltage is applied to the resistors that are connected to base 2 and base 1. The resistors connected to base 2 and base 1 create a voltage drop with the internal resistance of the UJT. When the charge in the capacitor grows to a value that is larger than the voltage drop across the UJT and the resistor connected to the base 1 lead, a current path is developed through the emitter and through the resistor that is connected to base 1. When current flows through this path, it creates a voltage pulse that increases to its maximum voltage very quickly. The waveform of this pulse is shown in the diagram, and you can see that it turns on and increases to its maximum value immediately. The duration of the pulse depends on the internal resistance of the UJT and the ratio of the size of the base 2 and base 1 resistors.

When the UJT is producing the pulse, the capacitor is discharging, and it begins to charge up immediately and repeats the process as its voltage charge increases to a point where it is larger than the voltage across the UJT. At this point the UJT produces another pulse. The time between pulses is determined by the size of the resistance and capacitance of the resistor and capacitor that are connected to the emitter. If the amount of resistance is increased, the time for the capacitor to charge increases and the time between pulses increases. If the amount of resistance at the emitter is reduced, the capacitor charges more quickly and creates pulses that are grouped closer together. This type of circuit is called an *oscillator*, and the number of pulses the oscillator produces in 1 s is called the *oscillator frequency*.

16.20
Testing the UJT

The unijunction transistor can be tested in the same way as a PN junction. The UJT must be isolated from its circuit, and you can test the PN junction between base 2 and emitter and then test the PN junction between base 1 and emitter. Use an ohmmeter and switch the polarity of its probes so that you forward bias and reverse bias each junction. Remember that when you forward bias a PN junction, it should have low resistance, and when you reverse bias a PN junction, it should have high resistance.

16.21
The Silicon-Controlled Rectifier

The silicon-controlled rectifier is called an SCR, and it is made by combining four PN sections of material. Figure 16–24a shows the electronic symbol for the SCR, and Fig. 16–24b shows the PN material for the SCR. The terminals on the SCR are identified as the anode, cathode, and gate. Because the SCR is basically a diode that is controlled by a gate, its symbol uses the arrow from the basic rectifier diode that you studied at the beginning of this chapter. When the SCR is turned on, it can conduct large amounts of DC voltage and current (more than 1,000 V and 1,000 A) through its anode and cathode. The major difference between the SCR and the junction diode you learned about in Section 16–5 is that the junction diode is always able to pass current in one direction when the diode is forward biased. The SCR is forward biased by applying positive voltage to its anode and negative voltage to its cathode. At this point the SCR still has high resistance between its anode-cathode junction. If a positive voltage pulse is applied to the SCR gate, the SCR's anode-cathode junction will have low resistance and the SCR will be in conduction. When the pulse is removed from the gate of the SCR, it will remain in conduction because the positive current that comes through the anode replaces the voltage the gate provided. The only way to turn the SCR off is to provide reverse-bias voltage to the anode-cathode or reduce the current flowing through the anode-cathode to zero. You should remember that the AC sine wave has zero voltage right before it provides the negative half of its waveform. This means that if the SCR is powered with AC

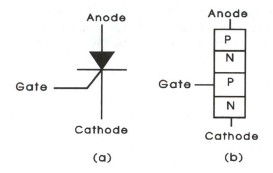

Figure 16–24 (a) Electronic symbol for the silicon-controlled rectifier. The terminals of the SCR are the anode, cathode, and gate. (b) P- and N-material in a silicon-controlled rectifier (SCR). The P and N materials are combined to make a PNPN junction.

voltage, the SCR is turned off when the AC waveform goes through 0 V and then to its negative half cycle. When the AC voltage waveform goes positive again, a gate pulse can be provided, and the SCR can go into conduction again. The gate is used to provide a pulse that is used to cause the SCR to go into conduction.

16.22
Operation of a
Silicon-Controlled Rectifier

Figure 16–25 shows an SCR connected in a circuit to control voltage to a DC load. The source voltage for this circuit is AC voltage. The main advantage of the SCR is that it will not go into conduction until it receives a pulse of voltage to its gate. The timing of the pulse can be controlled so that it can be delivered any time during the half-cycle that controls the amount of time the SCR is in conduction. The amount of time the SCR is in conduction controls the amount of current that flows through the SCR to its load. If the SCR is turned on immediately during each half-cycle, it conducts all the half-wave DC voltage just like a normal diode rectifier. If the gate delays the point at which the SCR turns on and goes into conduction at the 45° point of the half-wave, the amount of voltage and current the SCR conducts is 50% of the full applied voltage.

The other important feature of the SCR is that it will go into conduction only when its anode and cathode are forward biased. This means that if the supply voltage is AC, the SCR can go into conduction only during the positive half-cycle of the AC voltage. When the negative half-cycle occurs, the anode and cathode are reversed biased, and no current flows. This means the SCR is automatically turned off when the negative half of the AC sine wave occurs. Because the positive half of the AC sine wave occurs for 180°, the SCR can provide control of only 0 to 180° of the total 360° AC sine wave.

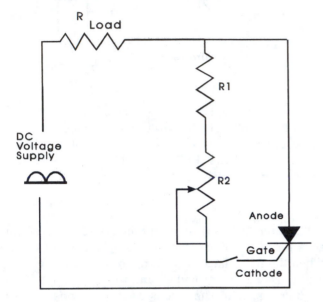

Figure 16–25 A silicon-controlled rectifier (SCR) shown in a circuit controlling DC voltage and current to its load.

It is also important to understand that because the turn-off point of the SCR is fixed to the point where the sine wave begins to go negative, the SCR can be controlled only by adjusting the point where it is turned on. The point where the SCR is turned on and goes into conduction is called the *firing angle*. If the SCR is turned on at the 10° point in the AC sine wave, its firing angle is 10°. If the SCR is turned on at the 45° point, its firing angle is 45°. The number of degrees the SCR remains in conduction is called the *conduction angle*. If the SCR is turned on at the 10° point, its conduction angle is 170°, which is the remainder of the 180° of the positive half of the AC sine wave.

16.23
Controlling an SCR

Figure 16–26 shows an SCR in a circuit with a UJT connected to its gate. The load in this circuit is a DC motor. This type of DC motor is often used as a damper-control motor. The circuit is powered by AC voltage, and the variable resistor in the oscillator (capacitor-resistor) circuit sets the timing for the pulse that is used to energize the gate of the SCR. Notice that a diode rectifier provides pulsing DC voltage for the capacitor, which

Figure 16–26 An SCR used to control a DC motor. A UJT is connected to the gate of the SCR to control its firing angle.

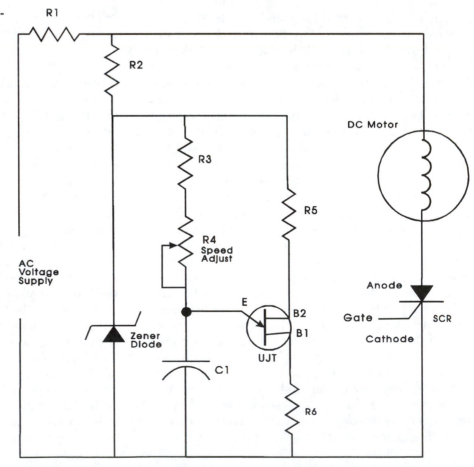

charges to set the timing for the pulse that comes from the UJT. Because this DC voltage comes from the original AC supply voltage, it has the same timing relationship of the original sine wave. This makes it easier to adjust the pulse from the UJT to turn on the SCR gate at just the right time to control the firing angle of the SCR from 0 to 180°. In reality the firing angle is usually controlled from 0–90°, which gives sufficient range of control to adjust the output DC voltage that is sent to the DC motor. You should remember that the speed of the DC motor can be controlled by adjusting the voltage sent to the armature and field. This diagram shows the waveform for the voltage at each point in this circuit. The load in this circuit could also be any other DC-powered load.

16.24
Testing an SCR

You will need to test an SCR to determine if it is faulty. Because the SCR is made of PN junctions, you can use forward bias and reverse bias tests to determine if it is faulty. In this test you should put the positive probe on the anode and the negative probe on the cathode. At this point the ohmmeter will still indicate the SCR has high resistance. If you use a jumper wire and connect positive voltage from the anode to the gate, you should notice the SCR will go into conduction and have low resistance. The SCR remains in conduction until the voltage applied to its anode and cathode is reverse biased or until the voltage applied to the anode and cathode is reduced to zero. This means that you can turn the ohmmeter polarity switch to the opposite setting or you can remove one of the probes, and the SCR will stop conducting. It is important to understand that the amount of current needed to keep the SCR in conduction is approximately 4 to 6 mA. This means that some high-impedance digital volt-ohmmeters will not have enough current when set to the ohms range to keep the SCR in conduction. If this is the case, you may need to test the SCR with an analog ohmmeter. The analog ohmmeter is a type of ohmmeter that has a needle and scale. You should also test the SCR for reverse bias to ensure that it has high resistance. Some times an SCR will not go into conduction because it has developed an open in its anode-cathode circuit. Other SCRs may stay in conduction at all times, which means the SCR is shorted.

16.25
Typical SCRs

SCRs are available in a variety of packages and case styles that include three terminals and larger types that have threads so that they can be mounted directly into

heat-sink material. The heat-sink material is made from metal and in some cases includes a fan that moves air over fins that are molded into the material to help remove the large amounts of heat that build up into the devices. Figure 16–27 shows examples of different types of SCRs.

16.26
Diacs

A unijunction transistor provides a positive pulse for the gate of the SCR, which allows the SCR to control large DC voltages and currents. A *diac* is an electronic component that provides a positive and negative pulse used as trigger signal for a device called a *triac*. A triac is similar to an SCR, except it can conduct voltage and current in both the positive and negative direction. Thus a triac provides a controlling function for AC voltage and current that is similar to the SCR. As a maintenance technician you may not notice the diac, because it is used as a device to produce a firing pulse for a triac. You learn in the next section how the triac is used in speed controls and other AC voltage-control circuits. Figure 16–28 shows the electronic symbol for a diac. Notice that the diac symbol consists of two arrows that show voltage pulses can be positive or negative. Because the diac symbol has two arrows, it can be shown in two different styles.

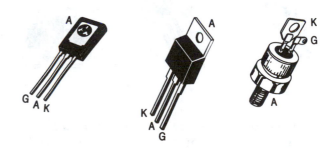

Figure 16–27 Typical SCRs shown in a variety of packages and case styles.
(Courtesy of Motorola, used by permission.)

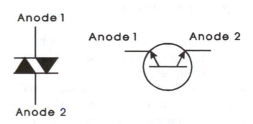

Figure 16–28 Electronic symbols for the diac. Notice the diac has two arrows, and it can be represented by either symbol.

Figure 16–29 (a) Diac in a circuit; (b) the pulse the diac produces.

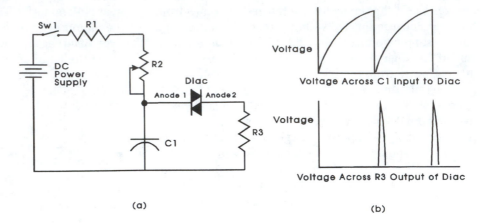

(a)

(b)

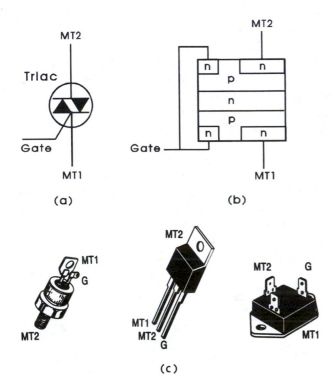

(a)

(b)

(c)

Figure 16–30 (a) Electronic symbol for the triac. (b) P-type and N-type material for the triac; (c) typical triac semiconductor devices.

(Copyright of Motorola, used by permission)

16.27
Operation of a Diac

When voltage is applied to a diac in a circuit, its PN junction remains in a high-resistance state and blocks the voltage until the voltage level reaches the breakover level. Thus when AC sine-wave voltage is applied to the circuit, the sine-wave voltage will increase to its peak level, return to zero, and then increase to negative peak voltage. Each time the voltage increases toward its peak and exceeds the breakover level of the diac, the diac provides a sharp

output pulse of the same polarity as the voltage that is supplied to it. For example if the breakover voltage level is 18 V, the diac blocks voltage until the AC voltage exceeds the 18-V level. At this point, the diac produces a pulse that is approximately 18 V. Figure 16–29 shows the diac in its circuit and the pulse that it creates.

16.28
Triacs

A *triac* is basically two SCRs that have been connected back to back in parallel so that one of the SCRs will conduct the positive part of an AC signal and the other will conduct the negative part of an AC signal. As you know, the SCR can control voltage and current in only one direction, which means it is limited to DC circuits when it is used by itself. Because the triac acts like two SCRs that are connected inverse parallel, one section of the triac can control the positive half of the AC voltage and the other section of the triac can control the negative half of AC voltage. Figure 16–30a shows the electronic symbol of the triac, and Fig. 16–30b shows the arrangement of its P-type and N-type materials. The terminals of the triac are called *main terminal* 1 (MT1), *main terminal* 2 (MT2), and *gate*. Because the triac is basically two SCRs that are connected in an inverse parallel configuration, MT1 and MT2 do not have any particular polarity.

16.29
Using a Triac as a Switch

A triac can be used in an industrial circuit as a simple on-off-type switch. In this type of application, the MT1 and MT2 terminals are connected in series with the AC load. When the gate gets a positive pulse signal, the triac turns on for the positive half of the AC cycle. When the AC voltage waveform returns from its positive peak to 0 V, the

Figure 16–31 A triac uses a switch to turn on voltage to a gas valve for a furnace. The triac receives its gate signal from a small amount of voltage that moves through the temperature-sensing element.

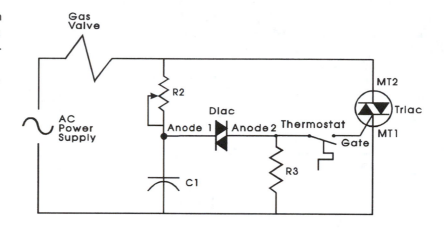

Figure 16–32 A triac used to control variable voltage to an AC electric heating element. Notice a diac is used to provide a positive and negative pulse to the triac gate.

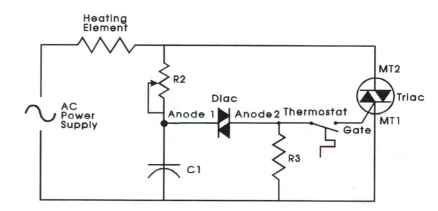

triac turns off. Next the negative half-cycle of the AC voltage waveform reaches the triac, and it receives a negative pulse on its gate and goes into conduction again.

This means that the triac will look as though it turns on and stays on when AC voltage is applied. The load that is connected to the triac will receive the full AC sine wave, just as if it were connected to a simple single-pole switch. The major difference is the triac switch can be used for millions of on-off switching cycles. The other advantage of the triac switch is that the gate pulse can be a very small amount of voltage and current. This allows the triac to be used in temperature control where the temperature-sensing part of a thermostat can be a small solid-state sensing element called a *thermistor*. The sensing element can also be a very narrow strip of mercury in a glass bulb that is very accurate but can carry only a small amount of voltage or current.

Another useful switching application for a triac is the solid-state relay. Section 16.32 explains AC and DC solid-state relays. Figure 16–31 shows a triac in a thermostat connected to a 24-V AC gas valve in a furnace. The temperature-sensing element is connected to the gate terminal of the triac. When the temperature increases, it makes the mercury in the temperature-sensing element expand and make contact between the two metal terminals. The small amount of voltage flowing through the sensing element is sufficient to cause the triac to go into conduction and provide voltage to the gas valve.

16.30
Using a Triac for Variable Voltage Control

A triac can also be used in variable-voltage control circuits, because it can be turned on any time during the positive or negative half-cycle, similar to the way the SCR is controlled for DC-circuit applications. In this type of application, a diac is used to provide a positive and negative pulse that can be delayed from 0° to 180° to control the amount of current flowing through the triac. This type of circuit can be used to control the amount of current and voltage supplied to electric heating elements. This allows the amount of current and voltage to be controlled from zero to maximum, which in turn allows the temperature to be controlled very accurately. Figure 16–32 shows a diagram of a triac used to control an electric heating element that is powered by an AC voltage source. Notice that a variable resistor is connected with a capacitor to provide an oscillator pulse to the diac. Because the resistor and capacitor are connected to an AC voltage source, the pulse will be both positive and negative as the AC sine wave changes polarity. The triac is connected to the same AC voltage source, so the timing of the pulse from the diac to its gate will always be synchronized with the polarity of the voltage arriving at the main terminals of the triac.

16.31
Testing a Triac

Because the triac is made from P-type material and N-type material, it can be tested like other junction devices. The only point to remember is that because the triac is essentially two SCRs mounted inversely parallel to each other, some of the ohmmeter tests are not affected by the polarity of the ohmmeter leads. In the first test for a triac an ohmmeter should be used to test the continuity between MT1 and MT2. When no gate pulse is present, the resistance between these terminals should be infinite regardless of which ohmmeter probe is placed on each terminal. Figure 16–33 shows how the ohmmeter should be connected to the triac. In Fig. 16–33a you can see the positive ohmmeter probe is connected to MT1 and the negative probe is connected to MT2. Because no voltage is applied to the gate, the resistance should be infinite. When voltage from MT2 is jumped to the gate, the triac will go into conduction and the ohmmeter will indicate low resistance.

In Fig. 16–33b you can see that the positive ohmmeter probe is connected to MT2, and the negative probe is connected to MT1. Because no voltage is applied to the gate, the triac is not in conduction, and the ohmmeter will indicate the resistance at this time is high. When voltage from MT2 is jumped to the gate, the triac will go into conduction, and the ohmmeter will indicate the resistance is low. It is important to remember that the triac will stay in conduction only while the gate signal is applied from the same voltage source as MT2. As soon as the voltage source is removed, the triac will turn off.

16.32
Solid-State Relays

Another solid-state device that you will see frequently used in industrial control systems is called a *solid-state relay (SSR)*. The solid state relay uses DC voltages for the control part of the circuit, which acts like the coil of a regular relay, and it can switch AC voltages through the part of the circuit that acts like the contacts. The solid-state relay uses SCRs, transistors, or triacs to control various loads. It operates very much like an electromechanical relay in that it is energized by a small voltage that controls a larger voltage and current.

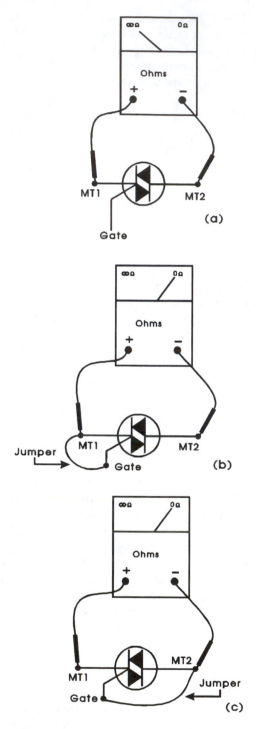

Figure 16–33   (a) Testing a triac by placing the ohmmeter's positive probe on MT2 and its negative probe on MT1. When gate voltage is applied from the MT2 probe, the triac goes into conduction. (b) Testing a triac by placing the ohmmeter's positive probe on MT1 and its negative probe on MT2. When gate voltage is applied from the MT2 probe, the triac goes into conduction.

Figure 16–34 shows a picture of a typical solid-state relay and provides a block diagram of the operation of the relay. Figure 16–35 shows the electrical diagram for

Figure 16–34 A solid-state relay rated for DC input and 120 VAC output.

a solid-state relay that uses a light-emitting diode (LED) and a photo transistor to accept the DC input and a triac to control the AC voltage and load in the output. The LED and phototransistor are called an *optoisolation device,* or an *optocoupler.* When a DC input signal is received at the SSR, it is converted to light through an LED. The LED and phototransistor are manufactured in an integrated circuit (IC) so that the light from the LED shines directly on a phototransistor. When the light strikes the phototransistor, its collector-emitter circuit goes to low resistance and the collector current flows through it to the base of the triac. The AC load is controlled directly by the triac. Notice that the power source for the load must be isolated from the input signal.

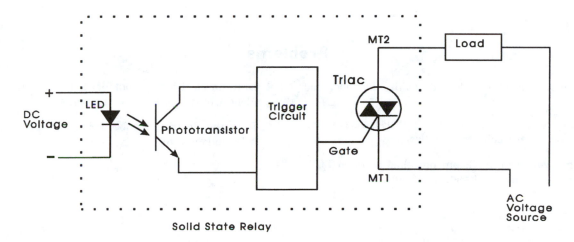

Figure 16–35 An electronic diagram of a solid-state relay that has a DC input signal and 120 VAC output.

Questions for This Chapter

1. Explain P-type and N-type material.

2. Explain the operation of a diode (PN) junction and show the input AC waveform and the output DC waveform.

3. Identify the terminals of a transistor and explain its operation.

4. Explain the operation of the SCR and explain the type of circuit where you would find one.

5. Explain the function of the unijunction transistor when it is used with an SCR.

True or False

1. The function of the triac is to provide switching similar an SCR, except the triac operates in an AC circuit.

2. The light-emitting diode (LED) is similar to a DC lightbulb in that it has a very tiny filament.

3. The main function of diode and SCRs is to convert AC voltage to DC voltage.

4. The full-wave bridge rectifier uses one diode.

5. The UJT and diac are important solid-state devices that provide pulse signals to other components to use as a firing signal.

Multiple Choice

1. A PN junction is forward biased when _____ .
    a. positive voltage is applied to the N-material and negative voltage is applied to the P-material.
    b. negative voltage is applied to the N-material and negative voltage is applied to the P-material.
    c. its junction has high resistance.

2. The rectifier _____ .
  a. converts DC voltage to AC voltage.
  b. converts AC voltage to DC voltage.
  c. can be used as a seven-segment display for numbers.

3. A circuit that has one SCR in it can _____ .
  a. control AC voltage.
  b. control DC voltage.
  c. control both AC and DC voltage.

4. The transistor operates similar to a relay in that _____ .
  a. its emitter is like the coil and its base and collector are like a set of contacts.

  b. its collector is like the coil and its base and emitter are like a set of contacts.
  c. its base is like a coil and its emitter and collector are like a set of contacts.

5. The solid-state relay _____ .
  a. uses an LED to receive a signal and a transistor to switch current.
  b. uses a capacitor to receive a signal and an IC to switch current.
  c. uses a transistor to receive a signal and an IC to switch current.

Problems

1. Draw the symbol for a diode and identify the anode and cathode.

2. Draw the symbol for an NPN and a PNP transistor and identify the emitter, collector, and base.

3. Draw the symbol for a silicon-controlled rectifier (SCR) and identify the anode, cathode, and gate.

4. Draw the electrical symbol for a unijunction transistor and identify base 1, base 2, and the emitter.

5. Draw the electrical symbol for a triac and identify MT1, MT2, and the gate.

◀ Chapter 17 ▶

Basic Hydraulics and Pneumatics

Objectives

After reading this chapter you will be able to:

1. Identify the basic parts of a hydraulic system.
2. Explain the operation of a directional control device.
3. Explain the operation of a flow-control device.
4. Explain the operation of a pressure-control device.

17.0
Overview of Hydraulics and Pneumatics

As a maintenance technician, you will need to be able to identify the basic parts of any hydraulic or pneumatic system and understand their operation. This information will also help you to follow a basic hydraulic or pneumatic diagram and analyze the operation of a system during troubleshooting. We begin this chapter by discussing a hydraulic system; we discuss pneumatic systems later in the chapter.

17.1
Basic Parts of a Typical Hydraulic System

One of the simplest hydraulic systems is the control of a hydraulic cylinder that extends and retracts. This type of cylinder may be used to open and close a mold on a plastic injection molding machine or open and

close a fixture for holding a part in machining application. An example of this cylinder is shown in Fig. 17–1; you can see that it is a linear actuator. When the cylinder is extended, the rod moves out of the cylinder, and when the cylinder is retracted, the rod moves back into the cylinder. If the cylinder is used in the plastic injection molding machine, the mold will be moved to the open position when the cylinder is retracted and to the closed position when the cylinder is extended. The basic parts of this system include the *reservoir*, which holds the hydraulic fluid, the *pump*, which moves the fluid, the *pressure-control device*, which sets the pressure for the system, the *directional control device*, which switches the fluid to the proper cavity on the cylinder to cause it to open and close, the *cylinder*, which has a rod that extends and retracts, and all the *tubing*, which routes the fluid throughout the system. The system may also have a flow-control device, which controls the speed at which the cylinder extends or retracts by setting the amount of fluid that is flowing. The more fluid that is flowing, the faster the cylinder rod moves.

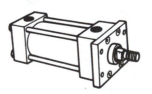

Figure 17–1 A typical hydraulic cylinder.
(Courtesy of Vickers, Inc.).

17.2
Common Terms Used in Hydraulic Systems: Work, Power, Horsepower, Pressure, Force, and Flow

If you want to fully understand the operation of a hydraulic system, you must also understand all the terms associated with it. The main purpose of any hydraulic system is to do work. *Work* is defined as exerting a force—such as a push or a pull—through some distance. This can be expressed as a formula: work = force × distance, or $W = f \times d$; the answer is expressed in foot-pounds. For example if a hydraulic cylinder pushes 100 lb of weight a distance of 2 ft, the hydraulic system needs to exert 200 ft-lb of work.

Force is defined as the pressure exerted per area. For example, pressure is defined as force per square inch.

Power is defined as the rate of doing work. This means that you can determine the power a hydraulic system produces if you can measure the amount of work it produces over a measured amount of time. The formula for power is shown here; the units for power are foot-pounds per time (ft-lb/time):

$$\text{Power} = \frac{\text{work}}{\text{time}}$$

Remember, work is force × distance.

The common unit for power in hydraulic systems is the *horsepower* (hp). One horsepower is equal to 33,000 ft-lb/min, or 550 ft-lb/s. For example, if the hydraulic system produces 99,000 ft-lb in 1 min, you can calculate the horsepower in the following manner:

$$\frac{99,000 \text{ ft} = \text{lb/min}}{33,000} = 3\text{hp}$$

To express 99,000 ft-lb/min in terms of seconds, divide 99,000 by 60 s/min, which gives 1,650 ft-lb/s. When this value is divided by 55 ft-lb/s, the answer still remains 3 hp.

In most cases when you are working around hydraulic systems, the work the system does is defined by the number of gallons of fluid that are pumped in 1 min and the amount of pounds of force exerted per square inch. The formula for this type of calculation is

$$\text{hp} = \frac{\text{gpm} \times \text{psi}}{1,714}$$

This formula can also be expressed as

$$\text{hp} = \text{gpm} \times \text{psi} \times 0.0007$$

It is also important to understand that horsepower can also be expressed in electrical units: 1 hp = 746 W, and it can be expressed in units of heat as 1 hp = 42.4 BTU.

When you first learn about hydraulic systems and encounter formulas, you may become discouraged, but you should be reminded that when you are working in a factory with hydraulic systems you will generally use printed charts or look-up tables printed on sliding charts to determine force, flow, and horsepower. You may also be able to obtain software for a computer to perform essential calculations and provide information about the system on which you are working. Another good resource is the technical staff at most hydraulic sales offices, who have a number of ways to help you to determine necessary elements of the system. The main key is to remember where the formulas are printed and not to worry about memorizing them.

17.2.1
Fluid Dynamics

Fluid dynamics is the science that controls the flow and pressure of hydraulic fluid. *Pressure* is defined as the force per unit area. An example of pressure is 100 lb/in.^2, which is also written as 100 psi. This means that 100 lb of force is exerted on every square inch, which is like placing a 100-lb weight on every square inch.

Flow is the movement of fluid through piping, controls, and actuators. Flow can be described in two ways. The first way is by its *velocity*, which is how fast the fluid particles move past a given point. The second way is to measure the amount of flow, which is called the *volume*. *Volume* can be described as the number of gallons of fluid or the number of cubic inches. The conversion of gallons per minute (gpm) to cubic inches per minute (in.^3/min) is shown next. Note that in the term cubic inches per minute, the word *per* is replaced with a slash (/). We see that gpm = 231 in.^3/min. This means that if you know the flow is 693 cu in./min and you are trying to find the number of gallons per minute, you would use the formula

$$\text{gpm} = \frac{\text{cubic inches/minute}}{231}$$

$$\text{gpm} = \frac{693 \text{ in.}^3/\text{min}}{231}$$

$$= 3 \text{ gpm}$$

If you know the number of gpm, you can calculate the cubic inches per minute by multiplying the number of gpm by 231. For example if you have 4 gpm, the formula would look like this:

$$\text{cubic inches/minute} = 4 \text{ gpm} \times 231 = 924 \text{ in.}^3/\text{min}$$

You may need to calculate cubic inches per minute or gpm when you are trying to figure what size of pump you need or the size of tubing you need. The term *viscosity* is the measure of a fluid's resistance to flow. If a fluid is thin and flows easily, it has a low viscosity. If the fluid is thick and flows poorly, it has a high viscosity.

17.3
The Reservoir

The reservoir, which is also called the *tank*, serves several functions in a hydraulic system. Figure 17–2 shows an example of a typical reservoir. The first and most obvious function is to provide a place to store sufficient fluid for all applications of the system. For example, if the system has multiple cylinders, they may not all be moving at the same time, so only a small portion of the fluid is used. If all the cylinders are moving at the same time, the amount of fluid being used is greater. The reservoir holds the extra fluid the system will need during peak demand times.

The second function of the reservoir is to provide a means of helping to dissipate heat that the fluid picks up as it moves through the system to the surrounding air. If the fluid picks up too much heat and the reservoir cannot pass it all to the air, a heat exchanger may be added to the system that uses water to remove additional heat. The reservoir has a large surface area to help allow heat to transfer from the fluid to the outside air. The tank may have baffles to help move the heat better.

Another function of the reservoir is to provide a platform for mounting the pump and motor. Because the reservoir is made of heavy steel, it provides an excellent place to mount the pump and motor, which also saves space by keeping these components off the floor. The reservoir has one or more drains that are used to empty the tank if the fluid becomes contaminated or needs to be changed. Access plates are mounted on the top or sides of the tank. These plates are bolted onto the tank so that they can be removed if maintenance personnel need to look into the tank or get inside it for any reason.

The tank will also have an *oil-level gage* called a *sight gage*. The sight gage is mounted on the side of the tank, where it is very visible and allows the level of the

Figure 17–2 Reservoir for hydraulic system.
(Courtesy of Vickers, Inc.).

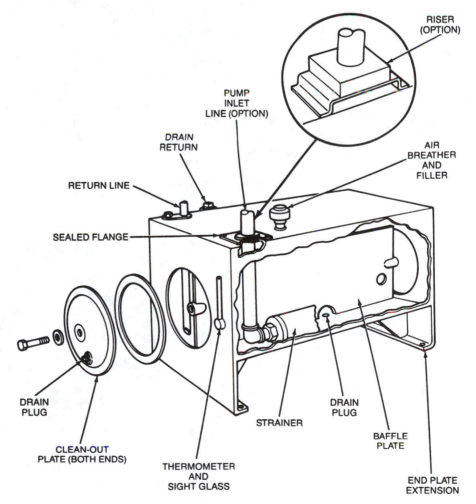

fluid to be checked without opening the tank. A filler access hole is provided to allow a place to add oil. The filler also has a breather that allows clean air to enter the tank. Typically the tank is not pressurized, so the breather allows outside air to enter the tank at all times. A drain plug is provided in the bottom of the tank to allow it to be drained for maintenance or to change the oil.

17.4
Pumps

The *hydraulic pump* converts electrical energy from a motor to fluid energy when its shaft is rotated. The output of the pump is defined by its rate of flow, which is called the *pump's displacement*. Pumps are rated by the amount of output flow and maximum operating pressure when its shaft is turning at its rated speed.

One of the classifications for pumps is *positive displacement pumps*. A positive displacement pump delivers a specific amount of fluid for each revolution of its shaft or for each cycle of the pump. The positive displacement pump can have a fixed or variable displacement. A second classification for hydraulic pumps is a *nonpositive displacement pump*. The output of a nonpositive displacement pump varies with the size of the input and output ports, the type of impeller, and the back pressure of the system. Figure 17–3 shows several examples of positive displacement–type pumps, such as a piston pump, a gear pump, a vane pump, and a lobe

Figure 17–3 Examples of positive displacement–type hydraulic pumps. (a) A double-gear-type pump; (b) a single-gear-type pump; (c) a vane-type pump; (d) a lobe-type pump; (e) a piston-type pump.
(Courtesy of Vickers, Inc.)
(continued)

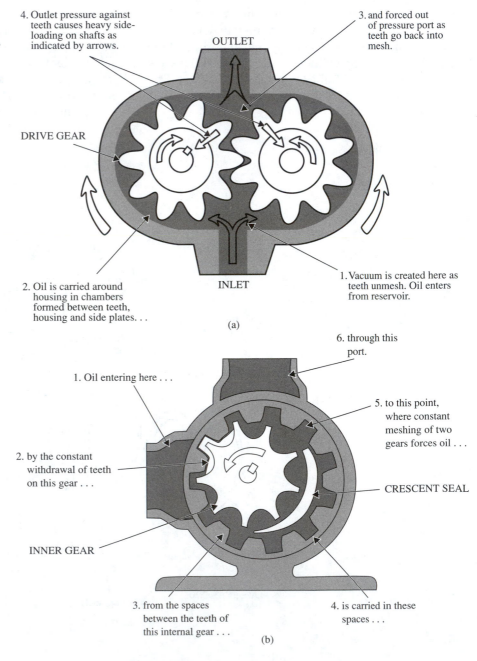

4. Outlet pressure against teeth causes heavy side-loading on shafts as indicated by arrows.

OUTLET

3. and forced out of pressure port as teeth go back into mesh.

DRIVE GEAR

1. Vacuum is created here as teeth unmesh. Oil enters from reservoir.

2. Oil is carried around housing in chambers formed between teeth, housing and side plates. . .

INLET

(a)

6. through this port.

1. Oil entering here . . .

5. to this point, where constant meshing of two gears forces oil . . .

2. by the constant withdrawal of teeth on this gear . . .

CRESCENT SEAL

INNER GEAR

3. from the spaces between the teeth of this internal gear . . .

4. is carried in these spaces . . .

(b)

Figure 17–3   *(continued)*

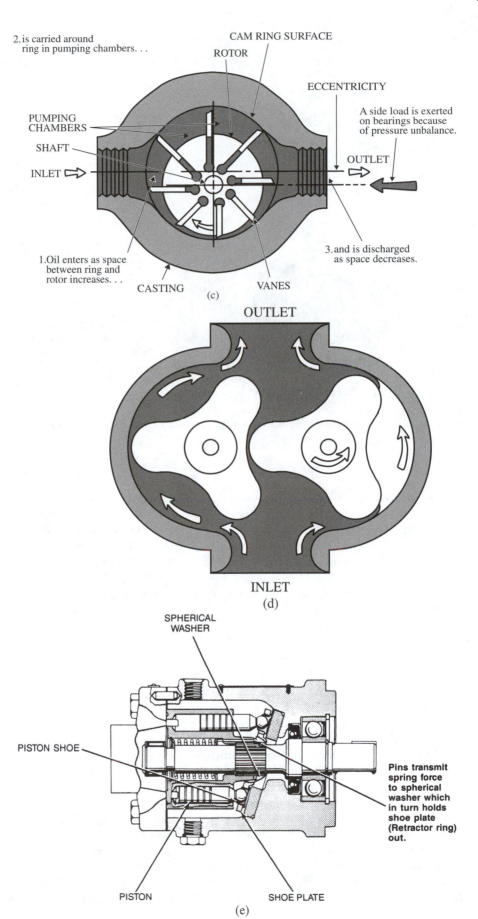

2. is carried around ring in pumping chambers. . .

CAM RING SURFACE

ROTOR

ECCENTRICITY

A side load is exerted on bearings because of pressure unbalance.

PUMPING CHAMBERS

SHAFT

INLET

OUTLET

1. Oil enters as space between ring and rotor increases. . .

CASTING

VANES

3. and is discharged as space decreases.

(c)

OUTLET

INLET

(d)

SPHERICAL WASHER

PISTON SHOE

Pins transmit spring force to spherical washer which in turn holds shoe plate (Retractor ring) out.

PISTON

SHOE PLATE

(e)

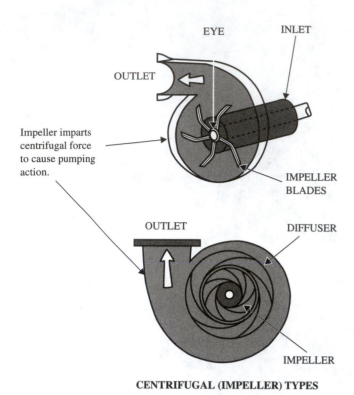

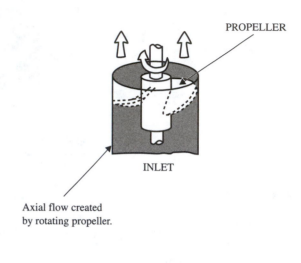

Figure 17–4   Example of nonpositive displacement–type pumps: a centrifugal- (impeller) type pump and an axial- (propeller) type pump.
(Courtesy of Vickers, Inc.)

pump. Figure 17–4 shows examples of nonpositive displacement–type pumps, such as impeller pumps with centrifugal impeller blades and axial (propeller) blades. You can see from the diagrams that the positive displacement pumps each have multiple cavities that fill with oil as the pump shaft turns. These cavities change shape and size as the gears, lobes, or vanes move oil through the pump. The main point to remember about positive displacement–type pumps is that they move the same amount of oil with each rotation of the shaft. This produces a constant flow for a specific pump speed.

A second point to remember is that hydraulic fluid cannot be compressed, so as long as the pump causes flow, the fluid will create pressure when it reaches restrictions throughout the system. When the amount of restriction is controlled, the pressure in the system can be controlled. Nonpositive displacement–type pumps do not pump the same amount of fluid with each rotation of the shaft. This means that some amount of fluid may bypass the impeller or propeller on the shaft when the flow of fluid is restricted. This characteristic is very useful in some systems, such as transfer pumps and sump pumps, because the pump will not become damaged if unwanted restrictions—such as dirty filters or foreign material—create blockages in a system. This characteristic is typically not useful in most hydraulic

systems, because the flow may become unpredictable. Most hydraulic systems in industrial applications use positive displacement-type pumps, because they will continue to try to pump the same amount of fluid under all kinds of changing conditions within the system. The main way the volume of fluid pumped through the system is changed with this type of pump is that one or more bypass routes back to the reservoir are provided. The flow through the bypass routes can be adjusted, which basically changes the volume of fluid moving through the system.

17.5
Variable-Volume Pumps

Because positive displacement pumps move the same volume of fluid with each rotation of their shafts, a system needs to vary the speed of the motor to change the displacement of the pump. The speed of older electrical motors could not be changed easily while maintaining their horsepower rating at slow speeds, so variable-volume pumps were perfected. Figure 17–5 shows a variable-volume pump called a radial piston pump or a bent-axis piston pump. The way these pumps change the volume of fluid that is moved with each rotation of their

---

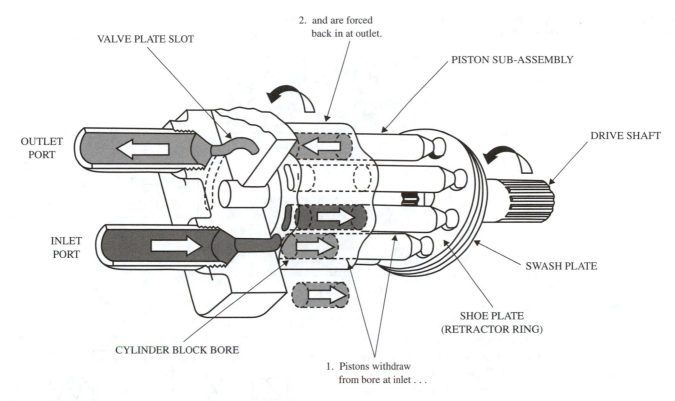

Figure 17–5 Example of a variable piston–type pump. (Courtesy of Vickers, Inc.)

shafts is to change the length of the piston stroke. From the diagram you can see the radial piston pump uses a *swash plate* that can be tilted to change the length of the piston's stroke. Figure 17–6 shows the different positions of the swash plate as it changes the amount of fluid the pistons will move. If the swash plate is tilted to its maximum position, it allows the piston to move full stroke and pump the maximum amount of fluid with each stroke. When the swash plate is tilted to its minimum position, the stroke of the piston is reduced to the point where the pump has no fluid flow. The position of the swash plate can be moved manually with a positioning arm, or it can be positioned automatically by hydraulic fluid.

The application for the hydraulic system is the determining factor in whether a fixed positive displacement pump or a variable positive displacement pump is used. For example many presses such as plastic presses and metal stamping presses may use a combination of fixed and variable pumps. In these types of applications, the system generally has one or more large fixed pumps up to 50 hp and one or more smaller variable-volume pumps. The larger pumps move the large volumes of fluid needed to move the large dies or

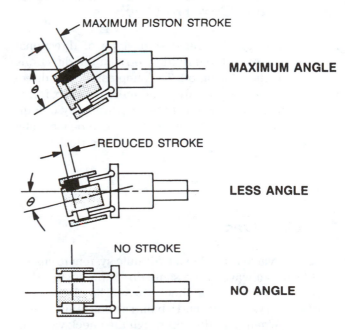

Figure 17–6 When the swash plate is at the maximum angle, the pistons will pump maximum volume. When the swash plate is at no angle, the pistons will pump minimum volume.

(Courtesy of Vickers, Inc.)

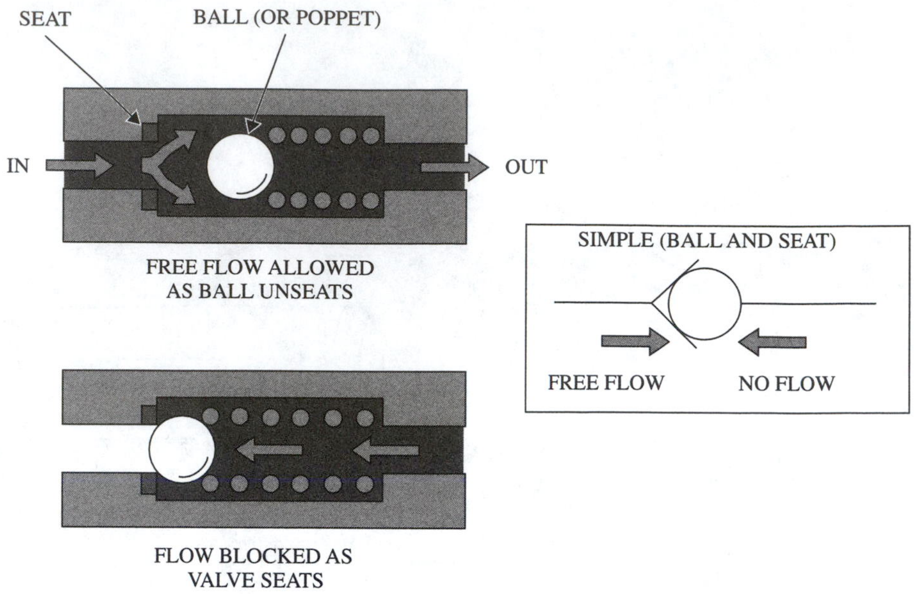

Figure 17–7 (a) A simple check valve includes a ball (poppet) that is forced against its seat to stop fluid flow when the direction of flow is wrong. (b) The hydraulic diagram symbol for a check valve.
(Courtesy of Vickers, Inc.)

SEAT BALL (OR POPPET)

IN OUT

FREE FLOW ALLOWED
AS BALL UNSEATS

SIMPLE (BALL AND SEAT)

FREE FLOW NO FLOW

FLOW BLOCKED AS
VALVE SEATS

molds in the system, and the variable pumps are used to provide the last amount of fluid to precisely control speed and pressure. The variable pumps are generally controlled by feedback sensors, such as pressure and flow sensors, to keep the system operation as accurate as possible.

17.6
Directional Control Devices

All hydraulic systems need some type of directional control devices to force fluid to flow in the directions required by the application. The simplest form of direction control is a check valve, and more complex forms of direction control include two-way, three-way, or four-way valves. Directional control valves can be manually actuated, hydraulically actuated, or electrically actuated by solenoids.

17.7
Check Valves

The check valve is used in a hydraulic system to ensure fluid flow is always in the same direction. Figure 17–7 shows an example of a check valve. The check valve consists of a valve body and a ball. The ball is also called a poppet. When fluid flows through the check valves in the correct direction, the ball moves away from its seat to allow maximum flow. When the flow is in the wrong direction, the ball is forced against its seat, and all flow is stopped. Some check valves also have a spring to hold the poppet (ball) in the proper location. Check valves

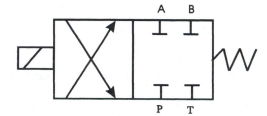

A B

P T

Figure 17–8 A two-position valve.

are used in hydraulic systems to ensure fluid flow is always in the correct direction.

17.8
Directional Control Valves

Directional control valves are used to control the direction in which fluid flows through a hydraulic system. In our first example, where the hydraulic cylinder was extended and retracted, a directional control valve is used to change the direction of flow from the rear of the cylinder, to make it extend, to the front of the cylinder, to make it retract. Control valves are categorized by the number of positions, two or three, and the number of ways (two-way, three-way, or four-way). Figure 17–8 shows a diagram of a two-position control valve.

The symbol for the two-position valve actually shows several important parts of the valve. For example this valve is actuated by an electric solenoid. The solenoid symbol is shown on the left side of the valve and also in Fig. 17–9. If the valve is manually operated, it has the symbol of the manual operator, which is shown in Fig. 17–10.

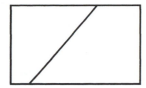

Figure 17–9   Symbol of solenoid actuator for valve.

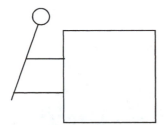

Figure 17–10   Symbol for manual actuator.

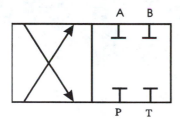

Figure 17–11   Diagram of two-position hydraulic valve.

Figure 17–12   Arrows show direction of fluid flow.

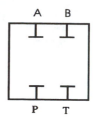

Figure 17–13   Ports are identified by letters.

You should notice that the diagram of a two-position valve also shows a second view of the valve. Figure 17–13 shows this second view of the valve diagram. In this part of the diagram all four ports are blocked when the valve is shifted to the right. The spring symbol on the right side of the valve indicates the valve will return to the spring-loaded position when no power is applied to the solenoid. When the valve is in the spring-loaded position, all ports are blocked.

Figure 17–14 shows a cutaway picture of the two-position valve, and you can see that it has a spool that moves back and forth between the two positions. The spool is designed so that parts of it are cut away to allow fluid to flow around them, and other parts are full size to block fluid. When the spool shifts so that a cutaway part is aligned with two ports, fluid flows through those ports. When the full-size part of the spool is aligned with two ports, fluid flow is blocked between those ports. Figure 17–15 shows a diagram of how the spool for a two-position, four-way valve allows fluid to flow between port B and the pressure port and port A and the tank port when it shifted to the right. Part b of the diagram shows the spool shifted to the left so that the pressure port and port A have flow and the tank port and port B have flow.

17.9
A Three-Position Valve

Figure 17–16 shows a three-position, four-way valve that is a manually operated valve. The three-position valve has three distinct positions to which its spool can be moved that cause the direction of fluid flow to change. The diagram for the three-position valve shows the connections of the ports in each of the three positions. The left part of the diagram shows what happens in the valve when the actuator on the left is activated. You can see from the arrows in this part of the diagram that the pressure port, P, is connected to port B, whereas port A is connected to the tank port. When the actuator on the right side of the valve is activated, the flow through the valve is reversed so that the pressure port is connected to the port A and port B is connected

Figure 17–11 shows the diagram of a two-position valve. Because this valve is a two-position valve, its diagram indicates the body of the valve as two parts. The part on the left shows two arrows and the part on the right shows four ports identified with the letters A, B, P, and T. P stands for pressure port, T stands for the port that is connected to the tank, and A and B identify the two ports connected to the hydraulic cylinder. Figure 17–12 shows the block with the arrows, which show the direction of fluid flow. You can see from the directions of the arrows in the valve body that when the valve is shifted to the left, port P provides pressurized fluid to port B and port A is connected to the tank port.

Figure 17–14 A cutaway picture of a two-position, four-way valve that shows the shape of the spool and how it aligns with the ports.
(Courtesy of Vickers, Inc.)

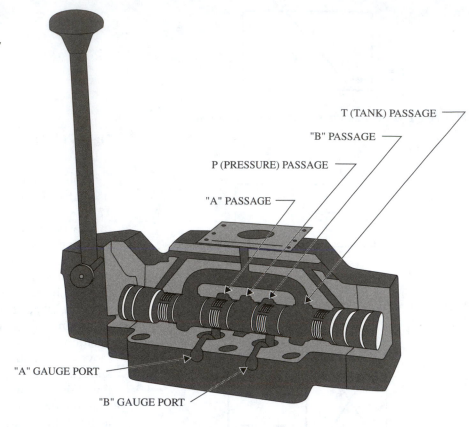

Figure 17–15 (a) Diagram of two-position, four-way valve that shows direction of fluid flow between the pressure port and port B and between the tank port and port A when the spool is shifted to the right. (b) When the spool is shifted to the left, port A is connected to pressure, and port B is connected to tank.

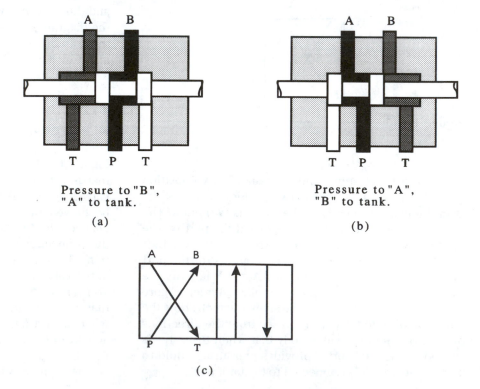

to the tank port. The middle diagram indicates that all ports are blocked when neither actuator is activated. The symbol for this valve shows that both ends of the valve have a spring return so that the valve will stay in the middle position, where all ports are blocked, when neither actuator is activated. You will find that hydraulic valves are available with a wide variety of options in regard to how the ports are connected.

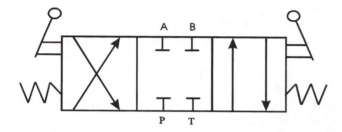

Figure 17–16   Three-position manually operated valve.

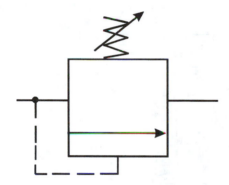

Figure 17–17   Symbol for pressure-relief valve. The arrow shows direction of fluid flow.

Figure 17–18   Cutaway diagram of pressure-relief valve. You can see the poppet and the handle that sets the valve's pressure-relief point.
(Courtesy of Vickers, Inc.)

17.10
Pressure-Control Devices (Relief Valves)

Hydraulic systems need to have a means of controlling the pressure in the system. As you know, the hydraulic pump provides fluid flow to the system. When the flow of the fluid is restricted, pressure will build in the system. When fluid is allowed to flow back to the tank, the pressure will drop because the tank is not pressurized. The function of pressure-control devices is to control system pressure by allowing more or less fluid to bypass the system and return to the tank. These valves are also called *relief valves* because they relieve the amount of pressure in the system.

Figure 17–17 shows the symbol for the pressure-relief valve. The symbol shows that the valve's relief pressure is controlled by spring tension and the direction of fluid flow through the valve when the pressure setting is exceeded. The arrow that is shown through the spring indicates this valve is a variable-type relief valve. Figure 17–18 shows a cutaway diagram of the relief valve, and you can see that the valve has a handle that allows the spring tension on the valve to be set manually. When the handle is turned clockwise, spring tension is increased, and the amount of hydraulic pressure must increase before the valve relieves pressure by allowing flow to bypass to the tank. When the valve handle is turned counterclockwise, the pressure setting is decreased.

Figure 17–19 shows how the fluid flow is blocked through the relief valve when the pressure setting is not exceeded. Figure 17–20 shows how fluid flows past the valve piston when the pressure setting is exceeded. You should remember that the outlet side of the pressure-relief valve is connected directly to the tank, so fluid can bypass the system to relieve pressure when

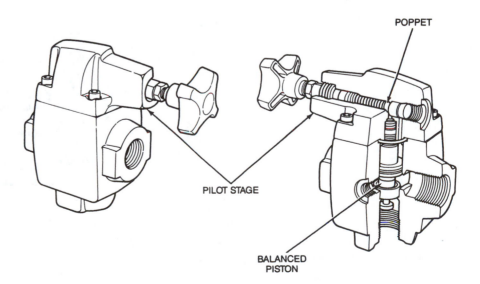

POPPET

PILOT STAGE

BALANCED PISTON

the pressure in the system exceeds the pressure setting of the valve.

17.11
Flow-Control Device

The fluid flow in a hydraulic system can be metered to control the speed at which a hydraulic cylinder extends or retracts. Flow control allows minor adjustments to be made in various parts of a system to balance the speed of its moving parts. Figure 17–21a shows the hydraulic symbol for a flow-control valve. From this figure you can see that the symbol looks like the hydraulic line is being squeezed down to restrict flow. The arrow through the valve indicates the valve is variable. The flow-control device is connected in series with the device they are controlling. This means that whatever the flow is in the flow-control device, the same flow will be moving through the cylinder or hydraulic motor with

which the device is in series because the fluid has only one path.

Figure 17–21b shows the symbol for a flow-control device with a check valve added in parallel. The check valve blocks fluid flow in one direction and forces all fluid flow to move through the flow-control valve. This causes the fluid flow to be controlled by the valve setting in this direction. When the fluid flow is reversed, the check valve will unseat and allow fluid flow at full flow to bypass the flow-control valve. This type of valve allows the flow to be controlled when a cylinder is extending so that its speed can be controlled. When the flow is in the opposite direction, it is allowed to bypass the flow-control valve by flowing through the check valve.

A system that is controlling the extending and retracting speed of a cylinder rod may have two flow-control valves. One will be in series with the circuit that causes the cylinder to extend, and the other will be in series with the circuit that causes the cylinder rod to retract. Each of these flow-control valves will have a check

Figure 17–19 When fluid pressure is below the setting on the valve, spring tension keeps the piston on its seat and no fluid flows.
(Courtesy of Vickers, Inc.)

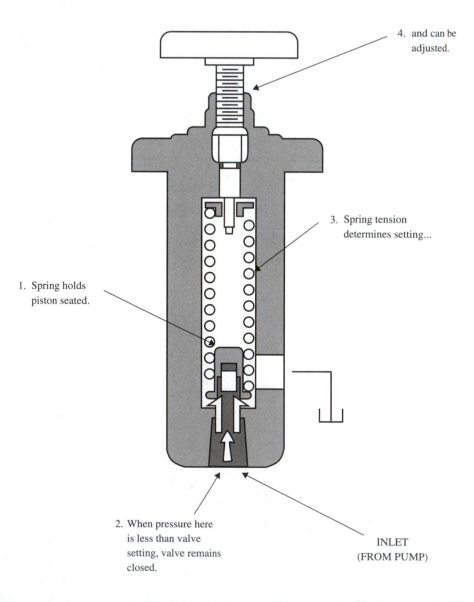

4. and can be adjusted.

3. Spring tension determines setting...

1. Spring holds piston seated.

2. When pressure here is less than valve setting, valve remains closed.

INLET
(FROM PUMP)

valve, which allows full flow in the opposite direction, which means each flow-control valve can be set at a different rate. You need to understand at this point that in some applications, such as opening and closing the mold (clamp section) on a plastic injection molding machine, opening and closing need to be controlled at different speeds, and the flow-control valve will accomplish this.

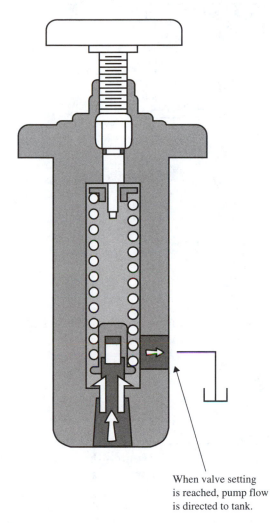

When valve setting is reached, pump flow is directed to tank.

Figure 17–20 When fluid pressure exceeds the valve setting, the piston is unseated, and fluid is allowed to flow to the tank.
(Courtesy of Vickers, Inc.)

Figure 17–21 (a) Symbol for flow-control valve; (b) symbol for flow-control valve with integral check valve.

17.12
Hydraulic Cylinders and Motors

In every hydraulic system, an actuator is provided to change the hydraulic energy to linear or rotary motion. Figure 17–22 shows an example of a typical cylinder, and Figure 17–23 shows an example of how the cylinder piston rod extends and retracts. In Figure 17–23a you can see that when fluid pressure is applied to port A, it forces the piston to extend. The fluid that is in front of the piston is allowed to flow through port B directly to the tank, so there is no resistance to the piston movement. When the fluid flow is reversed, the piston will retract, as shown in Figure 17–23b.

17.13
A Typical Hydraulic Diagram

As you work on hydraulic systems on machines, you will begin to notice several parts will be repetitive. Figure 17–24 shows a typical hydraulic diagram. In this diagram you can see the pump is identified on the bottom left side. The reservoir is identified by the symbol ⊥, which is shown four times in the diagram. The reason the reservoir is shown so many times is to make the drawing look cleaner by not having a large number of lines crossing each other. Each time a line returns to the reservoir, the symbol is shown.

After fluid leaves the pump in this diagram, you can see a pressure-relief valve is connected in the parallel in the circuit. The pressure-relief valve is set for the working pressure of the system. If the pressure exceeds this pressure, fluid will be allowed to flow directly to the reservoir.

The next component in the circuit is a directional control valve. From the symbol you can see that it is a two-position valve. This valve controls the direction of flow to the hydraulic load, which is a hydraulic motor. When the valve is switched one way, the motor shaft turns in a clockwise direction, and when the valve is switched the other way, the motor shaft turns in a counterclockwise direction. In some diagrams, a hydraulic

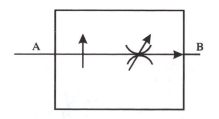

Symbol without check
(a)

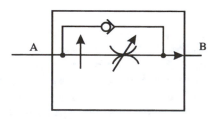

Symbol with integral check
(b)

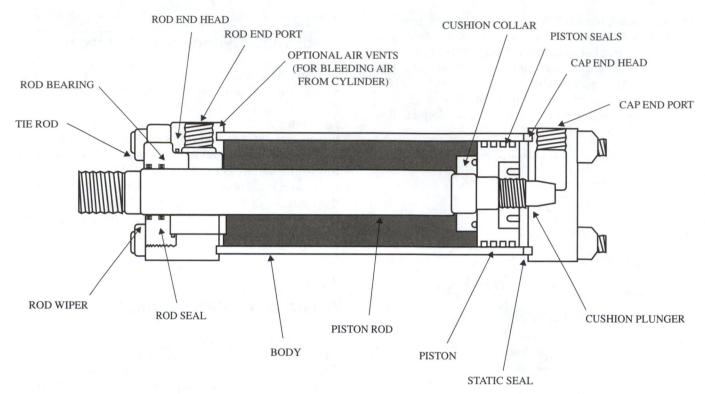

Figure 17–22 A typical hydraulic cylinder.

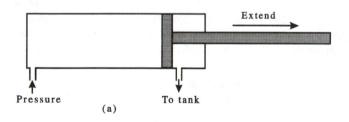

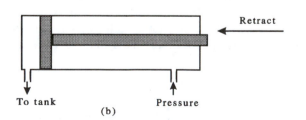

Figure 17–23 (a) Fluid pressure flows into the pressure port behind the piston and causes the cylinder to extend. Fluid in front of the piston moves freely back to the tank. (b) When the cylinder retracts, pressure flows in front of the piston and forces it to move to the back of the cylinder, causing the piston rod to retract.

cylinder is used instead of a hydraulic motor. If flow control is used, it is placed in the circuit in series with the directional control valve.

17.14
Pneumatics

A pneumatic system is very similar to a hydraulic system in that it has a power source, directional controls, pressure controls, flow controls, and actuators. The main difference between the two types of systems is that pneumatic systems use air for fluid instead of a liquid-type hydraulic fluid. One of the differences between the two is that air compresses, whereas hydraulic fluid will not. Another difference is that the pneumatic system does not send its return fluid back to the tank where it started; instead, the return fluid is called *exhaust*, and this air is just vented to atmosphere. Another difference is that when a pneumatic system has a leak, there is nothing that spills or drips as in the hydraulic system, where leaks are a constant worry.

Because air is compressible, it also contains a certain amount of moisture, which turns to water when it is allowed to condense. This water causes problems with corrosion, and it will wash away lubrication, so it must be controlled. A trap is used to collect the water and remove it from the system.

Figure 17–24 A typical hydraulic diagram.

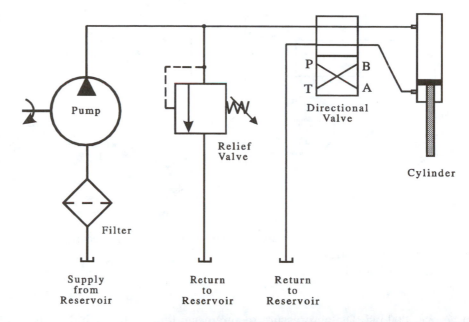

Figure 17–25 (a) Typical single-stage piston compressor; (b) typical single-stage, vane-type compressor. (Courtesy of Ingersoll-Rand)

17.15
Compressors and Tanks

The air compressor is the air pump for the pneumatic system. Typical compressors have pistons or vanes like the hydraulic pump. Figure 17–25 shows an example of a typical piston-type compressor and a typical vane-type compressor. The operation of the compressor begins as air enters the chamber and the pistons or vanes move. The amount of space the air is in is reduced sig-

nificantly, which causes the air to become compressed and increase its pressure. Because air is the fluid for the pneumatic system, it will begin to flow anytime a pressure difference is created.

Some compressors have several pistons to provide stages. The size of each additional piston is smaller, and as the air moves from piston to piston in the compressor, the air becomes more compressed and builds additional pressure. Typical pressure for a single-stage compressor can be up to 90 psi. Pressure in multistage compressors

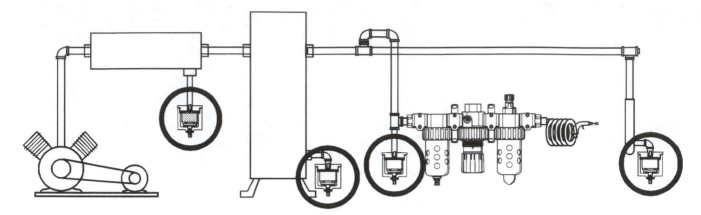

Figure 17–26 Typical pneumatic system, which includes two-stage compressor, intercooler/dryer, air tank, filter with drain, lubricator, and regulator.
(Courtesy of Parker Hannifin)

can be over 300 psi. The air-pressure requirements for most industrial applications are less than 90 psi.

17.16
Air Tanks

When air leaves the compressor, its temperature has increased significantly and it is usually stored temporarily in a large tank. The purpose of the tank is to provide a sufficient volume of air so that the equipment or tools that use air will have a sufficient supply at all times. Because the air stays in the tank for a time, it begins to cool down, and the moisture in the air begins to condense and drop out to the bottom of the tank as water. The tank has some type of drain to allow moisture to be removed on a continual basis. In some systems, a cooler called an *intercooler* is provided between the compressor and the tank to cause the temperature of the air to drop significantly, which causes the maximum amount of moisture to drop out of the air. The main purpose of the compressor and tank is to provide clean, dry air for the system.

17.17
Filters and Lubricators

Because the system needs clean air, a filter is also required. The filter is generally installed directly after the supply tank. Figure 17–26 shows a diagram of a typical system with the compressor, intercooler, tank, filters with moisture-removal capabilities, lubricator, and regulator. From this diagram you can see that the filter has a drain. The drain is needed because a small additional amount of moisture will drop out of the air as it passes

through the filter and its pressure drops slightly. A lubricator is provided to add a small amount of lubricant to the air as it passes through the lubricator into the system. Most active components in any pneumatic system need a small amount of continuous lubrication. The lubricator adds a small amount of oil to all of the air that moves from the tank. The final piece of the system is an in-line air regulator. The air regulator is similar to the pressure-control device. The regulator for the system is manually adjusted and it has a pressure gage attached to display the system pressure.

17.18
Filters, Regulators, and Lubricators

The air system needs a filter to keep the air clean. Usually the filter is incorporated with a disposable cartridge that can be exchanged when it is dirty. Figure 17–27 shows a picture of a typical air filter, regulator, and lubricator. From this picture you can see that the filter is the first piece in this unit. The filter consists of a filter element and a glass bowl that collects moisture. The bottom of the filter is a glass cup that has a drain mechanism in case this part of the system collects a small amount of moisture. Because the filter creates a small amount of pressure drop, any remaining moisture in the air will condense out at this point. The filter may have an automatic drain or a manual drain. If the drain valve is manual, someone will need to continually open the drain valve to drain the filter.

The next part of this unit is the regulator. The regulator is a pressure-control device that sets the maximum pressure for the system. The regulator is a manually operated control. When the regulator cap is turned clockwise, it increases the pressure setting, and when it

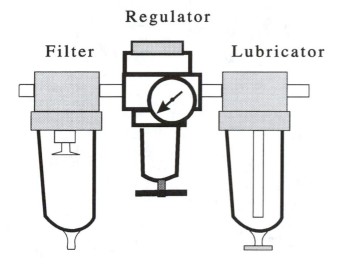

Figure 17–27 A filter, regulator, and lubricator for a pneumatic system.
(Courtesy of Parker Hannifin)

is turned counterclockwise, it reduces the pressure setting. A pressure gage is attached so that the pressure can be observed.

The third part of the unit is a lubricator. A lubricator provides a means of adding a small amount of lubrication for the system. Because the actuators in the air system have a number of moving parts, it is important that they are constantly lubricated. The lubricator is designed so that air moves through two paths. The majority of the air will move through a venturi section. The venturi section is a narrowing of the air passage that creates a pressure drop when air moves past it back to the full-size passage. A tube is connected to the venturi, and its other end is submersed in the oil in the bottom of the lubricator cup. At the point where the venturi creates a pressure drop, the oil changes into a fine mist, when it attaches itself to the air molecules and is carried to its destination, where it coats the moving parts in the actuator. The amount of lubrication that is allowed to be added to the air can be controlled by adjusting a small needle valve. Because this system depends on a minimum level of oil in the lubricator, it must be monitored and refilled as needed.

17.19
Control Valves

Pneumatic control valves are available in a variety of styles. These include everything from simple two-way, two-port, two-position, two-way, three-port, two-position, to four-way, four-port, two- and three-position valves. You will notice these valves are very similar to the hydraulic valves, and they provide similar functions. The next section identifies these valves.

Figure 17–28 Single-solenoid, two-way, three-port, two-position valve.
(Courtesy of Parker Hannifin)

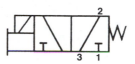

Figure 17–29 Symbol of a two-way, three-port, two-position valve.
(Courtesy of Parker Hannifin)

17.19.1
Two-Way, Three-Port, Two-Position Pneumatic Valves

Figure 17–28 shows the picture and Fig. 17–29 shows the symbol for the two-way, three-port, two-position pneumatic valve. From the diagram you can see that the ports on the valve are identified as 1, 2, and 3. Because the valve is a two-position valve, the symbol shows two boxes side by side with the connections between the ports identified in each box. The solenoid symbol is shown on the left side of the valve and the spring that returns the valve to its normal position is shown on the right side. In a typical application, air supply is connected to port 1 and port 3 is the exhaust port. The exhaust port is similar to the tank port on a hydraulic valve. In the pneumatic valve, the return air does not have to be returned to the system; instead it can simply be exhausted directly to the atmosphere right at the valve. Port 2 is connected to the pneumatic actuator, which could be some cylinder, motor, or pneumatic tool.

The valve operation is identified from the deenergized state. In the deenergized state, the spring positions the spool in the valve so that the air supply port (port 1) is blocked, and port 2—which is connected to the actuator—is connected to the exhaust port (port 3). This means that no supply air is being sent to the actuator. When the solenoid is energized, the spool in the valve is shifted by the magnetic energy developed by the solenoid

Figure 17–30 Six pneumatic valves mounted on a manifold. Air is supplied to all the valves through the large hole in the first valve.
(Courtesy of Parker Hannifin)

Figure 17–31 A typical hand-actuated valve.
(Courtesy of Parker Hannifin)

coil. When the spool shifts the air supplied to port 1, it is connected to the actuator through port 2. Basically you can view the diagrams in the two boxes as two conditions of the valve. The box on the left shows the valve connections when the solenoid is energized, and the box on the right shows the connections when the solenoid is deenergized and the spring has shifted the spool.

17.20
Valve Manifolds

In many applications, more than one pneumatic valve is used in a system. A way has been devised to supply air to all of the valves at one time through a manifold by connecting each of the valves together. This makes the plumbing for the system very simple, but it makes maintenance repair, or replacement of a valve much more

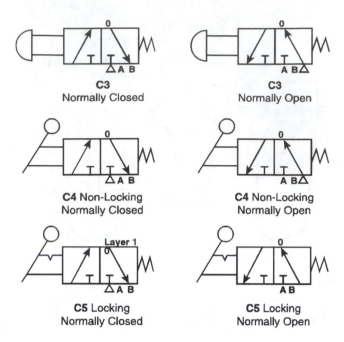

Figure 17–32 Symbols for manually actuated valves.

difficult. Figure 17–30 shows typical valves mounted on a manifold. Air is supplied to each valve through a large hole that is located directly through the middle of the valve. The access hole for the supply air on each individual valve lines up with the holes in the other valves so a constant supply of air is provided to all valves at a common point. The ports for each valve are located so that you can access each of them individually. It is important to ensure that no leaks occur between each valve and the manifold.

17.21
Hand-Operated Valves

Another type of pneumatic valve is a hand or manually operated valve. Figure 17–31 shows a picture and Fig. 17–32 shows examples of symbols for these valves. From this figure you can see that a wide variety of operators are provided, so that cylinders can be operated manually. Other types of valves, such as foot-operated valves, are also available for manual operation.

17.22
Four-Way, Three-Position Valve with Double Solenoids

A four-way, three-position valve is different from other valves in that it has two solenoids used to cause the valve to shift in both directions and a set of springs that

Figure 17–33 A typical double solenoid.
(Courtesy of Parker Hannifin)

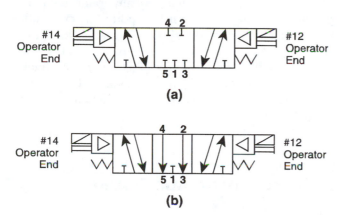

Figure 17–34 (a) A four-way, three-position valve with all ports blocked in the center position; (b) a four-way, three-position valve with all ports' exhaust in the center position.
(Courtesy of Parker Hannifin)

Figure 17–35 Typical single-ended and double-ended pneumatic cylinders.
(Courtesy of Parker Hannifin)

17.23
Actuators, Cylinders, and Pneumatic Motors

Pneumatic systems have a large variety of actuators, such as single-ended and double-ended pneumatic cylinders and pneumatic motors. The single-ended and double-ended cylinders are very similar to the hydraulic cylinders that you studied earlier. Figure 17–35 shows examples of single-ended and double-ended cylinders. The cylinders have two ports that allow the air to cause the cylinder rod to extend and retract. The pneumatic motor has a single port for air supply and an exhaust port. Pneumatic motors are used in a variety of hand tools and other types of rotary actuators.

An example of a rotary actuator that is not a motor is the rotary cylinder. This actuator looks like a typical rod-type cylinder, except its actuator is fixed in the middle of the cylinder. Figure 17–36 shows this type of actuator. When air is supplied to one end of the cylinder, it causes the actuator to rotate 180° in the clockwise direction. When air is applied to the other end of the cylinder, the

cause the valve to move to a third position called the *center position.* Any time both solenoids are deenergized, the spool is moved to the center position by springs that are mounted in each end of the valve. Figure 17–33 shows a picture of a double solenoid–type valve, and Fig. 17–34 shows a diagram of this type of valve. From the picture, you can see that solenoids are mounted on both ends of the valve. The diagram shows three boxes for the valve, indicating this valve is a three-position valve. The number of lines inside each box indicates this is a four-way valve that has five ports. The ports are numbered 1–5. Port 1 is the port where the air supply or pressure is connected. Ports 4 and 2 are connected to the actuator like the extend and retract ports of a cylinder. Ports 5 and 3 are generally allowed to exhaust the air directly to atmosphere.

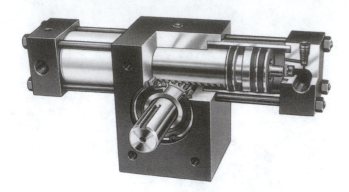

Figure 17–36 A rotary-type actuator. This actuator looks like a rod-type cylinder, but it has a fixed rotary actuator that rotates 180° in both clockwise and counterclockwise directions.
(Courtesy of Parker Hannifin)

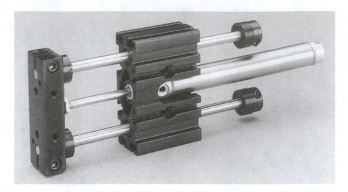

Figure 17–37 Linear slide.
(Courtesy of Parker Hannifin)

rotary actuator rotates 180° in the opposite direction. Rotary actuators are used in a variety of pick-and-place applications, where parts are picked up in one location, rotated 180°, and placed in a new location on a conveyor or assembly fixture.

17.24
Rodless Pneumatic Cylinders

One of the drawbacks of a traditional rod-type cylinder is that it takes up a little more than twice the amount of space taken up by the length of its rod. For example, if you have a 20-in. rod, you need a pneumatic cylinder that is slightly longer than 20 in. to hold the rod when it is retracted. When you take into account the length of the 20-in. rod when it is extended and the more than 20-in. cylinder, the total space needed is more than 40 in.

Rodless pneumatic cylinders were developed because they take up approximately half the space of traditional cylinders. The main feature of a rodless cylinder is the piston inside the air cylinder, which moves back and forth through the cylinder when air is applied to either end of the cylinder but does not have a rod attached to it. Instead, the piston has a strong magnet attached to it, and the load that is usually connected to end of the rod in a traditional cylinder is now connected to the magnetic coupler that is mounted on linear slide rails on the outside of the cylinder. When the piston inside the cylinder moves from one end of the cylinder to the other, the coupler on the outside cylinder follows this movement along the rails. Figure 17–37 shows a picture of a typical rodless cylinder called a linear slide. From this figure you can see that these cylinders look like they

have a sliding bed that moves along fixed linear rails. This is the coupler that is magnetically connected to the piston inside the cylinder. It is important to remember that rodless cylinders can also be very large (more than 20 ft long), and the advantage they provide is that they take up less than half the space of a traditional cylinder. This makes the rodless cylinder very useful in applications such as pick-and-place robots, which move in linear paths, and where space is important.

17.25
A Typical Pneumatic System

When you are working on the factory floor, you need to look at the pneumatic system as a total system rather than a group of unrelated parts. Figure 17–38 shows an example of a complete pneumatic system that is used in a pick-and-place robot application. In this application, three pneumatic cylinders control three axes of motion for the robot. You can see that the longest cylinder controls horizontal travel, and the vertical and depth axes are about the same size. Each cylinder is controlled by its own three-position, four-way, five-port directional control valve that is operated by two electric solenoid valves. The air supply for this system comes from an air compressor and is stored in a tank. The lubricator and filter provide a clean supply of air for the robot control. The major difference between the pneumatic diagram and the hydraulic diagram is that the pneumatic system does not try to return the air to the system after it is used to move a cylinder; instead air is exhausted to the atmosphere at the point where it is used.

When you are troubleshooting a pneumatic system, all you need to do is test to see that the proper amount of supply air is available at each control device and each operational device.

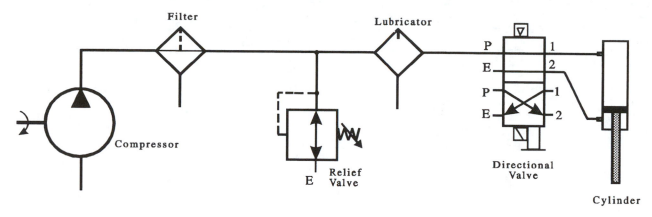

Figure 17–38 Typical pneumatic components.

Questions for This Chapter

1. Explain the operation of a hydraulic cylinder.

2. What is the difference between a positive displacement and a nonpositive displacement pump?

3. Explain the operation of a flow-control valve.

4. Explain the operation of a pressure-control valve.

5. Explain the difference between a hydraulic system and a pneumatic system.

True or False

1. Hydraulic fluid can be compressed like air in a pneumatic system.

2. Air in a pneumatic system is returned to a reservoir or tank after it is used in an air cylinder.

3. An impeller-type hydraulic pump is a positive displacement pump.

4. A check valve allows fluid flow in only one direction.

5. A directional control valve can be operated by hand or by an electric solenoid coil.

Multiple Choice

1. In a pneumatic system the filter is used to _____.
   a. filter air.
   b. filter air and provide a place to remove moisture.
   c. accumulate air.

2. The regulator in a pneumatic system is used to _____.
   a. control the flow in a system.
   b. set the maximum pressure in a system.
   c. control flow and set minimum pressure in a system.

3. The pressure-relief valve in a hydraulic system _____.
   a. sets the maximum pressure for the system.
   b. sets the maximum flow for the system.
   c. sets the maximum pressure and flow for a system.

4. The function of a reservoir (tank) in a hydraulic system is _____.
   a. to provide a place to store fluid.
   b. to provide a way to dissipate heat from the fluid.
   c. both a and b.

5. A positive displacement pump is a hydraulic pump that _____.
   a. can pump a variable amount of volume with each rotation of its shaft.
   b. can pump a constant amount of volume with each rotation of its shaft.
   c. can pump either variable volume or constant volume with each rotation of its shaft.

Problems

1. Calculate the horsepower of a hydraulic pump that moves 20 gpm at 500 psi.

2. Calculate the horsepower of a system that moves 132,000 ft-lb in 1 min.

3. Calculate the flow of a system that moves 1155 in.^{3} of fluid in 1 min.

4. Draw a sketch of a variable-volume hydraulic pump and explain how it varies the volume of fluid that it pumps.

5. Draw a sketch of a pneumatic system and identify the main parts.

◄ Chapter 18 ►

Basic Mechanics

Objectives

After reading this chapter you will be able to:

1. Identify the terms *mass*, *force*, and *acceleration*.
2. Explain the terms *radius*, *diameter*, and *circumference*.
3. Explain the terms *potential energy* and *kinetic energy*.
4. Calculate the force, distance, and speed of pulleys.

18.0
Overview of Basic Mechanics

In order to understand the basic information about mechanical systems, you will need to fully understand a variety of technical terms. You will also need to understand that all parts of every machine can be explained in terms of six basic, simple machines. These simple machines are the lever, the inclined plane (wedge), the wheel and axle, the screw thread, simple and compound gears, and pulleys. When these simple machine parts are put together, they are called *compound machines*. If you understand the basic theory involved with each simple machine and compound machine, you will be able to troubleshoot all the components in a power-transmission system. This chapter provides a review of the terms that are used to explain the applications of mechanical systems and power transmission and an explanation of the six simple machines.

18.1
Basic Systems of Measurements

When you work on power-transmission systems, you will encounter units of linear measurements, such as inches, feet, and yards, or units of weight, such as ounces and pounds. These units are part of a system that is called the English (or customary) system of measurement. The English system has been used extensively in the United States and is probably the one you used when you were younger. You may also encounter units of linear measurement such as centimeters and meters or units of weight such as kilograms that are part of a standard system called the metric system of measurement. In today's complex world, you will find equipment with one or both types of units used to identify measurements.

It is not the intent of this textbook to determine which system is the best or which system is more usable; instead this book presents material so that as a technician, you will be able to work accurately with the values that you are provided. The one point this book does make repeatedly is that neither system was designed to be converted easily to the other. In other words, if you locate a value on a machine such as its weight and it's listed in pounds, it is not necessary to convert this value to its metric equivalent. Figure 18–1 shows examples of weights and measures in both the English and metric units.

Figure 18–1 Table of weights and measures.

	English and SI (Metric) Units	
Dimension	*English Units*	*Metric Units*
Length	foot (ft)	meter (m)
Mass	pound (lb)	kilogram (kg)
Time	second (s)	second (s)
Energy	foot-pound (ft-lb)	kilogram-meter (kg-m)
Force	pound (lb)	kilogram (kg)

Figure 18–2 Prefixes for units of measure.

Prefix	*Word Value*	*Numerical Value*	*Exponential Value*
centi	1 hundred	100	1×10^2
kilo	1 thousand	1,000	1×10^3
mega	1 million	1,000,000	1×10^6
milli	1 thousandth	0.001	1×10^{-3}
micro	1 millionth	0.000001	1×10^{-6}

18.1.1
Prefixes

When you are working with units of measure, you will encounter prefix words that indicate how large or small a number is. For example the word *kilometer* uses the prefix *kilo*. A kilometer is equal to 1,000 m. A prefix such as *kilo* allows a measurement to be given with smaller numbers. For example, if we can give output of a nuclear power plant as 15 MW. This is the same as saying the output is 15 million watts. The measurement 15 MW is the same number as 15,000,000 W, but it is easier to write.

The most important point to remember is that you cannot control what values are provided, but you need to be able to write measurements in a more convenient form. The table in Fig. 18–2 shows the most widely used prefixes and their values. The exponential value of each prefix is also provided.

Exponents represent the number of zeros in the multiplier. For example, the value 15 km is the same as $15 \times 1,000$ m, because kilo means 1,000. Because 1,000 has 3 zeros, $1,000 = 10^3$. So, $15 \text{ km} = 15 \times 10^3$ m. (You can also remember that $1,000 = 10 \times 10 \times 10 = 10^3$.)

18.2
Basic Terms

18.2.1
Mass, Acceleration, Gravity, and Force

Mass is the term that identifies the quantity of matter in a body divided by the acceleration of the body. In simpler terms, mass is how much substance or how much matter a body has. If you want to increase the mass, you have to add substance or matter to the body. Because the mass of a body cannot be measured by its weight alone, a more precise means of measuring mass utilizes the quantity of matter in the body divided by the acceleration of the body due to gravity. The larger the mass of a body, the more it tends to resist any force that causes it to move.

Acceleration can be defined by a change in speed to a body. It can also include a change of direction with the change of speed. An example of acceleration is a car that speeds up. When you are in a car that speeds up, you will feel the affects of acceleration. For example when the car speeds up, you will be pushed back against the seat.

When the car decelerates as the brakes are applied, you will tend to lean forward as the car stops. When the car turns to the left, you will tend to continue moving in a straight line, which makes you lean to the outside of the car as it moves into the turn. All these forces act on you because you have mass and the car is accelerating (including changing directions). Acceleration is measured in terms of distance over time. For example you can increase the speed of a body at a rate of 10 ft/s or in international units. One of the most common types of acceleration you will encounter is the acceleration caused by gravity on a free-falling body.

Gravity is the invisible force that the earth exerts on all mass. An object that is raised in the air, such as a ball, will fall at a predetermined rate when it is allowed to fall freely. The rate of acceleration for all bodies on earth is 32 ft/s/s. This means that if you allow something to drop from a tall building, it will fall 32 ft the first second and it will increase its speed another 32 ft per second during the each additional second it is allowed to fall. For example if you could drop a ball from a building that is tall enough, the ball would fall starting at a velocity of 0, and it would accelerate to 32 ft/s during the first second. If the ball has not yet hit the ground, its velocity would continue to increase to 64 ft/s at the end of the second second. If the ball were allowed to fall for another second, its acceleration would reach 96 ft/s at the end of the third second.

You may not see any reason to know this information, but as a maintenance technician you will find a wide variety of parts in a machine that use gravity to cause certain parts to operate. The important point for you to understand is that the rates of speed of these parts were calculated when the machine was designed, and the actual speed during machine operation should be close to the calculated speed. If the actual speed is slower, you will know that the machine may have a problem, such as insufficient lubrication, or the machine parts may be binding.

Force is any push or pull or anything that causes a change of motion, such as pushing, pulling, lifting, or turning an object. Force can also be described as any action that causes mass to change velocity. It can be defined as mass multiplied by acceleration. It could also include speeding up an object, causing it to accelerate, or slowing down an object, called deceleration. Two of the simplest types of force are the force of gravity and the force of friction.

18.2.2
Relationship between Mass, Force, and Acceleration

When you study machine operation, you will begin to see that certain functions are predictable due to the laws of physics. If you understand the basic relationships, you will see that troubleshooting the mechanical part of a machine will be much easier. For example, the previous sections of this chapter explained the definitions of mass, force, and acceleration. There is a basic formula that you need to know to understand how they work together. If you understand this simple relationship, you will be able to observe the operation of a machine and determine whether it is operating correctly or not. The formula relating mass, force, and acceleration is $F = ma$:

$$\text{Force} = \text{mass} \times \text{acceleration}$$

This formula can also be written as

$$\text{Mass} = \frac{\text{force}}{\text{acceleration}}$$

In simple terms, this formula means that if the mass of an object increases, the amount of force must increase to make it accelerate at the same speed.

18.2.3
Other Basic Terms

Torque is defined as a force applied to a turning radius. An example of torque is the force that your arm applies to a wrench to loosen a nut. Because your arm is applying a force in a circular motion, the force is called torque. Another example of torque is the force that a motor applies to a shaft that has gears mounted on it. Because the shaft is turning in a rotational manner, the force applied to the shaft to make it rotate is called torque.

Work is defined as force applied to an object multiplied by the distance the object is moved. The units of work are foot-pounds (ft-lb). Work is usually not used in applications by itself; instead, it is used as part of a formula to determine horsepower.

Power is defined as the rate work is done. The unit of power is the horsepower. One horsepower is equal to 33,000 ft-lb/min. The formula for horsepower is

$$\text{hp} = \frac{\text{work (ft} - \text{lb)}}{33,000 \times \text{time (min)}}$$

Power can also be calculated from electrical energy. The unit of electrical energy is the watt, and 1 hp is equal to 746 W.

Velocity is the quantitative value of measure of the speed and direction of the movement of a body.

18.2.4
Forces

When you are studying mechanical systems, it is important to differentiate different types of forces. For example when you are discussing forces applied to a gear mounted to a shaft, any force that occurs in the direction the gear is moving is called *thrust*. Another way to think of thrust is the force that causes something to move directly against face of a bearing or gear. Axial force is a force against the side of a bearing or gear.

Stress is defined as the force per unit area that causes an object to compress or stretch. *Strain* is defined as the resulting amount of change a body goes through when a stress is applied to it.

Friction is the resistive force that opposes the motion of two bodies. Rolling friction is the friction that is caused when a circular body, or wheel, moves about a flat surface. Static friction is the force (opposition to movement) that is caused when a body is at rest. Static friction tends to hold a body at rest and cause it to remain at rest. This friction must be overcome to allow the body to begin to move. Sliding friction is the opposition of one body when it is sliding over another body. Sliding friction makes the moving body tend to stay at rest.

Centrifugal force is the force that tends to cause a body that is rotating to move in a straight line. The best example of centrifugal force is when you are riding in an automobile that turns a sharp corner. When the automobile turns the corner, as a passenger you will tend to continue moving in the same direction (in a straight line) the car was traveling. The centrifugal force tends to make your body pull to the outside of the corner. Centrifugal force plays a part in applications where belts or chains are moving around sheaves.

Centripetal force is the force that holds rotating bodies together. This is basically a force in opposition to centrifugal force.

18.2.5
Energy

Energy is the capacity for doing work. The units for measuring energy are the same as those used to measure work. Energy cannot be created or destroyed; it can only be converted from one form to another. The basic forms of energy that you will encounter in industrial applications are potential energy, kinetic energy, acoustical energy, thermal energy, chemical energy, electrical energy, and nuclear energy. Potential energy and kinetic energy can be in the form of mechanical energy or hydraulic energy; you will find many industrial applications that show how energy is converted to a usable form. An example of mechanical potential energy is a compressed spring, and an exam-ple of hydraulic potential energy is a hydraulic accumulator that is under pressure.

18.2.5.1 Potential Energy

Potential energy is energy in a stored form that has the ability (potential) to do work. For example, a heavy weight that is lifted above a machine on a wire has the potential to do work when it is allowed to fall. This is also the effect a flywheel provides. Other examples of potential energy include a compress or extended spring. When a spring is compressed, it is ready to release all the stored energy. Another form of potential energy is a hydraulic accumulator that has pressure built up in it or stored in it. It is important to recognize various forms of potential energy because it must always be released prior to working on a machine.

18.2.5.2 Kinetic Energy

Kinetic energy is energy in action. When a hydraulic cylinder is moving, it is exerting kinetic energy. When a motor shaft is rotating, it is also exerting kinetic energy. Any time a part of a machine is moving, it is exerting kinetic energy.

18.2.6
Center of Gravity

The *center of gravity* can be defined as the point in a body from which it can be suspended or on which it can be rested that allows the body to be in perfect balance in every direction. The center of gravity is important for many machine parts, because it is the point where the force of gravity is equal on all parts of the machine, causing the part to balance. When a part is in balance, it does not require additional energy to keep it located or positioned properly. In some machines you will be required to locate the center of gravity of a part when it is being replaced or put back on a machine. It is also possible to use two points to rest or suspend a large component such as a beam by locating the resting points or suspension points that are equally spaced from the center of gravity.

18.2.7
Momentum

Momentum is the result of multiplying the mass of an object by its velocity. An example of momentum is when you give a ball a push so it starts rolling on a flat surface. Once the force has been applied to the ball to get it rolling, the ball has momentum, which keeps it rolling until the forces of gravity and friction cause it to stop rolling. The same concept can be applied to all the parts of a machine that are in motion. Once a force has been applied to them and they are moving, they will continue to move. This is why a machine needs brakes to stop its different parts.

18.2.8
Terms Involving Circles

When you are working with round components in a power-transmission system, such as gears, wheels, pulleys, or shafts, you will need to understand a few basic terms. Figure 18–3 shows a circle with a number of parts identified. In this figure you can see that the *diameter* of a circle is the distance through the center of a circle. The *radius* is the distance from the center of a circle to its outside edge. The radius is also half the diameter of a circle. The distance around the outside of a circle is called the *circumference*. The important point you need to understand is that if a piece of information about the circle is provided in a diagram, such as radius, you can calculate the diameter and circumference. Figure 18–4 shows several useful formulas for calculating diameter, radius, and circumference of a circle.

In some applications, you will discuss a round object in terms of the number of degrees it has. For example, a circle contains 360°.

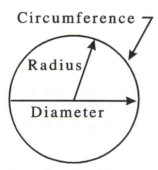

Figure 18–3   Example of the radius, diameter, and circumference of a circle.

$$A = \text{area}$$
$$d = \text{diameter}$$
$$r = \text{radius}$$
$$C = \text{circumference}$$
$$d = 2r$$
$$A = \pi r^2$$
$$C = \pi d$$
$$r = \frac{d}{2}$$

Figure 18–4   Formulas for calculating the area, radius, diameter, and circumference of a circle.

18.2.9
Newton's Laws

18.2.9.1   First Law of Motion
Bodies at rest tend to stay at rest, and bodies in motion tend to stay in motion unless external forces change the motion. An example of this law is when a force is applied to a ball that starts rolling across a flat surface. The ball continues to roll because of Newton's first law. We also call this *momentum*.

18.2.9.2   Second Law of Motion
A body in motion has a certain amount of motion called momentum. Momentum is controlled by the mass of a body and its velocity. The amount of momentum changes as the velocity of the body changes. For example, applications that have flywheels use a lot of momentum to get the flywheel to move at low speeds. When the flywheel reaches full speed, the amount of momentum needed to keep the flywheel turning is minimal.

18.2.9.3   Third Law of Motion
For every action there is an equal and opposite reaction. For example, if an automobile is traveling at a specific velocity, its brakes must apply an amount of braking force with friction that is equal to the force that is making the car move.

18.3
Six Simple Machines

All mechanical things can be explained in terms of six simple machines. These six simple machines are levers, inclined planes (wedges), the wheel and axle, screw threads, simple and compound gears, and pulleys. A machine can be defined as a mechanical device that needs an input of force and direction to achieve an output of work.

This section explains the mechanical advantage that each simple machine provides. When you are asked to troubleshoot a complex machine on the factory floor, you will begin to look at it in terms of the simple machines that makes up its basic functions, and you will find troubleshooting will be much simpler.

18.3.1
Levers

A *lever* is a simple bar that is placed against a fixed point so that it can be rotated. Figure 18–5 shows an example of a simple lever. From this example you can see the bar has a rock on one end and a fulcrum in the middle, and a force is applied to the opposite end. The fulcrum can be anything that is high enough and sturdy enough to allow

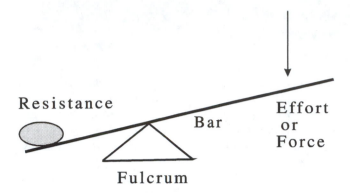

Figure 18–7   The amounts the resistance end moves upward and the force end moves downward must be measured for lever calculations.

Figure 18–5   A simple lever consists of a bar and fulcrum. A resistance (rock) is lifted when an effort (force) is applied to the raised end of the bar.

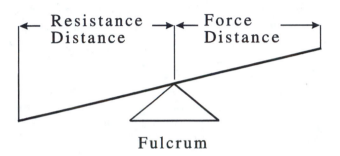

Figure 18–6   The resistance distance and force distance need to be measured to calculate the amount of load a lever can lift. Each distance is measured from the center of the fulcrum to the end of the bar.

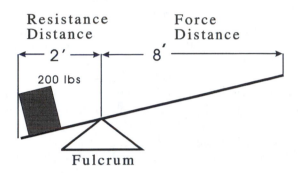

Figure 18–8   The resistance distance is 2 ft and the force distance is 8 ft. The weight to be lifted is 200 lb (Example 4–1).

the bar to be pulled against it so that the rock is lifted. The lever is a machine because you can gain mechanical advantage or speed by determining the proper length of the bar and the placement of the fulcrum.

The amount of weight the lever can lift can be calculated by knowing several basic measurements, such as the distance from the fulcrum to the end of the bar where the resistance rests and the distance between the fulcrum and the end of the bar where the force is exerted. Two other measurements are also needed. These are the distance the resistance will be moved upward when the lever is pressed on and the distance the force must move downward. The resistance distance and the force distance are shown in Fig. 18–6, and the resistance movement and the force movement are shown in Fig. 18–7.

18.3.1.1   Classes of Levers

The principal of the lever can be modified slightly to gain different advantages. These modified levers can be classified as *first-class levers*, *second-class levers*, and

third-class levers. The next part of the text explains the operation of each class of lever and provides examples of simple hand tools and machine parts that use these applications.

First-Class Levers.   The lever shown in Figure 18–8 is an example of a *first-class lever*. The definition of a first-class lever is a lever in which the fulcrum is placed between the resistance and the force. In this type of lever, the force must be applied downward to cause the load to move upward. If the fulcrum is placed nearer to the load end, the lever will be able to lift more weight, but the bar on the end where the force is applied will need to go through a longer distance to do this. The first-class lever increases force and reverses the direction of the motion. This means that a force of 100 lb pushing down on the end of the lever can lift a weight of 200 lb in the upward direction.

Mechanical Advantage of Force for First-Class Levers.   The mechanical advantage of force is determined by the position of the fulcrum under the bar. The placement of the fulcrum determines the ratio of the resistance distance and the force distance. The formula for calculating the advantage of force is

$$R_L \times R_D = E_F \times E_D$$

where R_L = the load resistance

R_D = the distance between the load and fulcrum

E_F = the effort force

E_D = the distance between the effort and the fulcrum

▶ **Example 18–1**

See Figure 18–8. If the bar is 10 ft long and the fulcrum is placed at 2 ft, the resistance load, R_L, is 2 ft from the fulcrum, and the effort force is applied to the bar at 8 ft from the fulcrum. If the resistance load (R_L) is 200 lb, how much force will need to be applied to lift the weight?

Solution:

$$E_F = \frac{R_L \times R_D}{E_D}$$

$$E_F = \frac{200 \text{ lb} \times 2 \text{ ft}}{8 \text{ ft}}$$

$$E_F = \frac{400 \text{ ft} - \text{lb}}{8 \text{ ft}}$$

$$E_F = 50 \text{ lb}$$

The mechanical advantage is based on the ratio of the two distances, 2 ft and 8 ft. Their ratio is 4:1. The important point about mechanical advantage is that this ratio is constant in this situation and you can use it to determine the amount of force needed to lift any weight with this lever and fulcrum. For example if the weight changes from 200 lb to 400 lb, you can use the ratio to determine the amount of force needed to raise 400 lb is 100 lb. ◀

Second-Class Levers. A *second-class lever* is a lever that has the load positioned between the fulcrum and the force end. Figure 18–9 shows an example of a second-class lever; you can see that the fulcrum is moved to the far left end of the bar, and the load is moved to the middle of the bar. An example of this type of lever in action is a wheelbarrow. Figure 18–10(a) shows the wheel on one end of the wheelbarrow is the fulcrum, and the handle on the opposite end is where you apply the force to lift the load, which is positioned in the middle. Another example of the second-class lever is shown in this figure, where a hydraulic jack is placed at one end of a bar and a load is lifted in the middle of the bar. A dump truck liftbed is another example of a second-class lever. In second-class levers, the load is moved in the same direction as the force.

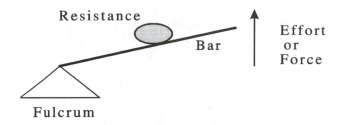

Resistance
Bar
Effort or Force
Fulcrum

Figure 18–9 Example of a second-class lever, where the load is placed between fulcrum and the force end.

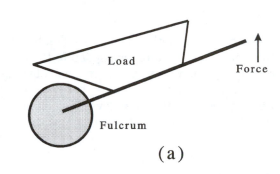

Load
Force
Fulcrum

(a)

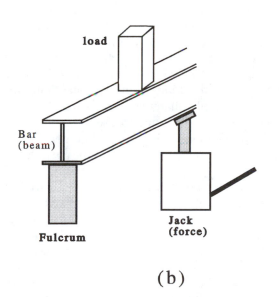

load
Bar (beam)
Fulcrum
Jack (force)

(b)

Figure 18–10 (a) A wheelbarrow is an example of a second-class lever. (b) A jack used to lift an I-beam, which lifts a load, is an example of a second-class lever.

Third-Class Levers. The third-class lever is the least used type of lever. In this type of lever, the fulcrum is located at one end, the load is at the other end and the effort (force) is applied in the middle. This type of lever is special in that it is not designed to increase force, rather it is designed to increase speed and distance that the load is moved. One example of a third class lever is a

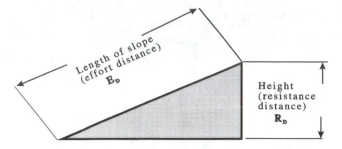

Figure 18–11 Example of an inclined plane. The effort distance and the resistance distance are shown on the plane.

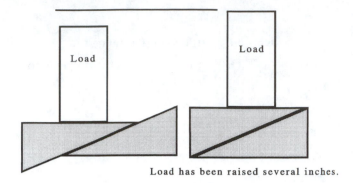

Load has been raised several inches.

Figure 18–12 The load on the right is raised by several inches after the incline planes (wedges) are moved together. This type of application is used to change the level of a machine.

fishing rod, where your hand on the end of the handle is the fulcrum point, the force is applied to the middle of the rod, by flipping your wrist, and the end of the fishing rod tip is the part of the machine that receives the increase of speed. In this type of lever, a small movement of the wrist, results in a large movement at the rod tip.

18.3.2
Inclined Plane (Wedge)

The second type of simple machine is called the inclined plane, or wedge. Figure 18–11 shows an example of an inclined plane. The incline provides a means of moving a weight that requires less effort than lifting the weight straight up. The inclined plane is integrated into a large number of machines in material-handling applications. For example, many conveyors that feed parts into a machine are positioned to use an incline rather than to lift parts straight up. The incline helps reduce the amount of force motors need to lift the parts into the machine.

Other uses of the inclined plane in industrial machines include using two inclined planes back to back. As the planes are forced together as a set of wedges, they cause the load that is placed on top of them to be raised. This type of application is used extensively in lifting and leveling machines that are being installed on the factory floor. Figure 18–12 shows an example of this application of inclined planes.

Another example of inclined planes is the *wedge.* Many machine tools, such as sawblades and other types of metal cutters, have teeth shaped as wedges. Because the teeth are in the shape of a wedge, the tool can remove material with less effort.

18.3.3
Wheel and Axle

The wheel and axle is a simple machine that consists of one or more wheels rigidly connected to a shaft. In typical applications, two or more sets of wheels are con-

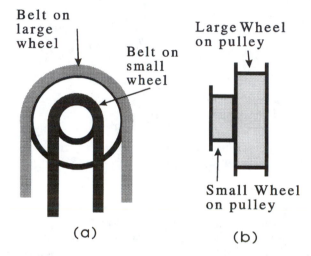

Figure 18–13 (a) Front view of a wheel and axle with belts. One of the belts is driven by a motor and the other belt is connected to a driveshaft of a machine. (b) Side view of a wheel and axle without its belts.

nected by belts to make a belt-and-pulley system. Figure 18–13 shows an example of two different-size pulleys that are connected to the same shaft. This figure shows a side view and a front view so that you can clearly see the difference of size. In an application like this one of the belts is driven by a motor, and the second belt is used to drive a machine shaft. The mechanical advantage of the wheel and axle comes from the size difference between any two pulleys. Figure 18–14 shows a set of multiple-size pulleys connected to a motor and a second set of pulleys connected to the driveshaft of a machine. In this figure you can see that the belt is around the largest pulley on the motor and on the smallest pulley on the drive shaft. This provides an advantage of speed to the driveshaft, so the driveshaft will run at a higher speed than the motor.

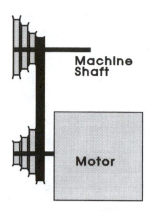

Figure 18–14 A multiple pulley is connected to a motor and a second one is connected to a machine shaft. The belt connects the largest pulley on the motor to the smallest pulley on the drive shaft. This provides an advantage of speed, where the machine drive shaft rotates faster than the motor.

Achieving an Advantage of Speed or an Advantage of Power. The exact increase in speed can be calculated from the size of each pulley. If a motor has a pulley that has a 4-in. diameter and a drive shaft has a pulley with a 1-in. diameter, the ratio of the two wheels is 4:1, which means that the drive shaft will run four times as fast as the motor shaft. For example, if the motor is turning at 500 rpm, the drive shaft will turn at 2000 rpm.

The pulley can be set for an advantage of power by placing the belt on the smallest pulley on the motor and using the largest pulley on the drive shaft. If the large pulley is 4 in. in diameter and the small pulley is 1 in. in diameter, the ratio is 4:1. This means that if the motor provides an effort of 100 lb, the drive shaft provides an effort of 400 lb.

The important point to remember about speed and power ratios is that they are inversely related. This means that when the motor has a small-diameter pulley and the drive shaft has a large pulley, the drive shaft gains an advantage of power, but it also loses the same amount of advantage of speed. In the previous example, where the motor provides an effort of 100 lb and the drive shaft gains an effort of 400 lb, if the motor is turning at 1,000 rpm, the drive shaft will be turning at a rate that is four times slower, which is 250 rpm.

18.3.4
Screw Threads

Screw threads are another type of simple machine. Screw threads are applied to round shafts, such as nuts and bolts. They are used to fasten parts together, transmit rotary motion to linear motion, and apply large forces with small amounts of effort. Screw threads are identified by the number of threads per inch and the

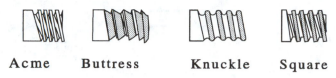

Figure 18–15 Examples of screw threads: acme thread, buttress thread, knuckle thread, and square thread.

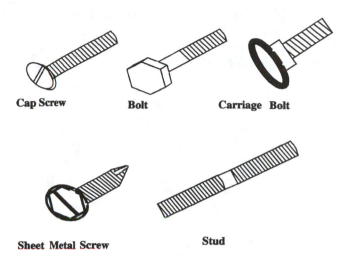

Figure 18–16 Types of fasteners that use threads.

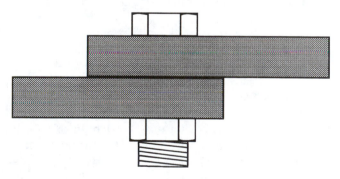

Figure 18–17 Example of a bolt and nut used as a fastener that connects two pieces of material together.

type of thread. Figure 18–15 shows examples of different types of screw threads.

18.3.4.1 Threads Used as Fasteners for Bolts and Screws

One use of threads in industrial equipment is as fasteners. Figure 18–16 shows examples of typical bolts and screws that are used as fasteners. If the threads are part of a bolt-type fastener, a nut with matching threads is needed. If the threads are part of a sheet metal screw or wood screw, the tapers on the threads are sufficient to hold two pieces of material together. Figure 18–17 shows an example of how a bolt and nut are used to hold two types of material together.

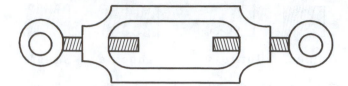

Figure 18–18 An example of a turnbuckle that is used to provide tension to wires or materials that are located a small distance apart. Because one end of the turnbuckle has right-hand threads and the other end has left-hand threads, the turnbuckle body can be turned to extend or retract the length of the turnbuckle.

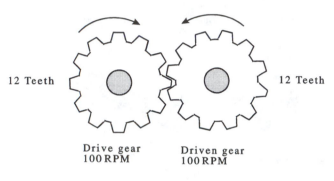

12 Teeth   12 Teeth

Drive gear   Driven gear
100 RPM   100 RPM

Gear ratio 1 : 1

Figure 18–19 Example of a drive gear and a driven gear.

18.3.5
Turnbuckle

A turnbuckle is a special application of threads. It is an apparatus that has the ability to extend or retract its length. This allows it to be used to tighten or loosen wires to provide exact tensioning. The turnbuckle has two eyebolts that match to a threaded body. The threads on one eyebolt are right-hand threads, and the threads in the other are left-hand threads. Because both right-hand and left-hand threads are used, the body of the turnbuckle can be turned in one direction to thread the eyebolts into the body and shorten the body. When the body is turned in the opposite direction, the bolts move out of the body and the turnbuckle becomes longer. Figure 18–18 shows a typical turnbuckle.

18.3.6
Simple and Compound Gears

Gears allow the smooth transmission of power and motion from one shaft (the driver) to a second shaft (the driven) by friction. In order for a gear to be effective, its teeth need to mesh with those of another gear. Figure 18–19 show an example of a driver gear that is meshed with a driven gear. In basic operation, gears have four

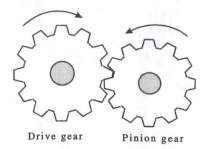

Drive gear   Pinion gear

Figure 18–20 Diagram of two gears that shows the direction of rotation of the driver and driven gear.

functions. First, they can provide motion to the driven shaft that is opposite in direction to the driveshaft; second, they can provide positive transfer of motion from the driversshaft to the driven shaft. Third, gears can provide exactly the same speed, a slower speed, or a higher speed from the driver to the drivenshaft; fourth, gears can be used to decrease or increase the amount of force from the driver to the driven shaft.

18.3.6.1 Changing Speed with Gears
Gears can be used individually or as part of a complex system in power-transmission systems. In most systems you will find a single *driver gear* and one or more *driven gears*. The driver gear is the gear that is connected to the shaft of the motor or source of power. If a single driven gear is used, it is connected to the part of the system that receives the power. If the gear is part of a transmission, it is combined with other gears, and they are placed in close proximity to each other so that their teeth will mesh with other gears.

Gears are used to change the speed of rotation of the driven shaft from that of the drive shaft. This is accomplished through gear ratios. For example, if the drive gear has 20 teeth and the driven gear has 10 teeth, the driven shaft will turn twice as fast as the driver gear. The formula for this is

$$\text{Output speed} = \omega \times \frac{N_1}{N_2}$$

where ω is the drive shaft speed, N_1 is number of teeth in the driver gear, and N_2 is the number of teeth in the driven gear.

18.3.6.2 Changing the Direction of Rotation with Gears
Another reason to use gears is to change the direction of rotation of the driven shaft. If the driven shaft needs to rotate in the same direction as the driver shaft, three gears must be used. Figure 18–20 shows a diagram that indicates the direction of rotation of a transmission

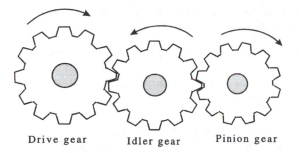

Figure 18–21 Diagram of three gears showing the direction of rotation of the driver, idler, and driven gears.

with two gears, and Fig. 18–21 shows a transmission with three gears. The drive gear is often called the *gear*. The middle gear is called the *idler*, and the driven gear is called the *pinion*. You should notice that if you keep adding gears to the transmission, if the total number of gears is even, the direction of rotation of the final shaft is opposite that of the driver gear, and if the total is odd, the direction of rotation for the final shaft is the same as that of the driver gear.

The speed and torque (measured in foot-pounds) transferred in a transmission of two or more gears are inversely related to each other. Thus if you select two gears that double the speed of the driven shaft, then the amount of torque of the driven gear is reduced by half. Also, if the gears are selected to provide an increase in torque at the driven shaft, the speed of the driven shaft will be decreased. You will find that the gears in a transmission are selected on the basis of the final speed for the driven shaft or the final torque the driven shaft must provide.

The remainder of this section explains the theory of operation of gears, introduces types of gears, provides strategies for analyzing problems with gears and transmissions, and explains methods of removing or repairing gears. You will also see products called gear reducers that are specifically designed to reduce the speed of the drive shaft by ratios of 5:1 up to 60:1 if slower speeds are needed.

18.3.6.3 Theory of Operation of Gears

Gears are used in machines to cause a positive transfer of power from one shaft to another, provide a change of speed, change direction of rotation of shafts, or gain an advantage of torque. For example if you have a driver gear that has 10 teeth and a driven gear that has 20 teeth, the driver shaft will turn twice as fast as the driven shaft, and the driven shaft will have twice the force as the drive shaft. The ratio of the number of teeth in gears is equal to the ratio of the speed and the ratio of force between the gears. Another important point to remember is that if two gears (the driver and driven) are used, the rotation of the drive gear and the driven gear will be in opposite directions. Each time another gear is added, the direction of rotation for that gear is changed. The driver shaft and the driven shaft can be mounted parallel to each other or at some angle to each other, depending on the application. Examples of these are shown in the next chapter. Gears are specifically designed for maximum power transfer with a minimum of friction.

Questions for This Chapter

1. Identify the term *mass*.

2. Explain the relationship between mass, force, and acceleration.

3. Explain the term *torque* and give an example of where you would find torque.

4. Explain the difference between potential energy and kinetic energy and give an example of each.

5. Explain the terms *radius, diameter*, and *circumference* in regard to a circle.

True or False

1. In the diagram in Figure 18–8 the amount of force to lift the weight will need to increase if the force distance is increased.

2. In Figure 18–14 the speed of the machine shaft increases when the belt is moved to the smallest pulley on the motor shaft.

3. In Figure 18–14 the force the motor provides to the machine shaft increases when the belt is moved to the smallest pulley.

4. A turnbuckle is a device with two ends that each have threads that allow the length of the turnbuckle to increase or decrease as the body is rotated.

5. When a transmission has three gears, the driven gear (pinion) will rotate in the same direction as the driver gear.

Multiple Choice

1. In the diagram in Figure 18–20, the force provided at the driven (pinion) gear will _____ .

a. be larger than the force of the drive gear.

b. be smaller than the force of the drive gear.

c. need to be measured to determine if it is larger or smaller than the drive gear.

2. In a transmission that has three gears, the driven gear and the driver gear _____ .

a. turn in the same direction.

b. turn in opposite directions.

c. need to be visually inspected because it is impossible to tell which way the driven gear will turn when the drive gear is turning in a clockwise direction.

3. When calculating mass, force, and acceleration, the amount of mass _____ .

a. increases when the amount of force increases and acceleration remains the same.

b. decreases when the amount of force increases and acceleration remains the same.

c. remains the same when the amount of force increases and acceleration remains the same.

4. When calculating horsepower, the amount of horsepower _____ .

a. increases when the amount of time decreases and the amount of work remains the same.

b. decreases when the amount of time decreases and the amount of work remains the same.

c. remains the same when the amount of time decreases and the amount of work remains the same.

5. The amount of weight a lever can lift in Figure 18–9 _____ .

a. increases when the length of the force distance increases.

b. decreases when the length of the force distance increases.

c. remains the same when the length of the force distance increases.

Problems

1. Determine the amount of force that is required to lift the weight shown in Figure 18–8 if the weight is 400 lb and the length of the force distance is 6 ft.

2. If the ratio of the pulleys where the belt is located in Figure 18–14 is 4:1, determine the speed of the driven pulley when the motor is running at 1,000 rpm.

3. Determine the amount of effort the drive shaft in Problem 2 will provide if the motor is providing an effort of 500 lb.

4. Determine the diameter and circumference of a circle that has a radius of 3 in.

5. Calculate the amount of horsepower a system produces when 99,000 ft-lb are moved in 1 s.

Chapter 19

Power-Transmission Systems

Objectives

After reading this chapter you will be able to:

1. Explain the operation of a coupling and identify several types that are used in power transmission.
2. Explain the operation of belts and identify several types that are used in power transmission.
3. Explain the operation of chains and identify several types that are used in power transmission.
4. Explain the operation of gears used in power transmission and identify several types.
5. Explain the operation of bearings used in machines and identify several types.

19.0
Overview of Power Transmission

The most important parts of the mechanical portion of a machine are the components that are used for power transmission. In many machines, an electric motor or hydraulic cylinder is used as a source of power. In the case of an electric motor, the power source is the rotating shaft. The power from the motor shaft must be transferred to other parts of the machine. The mechanical parts that transfer power on the machine are categorized as *power-transmission* components. This chapter explains the theory of operation and function of each of the components used in the transmission of power. These components include couplings, belts, chains, gears, pulleys, shafts, and bearings. You will also learn how these components fail and how to troubleshoot them. If a component is faulty, you will also learn how to remove and replace it.

19.1
Couplings

Couplings are used to transfer energy from a power source called the driver to the device being driven. In most cases an electric motor is the driver. Several different types of couplings are used in industrial applications. Figure 19–1 shows several examples of the types

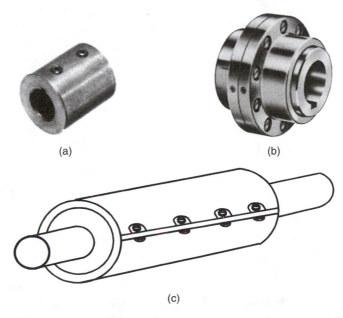

(a) (b)

(c)

Figure 19–1 (a) Example of sleeve-type couplings; (b) example of flange-type couplings; (c) example of ribbed-type couplings.

((a) and (b) courtesy of Lovejoy, Inc.)

of couplings you will find on the job. These include the flexible-type coupling and the rigid-type coupling. There are several types of flexible couplings, including the gear type, the slider type, the grid type, and the chain type. Rigid-type couplings are also called solid-type couplings, and they include the sleeve type, ribbed type, and the flange type. Several additional types of couplings are also used in industry; they include the fluid coupling and the clutch type coupling.

19.1.1
Flexible Couplings

Flexible couplings allow the connection of two shafts that are not aligned exactly. This may occur when one or both of the two shafts that are being connected move slightly as the machine functions. This may occur through normal movement of larger machines such as presses or it may occur through normal vibrations. A simple example that most everyone has seen is the drive-shaft on a large truck. If you are in traffic and a large truck pulls up beside you, you can see the exposed drive-shaft on the undercarriage of the truck. Because one end of the driveshaft is connected to the engine, which is connected to the front half of the truck, and the other end is connected to the differential gears mounted at the rear of the truck, the front and back of the truck may flex several inches as the truck moves over potholes or other bumps in the road. When the front tires and back tires are at different levels, the shafts from the transmission and the differential are not aligned exactly. If a fixed-shaft connection were used, the shaft would bind and damage the transmission or the differential. The answer to this problem is to use one or two universal joints, which provide flexible connections between the transmission and the shaft. There are many machines used in industry that also cause a substantial amount of flexing at the shaft coupling similar to the example of the truck. This section shows several examples of mechanical flexible couplings used in industrial applications today.

19.1.2
Types of Misalignment

There are four basic types of misalignment that you will encounter while working with shafts and couplings; examples of each are shown in Figure 19–2. The first type is called *angular misalignment*, and it occurs when the two shafts that are connected with the coupling do not meet at the same angle. From Fig. 19–2a you can see that one of the shafts is aligned correctly, and the second one is at a slight angle. The second type of misalignment is called *parallel misalignment*; an example

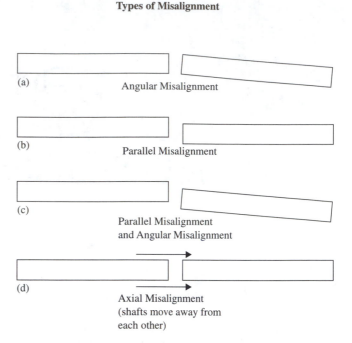

Types of Misalignment

(a) Angular Misalignment

(b) Parallel Misalignment

(c) Parallel Misalignment and Angular Misalignment

(d) Axial Misalignment (shafts move away from each other)

Figure 19–2 (a) Example of angular misalignment; (b) example of parallel misalignment; (c) example of parallel and angular misalignment; (d) example of axial misalignment.

of it is shown in Figure 19–2b. From this diagram you can see that two shafts meet at the same angle, so they are parallel. The problem with this type of misalignment is caused by the two shafts being offset to the left or right of each other by some distance. The third type of misalignment is a combination of the first two, where the shafts have both angular misalignment and parallel misalignment. An example of this type of misalignment is shown in Fig. 19–2c. The fourth type of misalignment is called *axial misalignment*, and it occurs when the two shafts move toward and away from each other, but the remainder of their alignment is correct. When this action occurs, the two shafts may pull out of a typical coupling, but a special type of coupling is designed to allow this type of motion.

19.1.3
Flexible Gear Coupling

A flexible gear coupling is shown in Fig. 19–3. This type of coupling is used in high-speed, high-torque applications. The coupling is made of two sections that are essentially identical, which allows them to be joined easily. The teeth in each gear mesh with the teeth of the other gear so that the transmission of power is smooth even at high speeds. This type of coupling also allows for considerable misalignment at higher speeds. How-

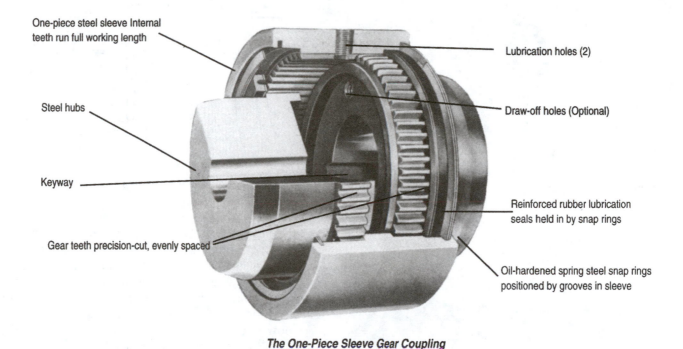

One-piece steel sleeve Internal
teeth run full working length

Lubrication holes (2)

Draw-off holes (Optional)

Steel hubs

Keyway

Reinforced rubber lubrication
seals held in by snap rings

Gear teeth precision-cut, evenly spaced

Oil-hardened spring steel snap rings
positioned by grooves in sleeve

The One-Piece Sleeve Gear Coupling

Figure 19–3   Example of flexible gear coupling.
(Courtesy of Lovejoy, Inc.)

ever, this type of coupling requires more maintenance
than other types of couplings.

19.1.4
Silent Chain Coupling

Figure 19–4 shows an example of a silent chain–type
coupling. This coupling is made of two identical sections wrapped by a chain. The chain is wrapped tightly
so that it covers both sections of the coupling. This configuration allows for maximum power and torque transfer for the most heavy-duty applications. A plastic cover
is used to cover the chain coupling so that the chain
cannot pull off as the shaft rotates at high speed. Another version of the chain coupling uses a plastic chain
made of nylon material. This type of chain coupling is
used in lighter loads where the torque and power requirements are smaller.

19.1.5
Roller Chain Coupling

Figure 19–5 shows a typical roller chain–type coupling.
This type of coupling is designed to provide high torque
at low speeds. You can see that each piece of the coupling is identical, and there is one row of teeth in each
gear. The chain is designed to cover both rows of teeth

Figure 19–4   Silent chain–type coupling.
(Courtesy of Power Transmission Distributors Association's Power
Transmission Handbook)

as it wraps around the gear teeth in the coupling. The
coupling can accommodate small amounts of misalignment and is easy to remove and replace. This type of
coupling needs lubrication and a cover so that the lubrication is not thrown off when the shaft rotates.

Figure 19–5   Roller chain–type coupling.
(Courtesy of Power Transmission Distributors Association's Power Transmission Handbook)

Figure 19–6   A grid coupling uses a metal strip of steel that fits in the slots when the two faces of the coupling are aligned. A covering keeps the metal strip in place.
(Courtesy of Lovejoy, Inc.)

19.1.6
Metallic Grid Coupling

Figure 19–6 shows a typical metallic grid coupling. This type of coupling is designed for low to medium speed with high torque transfer. It is made in two identical

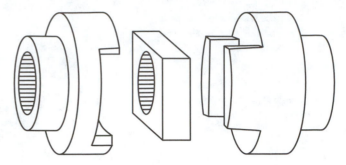

Figure 19–7   A slider-type coupling.

parts that have a series of slots that weave back and forth through the outside surface of the coupling. The slots are designed to receive a steel strip that fits tightly into them. The steel strip causes the two sections of the coupling to mate tightly and transmit power smoothly. The inside of each section of the coupling is keyed to accept power from the shaft. A cover is placed over the outside of the coupling so that when it rotates, the steel strip is not thrown out of the slot by centrifugal force. This type of coupling is rugged and can be used in applications such as a plastic extruder.

One advantage of the grid-type coupling is that it can accommodate a small amount of misalignment.

19.1.7
Slider Block Coupling

Figure 19–7 shows an example of slider block coupling. This type of coupling is also called an Oldham coupling, and you can see it has two identical pieces that are mated with a metal block. The two pieces of the coupling are connected with the block in the middle of them so that power and torque are transmitted between them. The important features of this type of coupling are its simplicity of design, its ability to transfer high-torque loads, and its ability to be changed quickly when it is worn. Another feature of this coupling is that it can accommodate shaft misalignments of up to 10°.

19.1.8
Jaw-type Coupling (Elastomeric Spider Coupling)

Figure 19–8 shows an example of the jaw-type coupling that has a rubber (elastomeric) spider element that mounts between the two pieces of the coupling. The rubber piece is used to absorb excessive amounts of vibration. This type of coupling is also used in applications where the direction of rotation is changed frequently. When the direction of rotation is changed, the rubber piece absorbs some of the torque so that it is not passed

Figure 19–8 A jaw-type coupling.
(Courtesy of Lovejoy, Inc.)

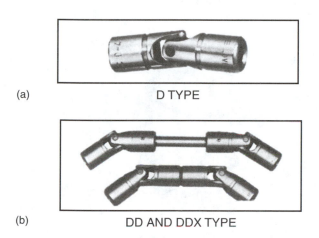

(a)   D TYPE

(b)   DD AND DDX TYPE

Figure 19–9 (a) Single universal joint; (b) double universal joints.
(Courtesy of Lovejoy, Inc.)

directly to the shaft. If the application requires more vibration absorption, a larger version of this type of coupling, called a shear-type donut coupling, can be used.

19.1.9
Universal Joints

Figure 19–9 shows an example of single and double universal joints. From the diagram you can see that this type of coupling allows for the largest amounts of misalignment. In many applications involving misalignment in more than one direction or plane, a double universal joint is used. In some applications the natural movement of the machine causes the two shafts that are being coupled to flex in such a way that more traditional couplings will wear out prematurely. The universal joint is made to be disassembled quickly by removing pins that hold the joint together. When you are troubleshooting this type of coupling, you will need to inspect the flexibility of both halves. It is normal that one

Figure 19–10 Example of sleeve-type bearing.
(Courtesy of Lovejoy, Inc.)

side of the coupling will wear more than the other due to the load of the stresses on the couplings.

19.1.10
Rigid Couplings

Rigid couplings can be used when two shafts are aligned precisely. Rigid couplings can generally transmit more power than flexible couplings. There are several types of rigid couplings available for use, including the sleeve coupling, the flange coupling, and the ribbed coupling. This section explains the advantage of each. Some couplings are sized much larger than their load rating because they tend to wear longer. When you select a coupling for an application or as a replacement, you need to determine its size and the amount of horsepower and torque it can transmit.

19.1.10.1 Sleeve Couplings
Figure 19–10 shows an example of a typical sleeve coupling. From this figure you can see that the coupling is made of a solid metal outer sleeve; the inner sleeve can be different sizes to accommodate a variety of shaft sizes. Two set screws are provided to connect the sleeve coupling to each end of the shaft. If the shafts are turning larger loads that require a larger amount of torque, keystock may be used with the set screws. The sleeve coupling can be installed or removed only when one of the shafts is loosened from its mounting screws and pulled back. This type of coupling is used for light-duty and medium-duty applications that may have up to 7.5° of misalignment.

19.1.10.2 Flange Couplings
Figure 19–11 shows an example of a flanged coupling. This type of coupling has one half mounted on each shaft. The two halves are then drawn together and bolted together so that the two shafts move as one. The coupling generally has a keyway to transfer maximum power. Some types have a bushing available to compensate for differing shaft sizes. This type of coupling allows shafts of different diameters to be mated together and can be used where couplings must be changed quickly with a minimum amount of work.

Figure 19–11 Example of a flange coupling.
(Courtesy of Lovejoy, Inc.)

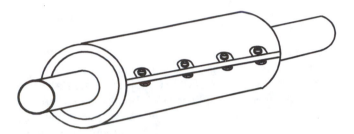

Figure 19–12 Example of a ribbed-type bearing.

19.1.10.3 Ribbed Couplings

Figure 19–12 shows an example of a typical ribbed coupling. From this figure you can see that the coupling is made of two halves that are bolted together. This feature allows you to remove and replace the coupling without removing the mounting bolts on one or both of the loads. This type of coupling is generally keyed to provide maximum torque transfer and is preferred where the coupling tends to wear out often and must be changed quickly.

19.2
Variable-Speed Controls with Belts and Pulleys

In industry it is typical to find motors whose shaft runs at a constant speed but that drive a shaft of a machine that needs to turn at variable speeds. One of the simplest ways to achieve this application is to use a motor with a belt drive. The motor and the load both have a sheave, which is similar to a pulley, and a belt is used to transfer energy. Figure 19–13 shows an example of a typical belt drive. From this photo you can see the motor is connected to the load by a belt. The speed of the load is determined by the ratio of the sizes of the sheave on the motor and the sheave on the drive.

Figure 19–13 A constant-speed motor connected to a variable-speed load by belt and sheave. The speed of the load is determined by the ratio of the sizes of the sheaves on the motor and load.
(Courtesy of The Gates Rubber Company)

Figure 19–14 A sheave with V-groove for a belt.
(Courtesy of Lovejoy, Inc.)

19.2.1
Sheaves

A *sheave* is a grooved pulley that mates with the belt. Figure 19–14 shows a typical sheave. The sheave has a bearing surface that is sized to fit the shaft of the motor or load, and it has a V-groove channel for the belt to fit into. The size of the diameter of the sheave where the belt fits into the V-groove determines the size of the sheave for calculations. The speed of the driven shaft can be calculated by knowing the diameter of the driver sheave and the driven sheave. The following formula can be used to determine the speed of the driven shaft:

$$\frac{\text{driver speed}}{\text{driven speed}} = \frac{\text{driven diameter}}{\text{driver diameter}}$$

▶ **Example 19–1**

Determine the speed of the driven shaft if the diameter of the driver sheave is 3 in. and the diameter of the

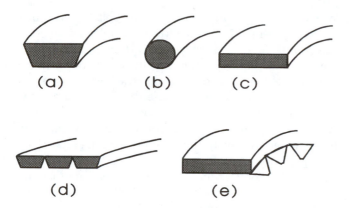

Figure 19–15 Types of belts: (a) V, (b) round, (c) flat, (d) banded or ribbed, (e) timing.

driven sheave is 6 in. The speed of the driver motor is 1,800 rpm.

Solution:

The first step is to rearrange the formula to calculate speed of the driven sheave. Use the following designations: S_1 = driver speed, S_2 = driven speed, D_1 = driver diameter, and D_2 = driven diameter.

$$\text{Driven speed} = \frac{\text{driver diameter} \times \text{driver speed}}{\text{driven diameter}}$$

$$S_2 = \frac{D_1 \times S_1}{D_2}$$

$$S_2 = \frac{3 \text{ in.} \times 1,800 \text{ rpm}}{6 \text{ in.}}$$

$$S_2 = \frac{5,400}{6} \text{ rpm}$$

$$S_2 = 900 \text{ rpm} \qquad \blacktriangleleft$$

19.2.2
Types of Belts

In modern industrial applications you will encounter a number of different types of belts. These include the round belt, the flat belt, the V-belt, the banded V-belt, the link belt, the timing belt, and the ribbed V-belt. Figure 19–15 shows examples of each of these types of belts. The most common type of belt is the V-belt, and you will find it in single-belt applications or in applications where multiple belts are used. If the load is so large that a single belt will not transfer sufficient power, multiple belts can be used to increase the power. Figure 19–16 shows an example of multiple belts.

Figure 19–16 An application where multiple belts are used.

(Courtesy of The Gates Rubber Company)

19.2.3
Adjustable-size Sheaves

In some applications, such as moving air with a belt-driven fan, the final speed of the fan must be adjustable. For this reason, the variable-size sheave is provided. The theory of operation for this application is that the smaller the drive sheave, the faster the driven shaft will turn. The variable sheave is designed so that one or both sides of the sheave are threaded so that it can move to create a larger- or smaller-diameter drive pulley. Figure 19–17 shows an example of a variable sheave. In Fig. 19–17a you can see that the movable part of the sheave is screwed in so that the sheave is at its narrowest condition. This causes the diameter of the sheave to be as large as possible. In this condition the motor speed remains constant, and because the driver sheave is large, the driven shaft will be at its fastest rate.

In Fig. 19–17b, the movable part of the sheave is unscrewed to make the groove in the sheave as wide as possible. This causes the belt to move toward the center of the sheave, which makes the diameter of the sheave as small as possible. In this case the motor speed is constant, but the driven shaft will turn at its slowest rate.

19.2.4
Timing Belts

Some industrial applications require the driver shaft and the driven shaft to be synchronized so that they are timed. The timing belt has teeth that match to the driver and driven sheaves. The teeth ensure that the two shafts retain their timing. Figure 19–18 shows an example of a timing belt and matching sheave. If you remove and replace a timing belt or try to replace a broken timing belt, it is very important that you ensure that the two sheaves are in the correct position (timed) when the belt is reinstalled.

Positive drive belts are similar to timing belts in that they have teeth in the belts and matching grooves in the sheave. The teeth and grooves allow for maximum power transfer between the belt and the sheave. The main problem with these types of belts is that the teeth begin to wear as the belt ages.

19.2.5
Maintaining and Troubleshooting Belts and Sheaves

At times you will be called to a machine to troubleshoot a problem with a belt and sheave. The problem may include the belt being broken or thrown from its sheave.

When this occurs you can replace the belt, but you must test the system before you allow it to go back into commission. If the belt is destroyed so that you can determine its size, you will need to measure the distance around the sheaves and determine the belt size that you need. You may be able to locate the specifications for the machine and determine the belt size from the data. If the data are not available you can try several belts from the measurement that you make. You can use the trial-and-error method until you have the proper-size belt for the application. After you determine that the proper-size belt is installed, you need to write the size of the belt into the maintenance file for the machine.

You may encounter a problem where the belt is loose and slipping rather than broken. If this is the case, you can loosen the tensioning mechanism or the bolts that secure the drive motor or the driven shaft and move them until the proper amount of tension is applied to the belt. Figure 19–19 shows an example of the proper method to tension a belt. In this figure you can see that a belt-tensioning tool is used to measure the proper amount of tension. Some maintenance personnel used to check the tension of belts by hand, but with modern belt drives, it is important that a tensioning tool be used. Figure 19–20 shows a method of adjusting belt tension.

Sheave set for 3 inch diameter

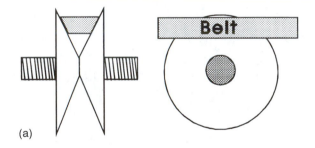

(a)

Sheave set for 2 inch diameter

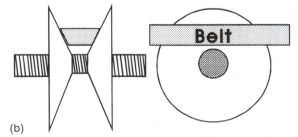

(b)

Figure 19–17 (a) The movable side of the sheave is screwed in so that the sheave is as tight as possible. This causes the belt to ride out on the sheave, making the diameter of the sheave as large as possible. (b) The movable side of the sheave is screwed out, making the V-groove as wide as possible. This causes the sheave to be as small as possible.

Figure 19–18 A timing belt with a matching timing sheave.

(Courtesy of The Gates Rubber Company)

19.2.6
Misalignment of Belts and Pulleys

Misalignment between two sheaves is the most common reason that belts fail prematurely. When the sheaves are not aligned correctly, the belt wears at a

Figure 19–19 Example of a belt-tensioning tool used to measure the proper amount of tension.
(Courtesy of The Gates Rubber Company)

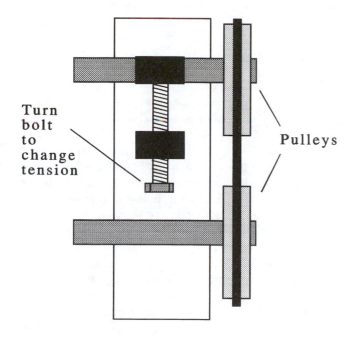

Turn
bolt
to
change
tension

Pulleys

Figure 19–20 One method of adjusting belt tension.

much faster rate and the amount of power that the belt can transfer is diminished. Misalignment results in belt noise, kinks (which increase belt fatigue), sidewall scouring, accelerated groove wear, and excessive overall wear. Figure 19–21 shows examples of how belts and sheaves can become misaligned. These include angular misalignment, where the two sheaves are not aligned in the same plane, and parallel misalignment, where the two shafts are not parallel with each other.

19.2.7
Aligning Sheaves

At times you will be troubleshooting and will find belts that are misaligned and wearing prematurely. Upon closer inspection you will realize you will need to realign the sheaves so that the belts will run true. The first step in this procedure is to use a straightedge such as a metal ruler or straight metal bar and place it against the edge of each sheave. Figure 19–22 shows an example of how a straightedge should be placed against the edge of the first sheave. The mounting bolts of the second sheave should be loosened so that it can be adjusted to fit tightly against the straight edge. After you have both sheaves aligned against the straightedge, you can tighten all mounting bolts.

19.3
Chain Drives

Some power-transfer applications in industry require more torque to be transferred from the driver shaft to the driven shaft than a belt can transfer. In these types of applications chains are used instead of belts, and sprockets that match the chain are used instead of a sheave. Figure 19–23 shows an example of a typical chain and sprocket. From this figure you can see that the links in the chain fit tightly to the teeth of the sprocket, which ensures that maximum power transfer occurs without the slippage that may occur with a belt. The size of the sprocket on the driver shaft can be compared to the size of the sprocket on the driven shaft to provide a ratio that can be used to determine the speed

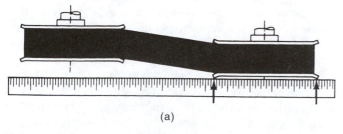

(a)

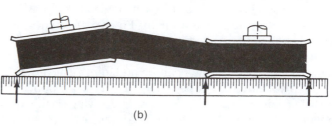

(b)

Figure 19–21 Types of misalignment.
(Courtesy of The Gates Rubber Company)

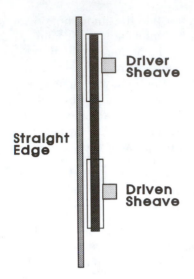

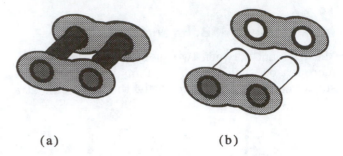

(a)   (b)

Figure 19–24 Example of (a) a typical chain link and (b) a master link.

Figure 19–22 A straightedge used to ensure proper alignment of driver and driven sheaves.

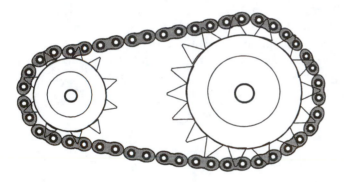

Figure 19–23 A typical chain and sprocket.

ratio similar to the way the speed ratio for belts can be determined. The chain and sprockets are made from high-quality steel if the amount of power required is high. You may also encounter some chains and sprockets that are made of plastic or high-quality nylon. In these applications the amount of power that is transferred may be smaller, but the application requires that the driver and driven sprockets maintain a timing function. The operation of chains and sprockets made of both materials is similar. The remainder of this section explains the types of chains and methods of troubleshooting and maintaining chain applications.

The chain is made of a series of single links that are joined together to make a chain. Figure 19–24 shows typical links that are used in a chain. In most chains the links are connected permanently when the chain is made, but additional links can be added or removed in the field with use of proper tools. In a typical application the chain comes in a very long piece, which is much longer than required for the job. The length of chain that is required is determined through measurements, and the chain is cut to length. The two ends of the chain are

joined in the field by a special link called a *master link*, which is used as the final link in the chain. The master link can be easily removed and replaced, allowing the chain to be installed on the sprocket with little effort. If the length of the chain is changed or if the chain breaks, extra links can be installed in the field. The other point to remember is that there are a number of types of chains whose links are specially designed for a specific application. The next section discusses the various types of chains you will encounter.

19.3.1
Types of Chains

There are several types of chains that are commonly used in industrial applications, and you will need to understand the features of each of them and why they are selected for use in a specific application. The first type of chain is called a *roller chain;* Figure 19–25a shows links of this chain. This type of chain is made of roller links and pin links. The roller chain can very efficiently transfer heavy loads at medium to high speeds (up to 50 ft/s). The chain links have the ability to roll as they move over the sprocket, which helps reduce the amount of friction between the chain and sprocket. When you are trying to purchase new chain for an application, you must identify the *pitch, chain width,* and *roller diameter* of the original chain and sprocket. The length can be determined by counting the links or making a precise measurement. The pitch of a chain is the distance between a specified point on a link to that exact point on the next link. The width of the chain is measured across the minimum distance between inside points in the link plates. The roller diameter is measured as the outside diameter of the link. Figure 19–25b shows a pintle-type chain. This type of chain is very similar to the roller-type chain. The silent-type chain is shown in Figure 19–25c. The main feature of the silent-type chain is that it is constructed so that the teeth of the chain are inverted. This means that when the chain fits over the sprocket, the sprocket comes into contact only with the links, which make the chain extremely quiet. This type of chain can

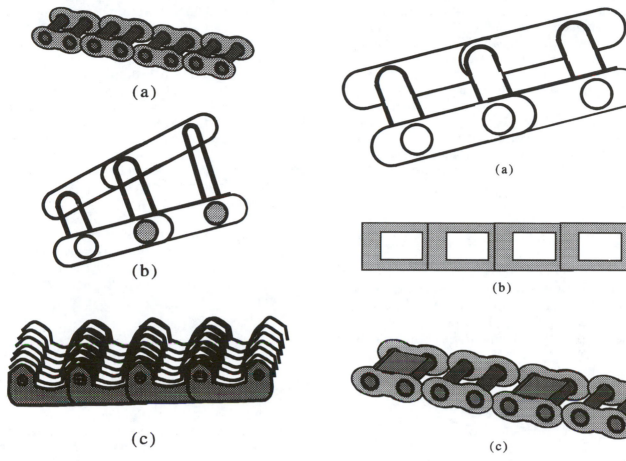

Figure 19–25 (a) Example of a roller-type chain; (b) pintle-type chain; (c) silent-type chain.

Figure 19–26 (a) Example of detachable link–type chain; (b) example of rollerless-type chain; (c) example of block-type chain.

transfer the same amount of energy as other types of chains, and it is usually found with multiple sections of chains rather than single links.

There are also several other types of chains that are used in industrial applications, which includes the detachable link, the rollerless chain, and the block chain. Examples of these chains are shown in Fig. 19–26. The detachable chain is used in applications where the chain speed is much slower. The rollerless-type chain has each link secured with a cotter pin, which can be removed to allow the chain to be disconnected and opened at any link. This is similar to having every link in the chain as a master link, which allows for changing one or more worn links anywhere in the chain with little trouble.

19.3.2
Removing and Replacing a Chain

When you are troubleshooting a chain drive, you may find that the chain needs to be removed to be repaired or replaced all together. When you must change the chain, you will need to locate a master link or a link that

can be opened so that the chain can be disconnected. After the master link is disassembled, you can remove the chain and move it to a workbench, where you can examine each link closely. You should flex the chain at each link as you inspect its operation. Each link should flex individually without binding. If you find a worn link or a link that is bent so that it cannot flex without binding, you will need to remove it from the chain. If you determine that a number of links are bad, you will need to make a decision about changing the entire chain. When you are ready to put the chain back on the sprocket, you will need to test its tension and its alignment.

19.3.3
Testing Chain Tension and Chain Alignment

When you are troubleshooting a chain drive, you may find that the chain tension is incorrect or that the alignment is incorrect. The chain can be tensioned by loosening the mounting bolt in the driver or driven part of the system and setting them to a location where the chain tension is correct. Figure 19–27 shows an example of proper chain tensioning.

Another problem you may encounter with chain is that the driver sprocket and the driven sprocket are not aligned properly. When you inspect the chain for wear, you may begin to suspect the alignment or the tension is incorrect, which is causing the chain to wear prematurely. Figure 19–28 shows how to align the two sprockets correctly. You may also find in order to provide proper alignment, you will need to take all tension off of the chain. After the chain is aligned, its tension will have to be set again.

19.4
Gears

Gears can be used individually or as part of a complex system in power-transmission systems. This section introduces types of gears, explains their operation and methods of installation, and discusses troubleshooting.

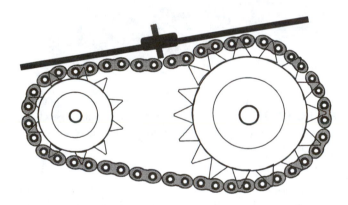

Figure 19–27 Example of methods of measuring chain tension with a table that is used to determine proper chain tension.

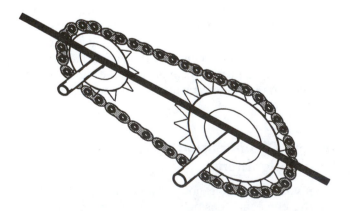

Figure 19–28 Using a straightedge to determine the alignment of driver and driven sprockets.

You will learn how these devices are a vital part of many machines; they are also subject to many problems. You will need to be able to observe gears in operation and determine what kind of problems that they have and prepare a plan of action to fix the problems. In most systems you will find a single *driver gear* and one or more *driven gears*, called *pinions*. The driver gear is the gear that is connected to the shaft of the motor or source of power. If a single driven gear is used, it is connected to the part of the system that receives the power. If the gear is part of a transmission, it is combined with other gears, and they are placed in close proximity to each other so that their teeth mesh with each other.

Gears are used to change the speed of rotation of the driven shaft from that of the drive shaft. This is accomplished through gear ratios. For example, if the drive gear has 20 teeth and the driven gear has 10 teeth, the driven shaft will turn twice as fast as the driver gear. The formula for this is

$$\text{Output speed} = \omega \times \frac{N_1}{N_2}$$

where ω is the driver shaft speed, and N_1 is the number of teeth in the driver gear, and N_2 is the number of teeth in the driven gear.

Another reason to use gears is to change the direction of rotation of the driven shaft. If the driven shaft needs to rotate in the same direction as the driver shaft, three gears need to be used, the driver, an idler, and a pinion. Figure 19–29 shows a diagram that indicates the direction of rotation of a transmission with two gears, and Fig. 19–30 shows a transmission with three gears. You should notice that if you keep adding gears to the transmission, if the total number of gears is even, the direction of rotation of the final shaft is opposite that of the driver gear, and if the total is odd, the direction of rotation for the final shaft is the same as that of the driver gear.

The speed and torque (measured in foot-pounds) transferred in a transmission of two or more gears are inversely related to each other. Thus if you select two gears

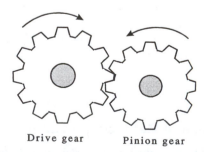

Drive gear   Pinion gear

Figure 19–29 Diagram of two gears that shows the direction of rotation of the driver and driven gear.

that double the speed of the driven shaft, then the amount of torque at the driven gear is reduced by half. Also, if the gears are selected to provide an increase in torque at the driven shaft, the speed of the driven shaft will be decreased. You will find that the gears in a transmission are selected on the basis of the final speed for the driven shaft or the final torque the driven shaft must provide.

The remainder of this section explains the theory of operation of gears, introduces types of gears, provides strategies for analyzing problems with gears and transmissions, and explains methods of removing or repairing gears. You will also see products called gear reducers that are specifically designed to reduce the speed of the drive shaft by ratios of 5:1 up to 60:1 if slower speeds are needed.

19.4.1
Theory of Operation of Gears

Gears are used in machines to cause a positive transfer of power from one shaft to another, provide a change of speed, change direction of rotation of shafts, or gain an advantage of torque. For example if you have a driver

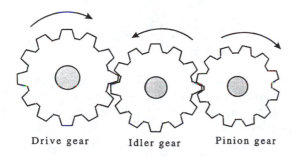

Drive gear   Idler gear   Pinion gear

Figure 19–30 Diagram of three gears that shows the direction of rotation of the driver, idler, and driven gears.

gear that has 10 teeth and a driven gear that has 20 teeth, the driver shaft will turn twice as fast as the driven shaft, and the driven shaft will have twice the force as the drive shaft. The ratio of the number of teeth in gears is equal to the ratio of the speed and the ratio of force between the gears. Another important point to remember is that if two gears (the driver and driven) are used, the rotation of the drive gear and the driven will be in opposite directions. Each time another gear is added, the direction of rotation for that gear is changed. The driver shaft and the driven shaft can be mounted parallel to each other or at some angle to each other, depending on the application. Examples of these are shown in the next section. Gears are specifically designed for maximum power transfer with a minimum of friction.

19.4.2
Categories of Gears

You can basically classify all the types of gears that you will find in industry today into three categories, each of which has a number of different types of gears. Figure 19–31 shows the three categories of gears as *parallel-axis gears*, where the drive shaft and driven shaft are mounted parallel to each other, *intersecting-axis gears*, where the drive shaft and driven shaft intersect at some angle (such as 90°), and *nonparallel and nonintersecting types of gears*. The figure shows examples of each of these types of gears so that you can recognize them when you are on the job.

19.4.3
Types of Gears

There are four common types of gears that you will find in industrial applications and machines. They include spur gears, helical gears, miter and bevel gears, and

Figure 19–31 Categories of gear types.

Categories of Gears	Types of Gears	Efficiency Percent
Parallel-axis gears	Spur gear	98–99.5%
	Rack	
	Internal gear	
	Helical gear	
	Helical rack	
	Double helical gear	
Intersecting-axis gears	Bevel gear	98–99%
	Spiral bevel	
	Spur bevel	
Nonparallel, nonintersecting gears	Worm gear	30–90%
	Screw gear	70–95%

worm gears. The teeth on the spur gear are cut straight and are parallel with the shaft. The number of teeth in the drive gear and the number of teeth in the driven gear are selected to create a gear ratio that will cause a specific speed or force. Another type of spur gear is used as a pinion with a rack. In this type of application the spur gear is rotated by the driver motor, and the rack moves in straight-line motion. This allows rotary motion to be converted to linear motion for use in applications such as motion control or robotic applications. The number of teeth in this type of application determines the speed or the force transferred to the rack.

The helical gear allows the driver and the driven shafts to meet parallel, like the spur gear, or at 90°. The teeth of the helical gear are cut at an angle. The number of teeth in the driver and driven gears again determine the relationship of speed and force. Helical gears provide improved tooth strength and provide increased contact ratio. Helical gears also can provide greater load-carrying capacity than spur gears.

The miter and bevel gears are the third type of gears used in industrial applications. The teeth on the gears are cut at an angle and they are beveled. This type of gear allows the teeth to mesh and transfer power with less wear on the teeth, and these gears are quieter than other types of gears. The bevel gear can run at much higher speeds than other gears with less wear.

The fourth type of gears is called a worm gear. This type of gear intersects at 90° and provides a high ratio of speed reduction in a small space. This type of gear should be the quietest type of gear.

19.4.3.1 Spur Gears

Spur gears are defined as gears that transmit power between two shafts that are parallel with each other. Figure 19–32 shows an example of two spur gears. You can see that the teeth for these gears, which are cut at 90° across the gear, must match with each other exactly so the teeth of each will mesh correctly with the other. The spur gear is a low-cost gear that is used for general-purpose loads at medium speeds (up to 4,000 rpm). If you increase the speed too high, this type of gear will become very noisy. Because it is low cost and requires little maintenance, the spur gear is the most widely used type of gear.

In some power-transmission applications a third gear, called an *idler gear*, is added to the transmission to cause the shaft rotation of the driven gear to be the same as the driver gear. From these diagrams you can see that if two gears are used, the rotation of the shafts is reversed. If a third gear is used, the first and last gear turn the same direction, and the idler gear in the middle of the set rotates the opposite direction.

Spur Gears with a Rack. Another type of spur gear that you will encounter in industrial applications is

Figure 19–32   Two spur gears. Notice the teeth are cut at exactly 90° across the face of the gear.
(Courtesy of Boston Gear)

Figure 19–33   A spur gear with a rack. When the gear turns (rotary motion), the rack moves linearly in a straight-line motion. The spur gear and rack convert rotary motion to linear motion.
(Courtesy of Quality Transmission Corporation)

called a *gear and rack*. The rack is used to convert rotary motion into linear rotation. When the shaft with the driver gear rotates, the teeth of the spur gear mesh with the teeth of the rack, which causes the rack to move in a linear motion. When the rotation of the drive shaft is reversed, the motion of the rack is also reduced. Figure 19–33 shows an example of a spur gear and rack.

19.4.3.2 Helical Gears

Helical gears are also used in applications where the drive shaft and the driven shaft are mounted parallel to each other. Figure 19–34 shows an example of helical gears. The main difference is that helical gears are cut on an angle across the gear face. The angle at which the teeth are cut on the gear is called the *gear pitch*. Helical gears are available with pitch (angle) from a few degrees to 45°, with the average being about 20°. Because the teeth are cut on an angle, they have the ability to transfer more load than the spur gears. The helical gears also can run much more quietly than spur gears.

Figure 19–34 Helical gear. Notice the teeth are cut at an angle across the face of the gear.
(Courtesy of Boston Gear)

The other important feature of helical gears is that they create a thrust load, which is a force along the length of the shaft. Special bearings are used with helical gears to counteract the thrust load. Another way to counteract the thrust load is to use double helical gears. Figure 19–35 shows an example of double helical gears. From this figure you can see that one set of helical gears is cut at one angle across the gear face, and the other set is cut in the opposite direction. The result is that the thrust created by one set of teeth is counteracted by the thrust in the opposite direction caused by the other set of teeth.

The helical gear can also be used with a rack to provide a means of transferring rotary motion to linear motion. The helical gears can provide more power transfer than typical spur gears with a rack.

19.4.3.3 Bevel Gears and Miter Gears

Figure 19–36 shows examples of *bevel-type* gears. These gears are used when the drive shaft and the driven shaft intersect at approximately 90°. The bevel gear provides the most efficient means of transmitting power of all the gear designs, but it may also be necessary to specify the bevel gear where space is limited and the shafts must be mounted at 90°. The intersection of the shafts in a bevel gear does not have to be exactly 90°. You may also find some applications where the shafts intersect at angles less than 90° and some where the shafts intersect at angles more than 90°. If the two gears in the bevel gear are of equal size and mounted at exactly 90°, the gear can be referred to as a *miter gear*. Thus you may encounter a transmission that some people call a miter gear and others call a bevel gear. If the

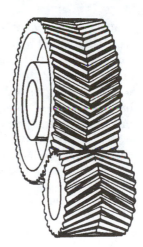

Figure 19–35 Example of double helical gears.
(Courtesy of Quality Transmission Corporation)

gears are of equal size and mounted at exactly 90°, they can be called either a bevel or miter gear.

19.4.3.4 Worm Gears

Figure 19–37 shows an example of a *worm gear*. The worm gear consists of cylindrical worm that meshes with a wheel gear. In this type of application the worm is generally the driver and the wheel gear is the driven gear. One application of the worm gear is called a *speed reducer*. Figure 19–38 shows an example of this type of speed reducer; its major features are the large amount of speed reduction (100:1) and the small package. The screw part of the transmission provides a larger mechanical advantage than other types of gears, but this

Figure 19–36 Example of bevel gears. Notice that the shafts of these two gears intersect at 90°.
(Courtesy of Boston Gear)

Figure 19–37 Example of worm gears.
(Courtesy of Boston Gear)

also creates the loss of efficiency. The other reason a worm gear is used is because it has more surface area that stays in contact between the worm and rotating gear, which makes it run quietly and allows it to absorb higher shock loads.

The efficiency of the worm gear is typically very low, which means you will lose some of the horsepower of the driver. For example if you need 1 hp to lift a load and you need to use a worm gear, the driver motor may need to be larger than 2 hp. Other problems with the worm gear are that it produces some axial loading forces, which require more expensive bearings to offset them, and that it has a tendency to produce large amounts of heat that must be carried away through additional lubrication.

19.4. 4
Gear Meshing and Backlash

One of the major problems you will encounter as a maintenance technician is a set of gears that will not mesh correctly or that have excessive backlash. Figure 19–39 shows a typical set of gears as they mesh. From this figure you can see how the gear teeth fit together. The depth of the gear must fit the mating gears exactly, with a small amount of clearance between the tip of the gear and the other gear. If there is not sufficient clearance, the gear teeth will wear down prematurely.

Figure 19–40 shows an example of how backlash is measured. From this figure you can see that a small amount of backlash is permissible, but as gears wear, the amount of backlash will increase and it will reach a point where it is unacceptable. As the amount of backlash increases, the gears will also begin to vibrate excessively when the gears change speed or change direction of rotation. If the gears are connected to a

motor that is controlled by servo system, the computer on the servo system will try to compensate for the excessive backlash and actually make the gears vibrate more excessively.

In some cases you can add shims or thrust washers to help reduce the amount of backlash. These devices help reduce the amount of distance of the backlash, but generally they are only temporary, and the gears will need to be replaced as soon as new gears can be located.

19.4.5
Troubleshooting Gears

When you are working with gears and transmission you will find problems with worn gears and noisy gears. When gears begin to wear, their teeth will not mesh correctly and they will not be able to transfer power smoothly. When this occurs, the shaft of the driver gear or the driven gear will begin to vibrate. This same type of problem can occur if the gears become dry or if lubrication is low. In some cases, you can add lubrication to the system and get the gears to return to their normal condition. If the gears continue to vibrate, they will need to be inspected and replaced. Excessive vibration can also be caused by misalignment or excessive end play.

If either of the two gears in the set have a worn bearing, broken teeth, or badly misaligned shafts, you will find that the gears tend to bind rather than vibrate. Your first indication that the gears are binding when you are called to equipment that has a gear drive is when you find the motor overloads are tripped. When you reset the overloads, you will find the motor may run for a few more hours or less and then the overloads will trip again. The reason the overloads are tripping is because gears are hitting. You will be able to detect the binding by observing the drive motor and the machine that is receiving the power and you will notice that the transfer of power is not smooth. If you try to turn the shaft by hand you will also find places where it turns smoothly,

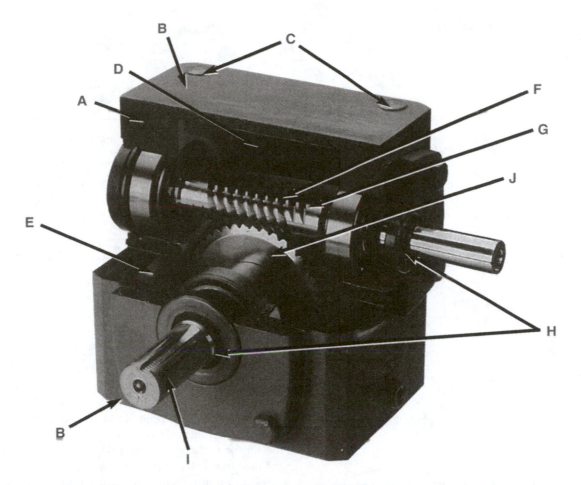

A. Rugged housing of fine-grained, gear-quality cast iron provides maximum strength and durability. Greater rigidity and one-piece construction ensure precise alignment of the worm and gear. This housing construction also provides superior resistance to caustic washdown solutions, plus high heat dissipation and reduced noise level. Pipe plugs allow easy fill, level and drain in any mounting position.

B. Housings are straddle-milled top and bottom for precise alignment of horizontal and vertical bases.

C. Multi-position mounting flexibility — threaded bolt holes let you install 700 Series speed reducers in almost any position.

D. Internal baffle assures positive leak-free venting.

E. Large oil reservoir provides highly efficient heat dissipation and lubrication for longer operating life.

F. High pressure angle on worm provides greater operating efficiency.

G. Integral input worm and shaft design made from high-strength case-hardened alloy steel. Reducer sizes 710 through 726 have pre-lubricant ball bearings; 732 through 760 have tapered roller bearings. Double lip oil seals are standard.

H. Super-finished oil seal diameters on both input and output shafts provide extended seal life.

I. High strength steel output shaft assures capacity for high torque and overhung loads.

J. High-strength bronze worm gear is straddle mounted between heavy-duty tapered roller bearings to increase thrust and overhung load capacities, sizes 713-760.

Figure 19–38 A typical worm gear.
(Courtesy of Boston Gear)

Figure 19–39 An example of how typical gears should mesh.

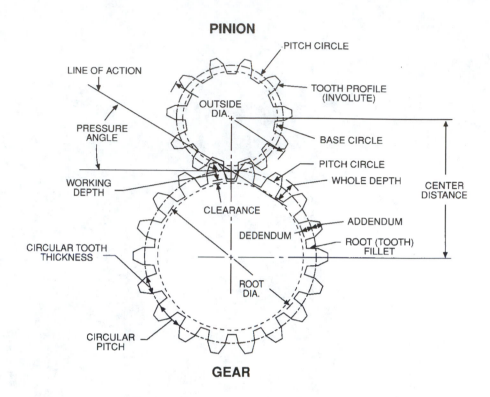

PINION

GEAR

Backlash

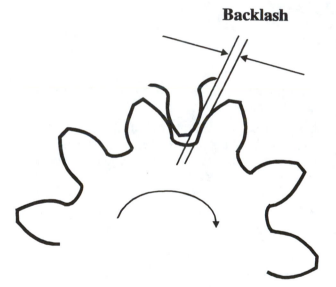

Figure 19–40 An example of how backlash is measured.

and you will find places where it binds excessively. Over time the binding will increase to the point that the shaft will not turn and the motor will stall at all times; then it will blow its fuses as quickly as you replace them. You may also locate the motor and gearbox and find one or both very hot, and you will find the fuses for the motor are blown. You may also notice that the motor's shaft will not rotate when you try to start the motor again. At this point the motor will be overheated and you may begin to think the problem is with the motor. You may also

find the motor is completely burned out at this point if it has the wrong size fuses and overloads. If you replace the motor without checking the gearbox, you will find the problem will return rather quickly.

One point to remember at this time. The first indication of problems with this type of condition is that the machine that the motor and gear are operating will stop functioning. When this occurs, usually the electrical maintenance personnel are called first. They will probably notice that the motor is overheating and the overloads are tripped. The first inclination at this point is to reset the overloads and walk away from the system. This may temporarily repair the problem, but if the real cause is worn or binding gears, the overheating of the motor will return. The second time the overloads must be reset, the inclination of the maintenance personnel is to increase the size of the overloads. If fuses are tripped, they may also want to increase the size of the fuses to allow it to run longer before the fuses blow again. The problem with increasing the overload or fuse size is that it will allow the motor to draw more current than it should; eventually the motor windings will be permanently damaged and the motor will need to be replaced.

19.5
Bearings

Bearings are required in industrial applications to minimize friction between moving parts and parts that ride

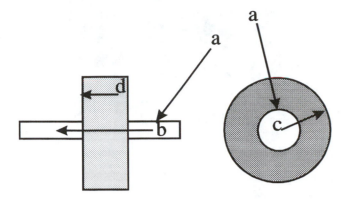

Figure 19–41   Journal (a) is the part of the bearing or shaft that supports the bearing. The axis (b) is an imaginary line that runs through the shaft. Radial force (c) is a force that emanates from the center of the bearing outward, like a spoke in a wheel. Thrust (d) is a force that acts against the side of the bearing.

Figure 19–42   Examples of a plain bearing, also called a bushing.
(Courtesy of Boston Gear)

on each other. The motion can be linear or rotary. When you are working in maintenance applications, you will encounter bearings; you must be able to identify each type of bearing as well as remove and replace bearings. This section introduces the various types of bearings you will find in industrial applications, explains methods of evaluating bearings to ensure they are operating correctly, and shows the proper way to remove and replace a bearing.

19.5.1
Important Terms Used with Bearings

There are several important terms that apply to bearings that need to be introduced so you will be able to understand some of the features of bearings that are designed to counteract the forces involved. Figure 19–41 shows a diagram of a bearing and shaft and defines several terms. The *journal* is shown at a. It is the part of the bearing or shaft where the bearing rides. In terms of the bearing, the journal is the inside part of the bearing that mates with the shaft. In terms of the shaft, the journal is the part of the shaft where the bearing mates with it. In some cases the journal on the shaft is polished to ensure the bearing fits tightly with the shaft. The *axis*, which is an imaginary line through the shaft, is shown at b. *Axial forces* refer to the direction forces are applied to a shaft. An axial force is a force that moves along the length of the shaft. At c is an illustration of *radial force*. A radial force is a force that originates at the center of a bearing or shaft and moves outward, like the spoke of a wheel, to the outside of the bearing. *Thrust*, which is a force that is applied against the side of the bearing, is shown at d. The arrow is pointing away from the bear-

ing in this example, but the force could also be an external force that is applied inward against the arrow to the side of the bearing.

19.5.2
Plain-type Bearings (Bushings)

As you work in industrial applications, you will find bearings that are designed for sliding applications and bearings that are designed for rolling applications. Bearings designed for sliding loads can be classified by the load they are designed for, such as axial (side) loads and loads that emanate out from the center of the bearing (radial loads). Bearings that are designed for rolling loads can be classified by the type of bearing device that is used in the bearing, such as ball bearings, roller bearings, and needle bearings.

The first type of bearing designed for sliding loads is called a *plain bearing*. Figure 19–42 shows an example of this type of bearing. From the figure you can see the bearing is a solid piece of metal that supports the shaft. This type of bearing is also called a *bushing*. The plain bearing is a rather simple design in that it has no moving parts. It uses a thin film of lubrication that exists between the bearing surface and the shaft surface. Because the lubricant is used to limit the amount of friction between the two surfaces, it is of the utmost importance that the proper amount of lubrication always be present. All plain bearings have some means of ensuring that lubrication is properly distributed across all bearing surfaces on a continual basis. Figure 19–43 shows an example of how the bearing has grooves to allow the lubrication to reach all areas of the bearing and shaft surface. The only way these types of bearings fail is when the lubrication is not present.

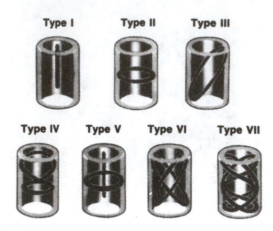

Figure 19–45   A split-type pillow block.
(Courtesy of Boston Gear)

Figure 19–43   Examples of grooves in a plain bearing, which ensure that lubrication is evenly distributed between the bearing and shaft.
(Courtesy of Boston Gear)

Figure 19–46   A flange-type block.
(Courtesy of Boston Gear)

Figure 19–44   A pillow block that supports a bushing.
(Courtesy of Boston Gear)

Figure 19–47   Examples of flange-type plain bearings.
(Courtesy of Boston Gear)

The plain bearing (bushing) is generally mounted in a pillow block. A typical pillow block is shown in Fig. 19–44. From this figure you can see the pillow block is designed to hold the bushing in place against all types of axial and radial loads. The pillow block has two mounting holes (or slots) that allow it to be mounted or secured. The mounting holes may be slotted to allow for proper tensioning of belts or chains that may be mounted to the shaft on which the bearing is mounted.

A pillow block can be designed as a solid unit or as a split unit. A split unit is shown in Fig. 19–45. From this example you can see a pillow block is made of a top section and a bottom section, which are connected together by bolts. This type of pillow block is used in applications where the pillow block is taken apart frequently to remove and replace the bearing. When you need to remove and replace the bearing, you can remove the mounting bolts from the pillow block and remove the top of the pillow block. Once the top half of the pillow block is removed, the shaft with the bearing

mounted on it is lifted out and a bearing puller is used to remove the bearing. When you have the new bearing installed on the shaft, you replace the shaft and bearing into the bottom half of the pillow block. After the bearing is properly located in the pillow block, the top half of the pillow block is replaced and the mounting bolts are reinstalled. If a sheave or sprocket is mounted on the shaft, the proper tension is set and the mounting bolts are tightened. Another type of mounting base for bearings is called a *flange-type mounting base*. This type of base is shown in Fig. 19–46.

Another type of plain bearing is called a *flange bearing*. A flange bearing provides support for both axial forces (forces from the side) and radial forces. The flange bearing looks similar to the bushing-type bearing, except it has a flange. Figure 19–47 shows a flange-

Figure 19–48   A picture of a typical ball bearing.
(Courtesy of Torrington)

type bearing. Another way to classify all bearings is to identify the bearings that are made of metal and those that are nonmetal, or plastic. You will find a large number of industrial bearings are made of metal, but an increasing number of plastic, nylon, and teflon bushings are also used in some applications.

19.5.3
Rolling Element–type Bearings

There are three basic types of *rolling element–type bearings*, the *ball bearing*, the *roller bearing*, and the *needle bearing*. Each of these types of bearings has an inner ring and outer ring, called a *race*. The inner race is designed to support the shaft on its inside and the bearing components, such as balls or rollers, on its outer edge. The outside race contains the balls or rollers against the inner race. The bearing race is generally made from hardened steel.

19.5.3.1   Ball Bearings
Figure 19–48 shows a picture of a typical *ball bearing*, and Fig. 19–49 shows a diagram of a typical ball bearing. From these figures you can see that the ball bearing consists of an inner race, an outer race, ball cage, and balls. The inside dimension of the race is called the *bore*. The bore of the bearing must match the outside diameter of the shaft. The balls are kept in place in the bearing by the *bearing cage*. The cage is designed to allow each ball to make contact with the inner race and the outer race. It also keeps the balls from coming loose and falling out of the bearing. If the bearing cage is damaged, it will allow one or more of the balls to come loose and fall out of the bearing. The outside race is used to keep the balls in contact with the inner race; its size determines the outside diameter of the bearing.

The side of the bearing is called the *face*. At times you will use a tool to apply force from a mallet to the face of the bearing to properly seat the bearing. You

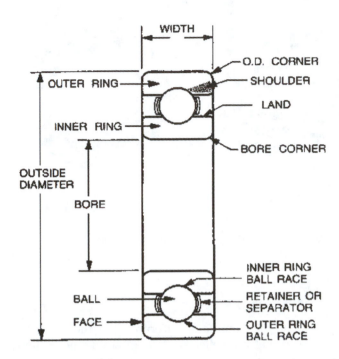

BALL BEARING

Figure 19–49   A diagram of a ball bearing that shows the inner race, outer race, ball cage, and balls.
(Courtesy of Torrington)

Figure 19–50   Double-row-type ball bearing.
(Courtesy of Torrington)

must be sure not to damage the bearing face or allow it to be mashed into the balls in the bearing cage when the bearing is installed on the shaft.

19.5.4
Double-Row Ball Bearings

A heavy-duty-type ball bearing is called a *double-row ball bearing*. This type of bearing uses a double row of balls that roll in two deep grooves. Figure 19–50 shows an example of a double-row ball bearing.

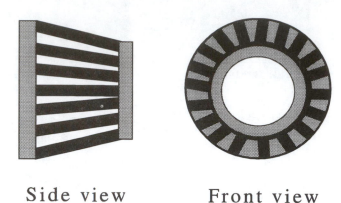

Side view   Front view

Figure 19–51 Tapered-type roller bearing.

Figure 19–52 Double-row tapered-type roller bearing.
(Courtesy of Torrington)

19.5.5
Roller Bearings

The roller-type bearing is different than the ball-type bearing in that the roller is made of solid-steel rollers that look like small steel tubes. The main feature of roller bearings is that they provide a larger rolling surface, which can support stronger forces. The roller bearing can also bear loads that have higher-impact loads or shock loads, which occur when shafts are started and stopped frequently. The roller bearing is made of an inner race, an outer race, and the rollers. A retainer ring ensures that the rollers stay on track inside the races. One problem with roller bearings is that they have problems with consistent loading and tracking of the rollers. Ball bearings, on the other hand, tend to track more consistently because their size and design are uniform. If the rollers are mounted in the race at a slight angle, they can support both axial and radial forces. When the rollers are mounted in the bearing at a slight angle, they are called *tapered roller bearings*. Figure 19–51 shows an example of a tapered roller bearing. Another type of roller bearing is the *double-row-type roller bearing*. This type of roller bearing allows the two separate rows of rollers to constantly align with load. This causes less wear on the bearing and allows it to support larger loads. Figure 19–52 shows an example of a double-row tapered-type roller bearing.

19.5.6
Needle Bearings

The third type of rolling element–type bearings is called the *needle bearing*. An example of a needle bearing is shown in Fig. 19–53. From this figure, you can see that the needle bearing has an outside race, but it does not have an inside race. This arrangement allows for the needles to be in contact with the shaft that it is supporting. In this type of bearing the rollers are called *needles* because they have a much smaller diameter than a

Figure 19–53 A heavy-duty needle roller bearing with a machined outer ring.
(Courtesy of Torrington)

typical roller. The needle bearing has the ability to carry much larger radial loads at much higher speeds than other types of rolling-element bearings. It is important that the needle bearing have adequate lubrication, which must be checked frequently.

19.5.7
Thrust Bearings (Washers)

Another type of bearing is called a *thrust bearing*. An example of this type of bearing is shown in Fig. 19–54. From this figure you can see that the thrust bearing is actually a special washer that can bear tremendous loads in the axial direction; it also limits side-to-side movement in bearings. The thrust bearing is generally used in conjunction with plain bearings, or bushings. The thrust bearing may be mounted on a shoulder that is cut into a shaft. In some applications, two or more thrust bearings are used to take up the required space between the shoulder of the shaft and the bushing-type bearing. If one of the thrust bearings (washers) begins

Figure 19–54 Examples of thrust bearings, which are also called thrust washers.
(Courtesy of Boston Gear)

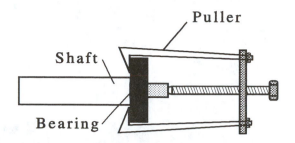

Figure 19–56 Example of a bearing puller mounted in place to remove a bearing.

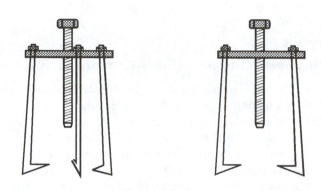

Figure 19–55 Types of bearing pullers.

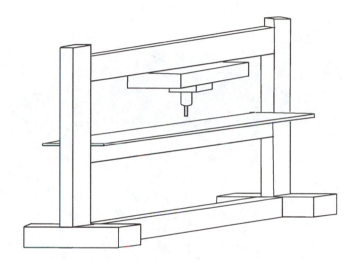

Figure 19–57 Example of a hydraulic-type press.

to wear, additional thrust washers can be added. If you disassemble a shaft with thrust washers, it is important to ensure that they are cleaned and well lubricated when they are replaced.

19.5.7.1 Maintaining Bearings

When you are working on a factory floor, you will be called to check bearings that are noisy or worn. If you need to remove and replace a bearing, you will need to use a bearing or gear puller. Figure 19–55 shows an example of several types of typical gear pullers. You will notice that each type of puller has a number of legs that are connected behind a bearing and a screw shaft in the middle. Figure 19–56 shows an example of a bearing puller mounted on a bearing ready to begin the removal process. Once the legs are in place, the screw shaft is turned clockwise until the bearing begins to move off the shaft. Different-size legs are available to allow the puller to be used on larger and shorter shafts.

19.5.8 Bearing Presses

In some applications you will need a press to remove a bearing from a race. The bearing press may need hy-

draulic pressure to provide enough force to remove the bearing. A typical hydraulic-type bearing press is shown in Fig. 19–57. If the bearing is smaller, an arbor press may be used to remove the bearing. An example of an arbor press is shown in Fig. 19–58. The press can also be used to reinstall the bearing in the race.

19.6
Clutches and Brakes

Many machines need a means of disengaging the driven shaft from that of the drive shaft of the motor that is providing the power for the machine. One simple way of doing this is with a clutch. A clutch provides a means of interrupting the power transmission from the shaft of the electric motor to the driven shaft of the machine. This allows the machine to be idled or disengaged while the electric motor continues to run at full speed. The clutch is especially useful when a machine must be idled frequently and it is impractical to start and stop the motor continually.

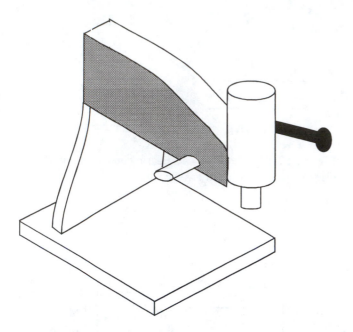

Figure 19–58 Example of an arbor-type press.

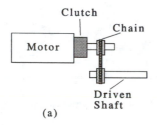

(a)

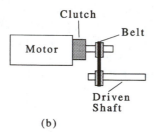

(b)

Figure 19–59 (a) Foot-mounted electromodule clutch used with a chain drive; (b) a belt-driven clutch.

Figure 19–59 shows examples of several types of clutches. The clutch is made of two parts, which can turn independent of each other. When the clutch is engaged, the two parts turn as a single unit. When the clutch is disengaged, the part of the clutch that is connected to the motor shaft can continue to spin, whereas the part of the clutch that is connected to the driven shaft on the machine can be stopped. From this diagram

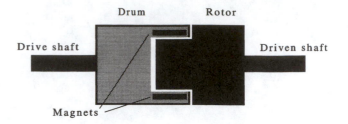

Figure 19–60 Example of an eddy current drive.

you can see that the clutch can be mounted directly to the motor shaft, or it can be mounted on an idler shaft between the motor shaft and the driven shaft. A clutch can also work with either chain or belt drives.

19.6.1
Magnetic Clutches (Eddy Current Drives)

The clutch can be designed to allow the driver shaft and the driven shaft to be coupled so that they can turn at different speeds. This type of clutch is called an *eddy current drive*. Figure 19–60 shows an example of an eddy current drive. The drive member of the eddy current drive is a steel drum driven by a constant-speed AC motor, and the driven member of the eddy current drive is a steel rotor that rotates freely inside the drum. Electromagnets allow a variable magnetic field between the drum and rotor. The speed of the rotor depends on the strength of the magnetic field. The clutch consists of a drum mounted on the motor shaft, a rotor mounted on the driven shaft, and a strong electromagnetic coil. The rotor fits inside the drum with only a small air gap separating them. A magnetic field is produced by the field coil and causes the drum to attract the rotor. When the magnetic field is at its strongest, the rotor and drum rotate at the same speed. When the magnetic field is weakened, the drum and rotor are allowed to rotate at different speeds. Because the drum is mounted directly on the electric motor shaft, it will continue to spin at the full rpm of the motor. The rotor, which is mounted on the driven shaft, will spin much more slowly. The magnetic field can be adjusted to fine-tune the speed of the driven shaft.

19.6.2
Brakes for Machines

Some machines need a means of stopping the shaft of a motor instantly when power is disconnected. Figure 19–61 shows a typical friction-type brake, which consists of a set of brake shoes and a brake drum attached to the machine shaft that is being controlled. The brake

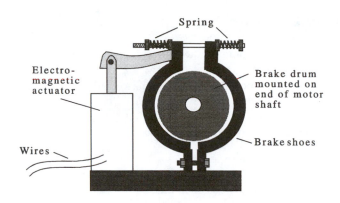

Figure 19–61 Example of a friction-type brake that consists of brake shoes and a brake drum connected to the shaft that is being controlled.

Figure 19–62 Example of friction-type brakes that use ceramic magnets and friction discs.
(Courtesy of Warner Electric, South Beloit, Ill.)

is shown with the magnet energized to keep the brakes off so that the motor can turn freely. When power is turned off, the springs will press the shoes tightly against the brake drum. The friction-type brake is similar to the brakes on the rear wheels of most automobiles. This type of brake uses heavy springs to make the brake shoes come into contact with the brake drum. When the brake shoes come into contact with the brake drum, friction builds up and stops the drum from turning.

When the brakes are not needed, an electric solenoid is energized, and its strong magnetic field pulls the shoes away from the drum. When power to the solenoid coil is deenergized, the strong springs pull the brake shoes tightly against the drum, and friction causes the drum to stop turning. Because the spring pressure makes the brake shoes press against the brake drum when electricity is not present, this type of brake is called a *fail-safe brake*.

Another version of the friction brake is shown in Fig. 19–62. This type of brake uses friction disks and ceramic magnets to cause the brake pads to operate. The ceramic magnet is a much stronger type of magnet, which allows this type of brake to last longer.

19.6.3
Clutch and Brake Modules

Some machine designs do not allow enough room for both a clutch and brakes to be mounted on a shaft. For this reason, a clutch and brake module is available that integrates both a clutch and a brake into the same unit. The clutch and brake module provides the function of both the clutch and brake, as discussed earlier in this section. Figure 19–63 shows examples of several clutch and brake modules.

Figure 19–63 Examples of a clutch and brake module.
(Courtesy of Warner Electric, South Beloit, Ill.)

Questions for This Chapter

1. Explain the difference between a flexible-type coupling and fixed-type coupling.

2. Explain the different types of coupling misalignment.

3. What is an arbor press used for?

4. Explain why couplings must be able to be changed quickly.

5. In a belt and sheave application, which sheave should be larger if the driven shaft is to turn faster than the drive shaft?

True or False

1. A rack-and-spur gear allows rotary motion of a motor shaft to be converted to linear motion in the rack.

2. A set of gears operates better when the backlash becomes larger.

3. A needle bearing is a rolling element–type bearing.

4. An eddy current drive (clutch) ensures the drive shaft will turn at the same speed as the driven shaft at all times.

5. A bearing puller is the correct tool for removing a bearing from a shaft.

Multiple Choice

1. A timing belt is _____ .
   a. a special type of V-belt that is reinforced with cable.
   b. a special type of belt that has teeth that mate with a pulley to ensure power transmission.
   c. a belt that is specially designed to be used with adjustable-size sheaves.

2. In the elastomeric spider coupling (jaw-type coupling), the rubber _____ .

   a. absorbs some of the vibration that develops in the coupling.
   b. makes a nonslip connection between the two pieces of the coupling.
   c. allows the two pieces to slip slightly so that the two shafts can run at differing speeds.

3. A clutch provides _____ .
   a. a means of connecting two shafts so that they can run at the same speed at all times.
   b. a means of allowing the drive shaft and driven shaft to run at different speeds.
   c. a means of completely disconnecting the drive shaft from the driven shaft.

4. A universal joint-type coupling _____ .
   a. is specifically designed for shafts that may have misalignment.
   b. is specifically designed for shafts that have misalignment.
   c. is designed to allow the drive shaft to completely disconnect from the driven shaft.

5. In a pulley application, an adjustable-size sheave _____ .
   a. allows the speed of the driven shaft to be faster than that of the drive shaft by making the size of the adjustable sheave that is mounted on the driven shaft slightly smaller.
   b. allows the speed of the driven shaft to be faster than that of the drive shaft by making the size of the adjustable sheave that is mounted on the driven shaft slightly larger.
   c. ensures the speeds of the driven shaft and drive shaft are always the same.

Problems

1. The sheave on a drive shaft is 4 in. in diameter, and the sheave on the driven shaft is 2 in. in diameter. If the motor shaft is turning at 1800 rpm, how fast will the driven shaft turn?

2. If the gear on a drive shaft has 30 teeth and the gear on the driven shaft has 10 teeth, how fast will the driven shaft turn if the drive shaft is turning at 100 rpm?

3. Select two gears and identify the number of teeth each will need to provide a 4:1 ratio of power.

4. Identify an application where you would select a bevel gear for a project.

5. Identify an application that would use a worm gear.

◀ **Chapter 20** ▶

Troubleshooting

Objectives

After reading this chapter you will be able to:

1. Develop a plan to methodically troubleshoot a machine and find faults.
2. Develop a plan to work safely when you are troubleshooting.
3. Find problems when power is lost on a machine.
4. Find problems in the control circuit of a machine.
5. Find problems in the load circuit of a machine.

20.0
Overview of Troubleshooting

When you are working on a factory floor, you will find that all the machines in the factory have a common goal: to produce as much product as possible for every hour that they are running. In most cases, the delivery of the product has been negotiated to occur within 24 h of when the product is manufactured. This means that if any machine in the factory breaks down, it is holding up this process, and some product orders are not going to be on time.

After you understand this concept, you will understand why it is so important to be able to troubleshoot a broken machine and quickly determine the problem and correct it. You will also learn the importance of finding the problem the first time you test the machine, rather than ending up changing several parts.

Troubleshooting can be described as designing tests whose answers will identify the problem or al-

low you to eliminate portions of the machine as potential sources of the problem. You will find that many times a troubleshooting test does not actually identify the exact problem; instead, it eliminates what cannot be the source of the problem. For example if you are troubleshooting an automobile that has one flat tire, you may find you can reach the conclusion by identifying which three tires are good, or you can get there by identifying which of the four is bad. What you will see is that it does not matter which approach you use; the result is still the same. This chapter explains how to approach troubleshooting and identify which parts of the machine to test. The tests explained in this chapter are based on the premise that you fully understand the theory of operation of all the parts of the machine. You may need to refer to information in previous chapters to review the theory of operation of some parts from time to time while you are making a troubleshooting test. It is very important that you be able to identify when a part is operating correctly and when it is not. You will also learn to use all your senses—such as sight, sound, touch, and smell—to locate problems.

20.1
When You Are Called to a Machine That Is Broken

When you are first called to a machine that is malfunctioning, you will need to follow a strategy that will ensure that you completely assess what parts of the machine are functioning and what parts are not. In some cases, all power will be lost and the machine will not be running. In other cases, the machine will be partially running, but it will not be completing its cycle correctly. In most shops, you will receive a work order that identifies the machine, the type of fault, and the location of the machine if the plant is a large facility.

One of the most important things that you need to do when you first arrive at a machine is to try to find the operator that called the problem in. The most important thing that the best troubleshooting technicians have learned over the years is to develop a longstanding relationship with each of the machine operators. This is important because no one knows a machine better than the operator. When the machine is malfunctioning, the operator can tell you how the machine is supposed to operate as well as tell you about the past history of the machine. If it is possible to locate the operator, begin the troubleshooting process with a short interview to determine what the exact nature of the problem is and what happened just before the machine stopped functioning. If the operator of the machine that is malfunctioning is not available, you may be able to find an operator of a similar machine who will provide some information.

The biggest mistake that most beginning troubleshooters make is to start making tests and replacing parts before they have a plan. The plan will help you think about how the machine is supposed to operate, and then you can verify how it is actually operating, which will help you determine what tests to make to find the problem.

20.2
Making a Plan for Safety

After you have talked to an operator, you also need to make a wide-ranging plan of all the safety concerns about this machine. You need to determine where the main power disconnect is located. You also need to determine if any other type of power is produced or used at this machine, including hydraulic, pneumatic, or steam. You should also determine if there are any potential energy sources that may be harmful, such as a mold or die that may drop due to gravity when you begin to work around the machine.

You should also locate the lock-out and tag-out procedure for this machine and follow the instructions. If you need to secure the power systems during any part of your troubleshooting activities, you can follow the instructions to do so.

20.3
Determining How Many Subsystems the Machine Has

The next step in the troubleshooting process is to make a mental note of how many subsystems are involved with the machine. For example, nearly all machines are powered with electricity. The electricity may be converted to hydraulic or pneumatic power. Some machines also have steam power. After you have determined the number of subsystems, you can begin to try the machine and see what will work and what is malfunctioning.

At this point you may be thinking that everything you have been requested to do to this point is a waste of time, because you are trying to get the machine back into production. In fact, if you do these first three steps—talk to the operator, review safety, and assess the machine for systems and subsystems—you should have spent all of 2 or 3 min. You will come to find that these are the best 3 min you can spend, because it will give you a good overall assessment of the situation. After you have been on the job for a while, you will find that many of the answers to the first three steps will be rather redundant, and you will be able to shorten the time.

20.4
Trying to Operate the Machine; Verifying the Problem

After you have completed the first three steps, you are ready to try to make the machine run so you can duplicate the fault that has been reported. You will find that this is the step that most novice troubleshooters forget. Sometimes troubleshooters will get a call that the machine will not run at all or has a hydraulic system failure, and they will assume the problem is the hydraulic pump and begin to change the pump without verifying the problem. You need to ask the operator to try to start and run the machine as though it is functioning properly. If the machine actually has a fault, it will continue to repeat the fault. In some cases the fault is intermittent and it will not occur every time. In this case, you will need to develop a longer-term strategy. In most cases the fault will occur again, as it did when the fault was called in.

If the machine does not respond at all or if it does not try to run, you must start your troubleshooting at the main power source for the machine. You will need to locate the electrical blueprints for this machine at this point. In most cases the main power source is the three-phase disconnect. The three-phase disconnect usually has three fuses that can blow. The first test of the power source should be a voltage test (line 1–2, line 2–3, and line 3–1). If the voltage is the same for each of these tests, you have proved that the system supply voltage is correct. If you do not get the same voltage at each of the three tests, you have proved that one of the legs of the power source is bad; then you should turn off the disconnect and remove and test the three fuses for continuity (ohms). A good fuse should show $0\ \Omega$, and an open fuse should show infinite resistance.

20.5
Designing Tests That Rule Out Problems

One thing that good troubleshooting technicians have learned is that they can develop troubleshooting tests that rule out as many parts of the machine as possible. The best way to explain this process is to think about a card trick, where you pull one card from the deck and someone tries to guess which card you pulled. Your only answer to them to any question they pose can be yes or no. For instance, Is the card a 6 of diamonds?

If you have a deck of 52 playing cards, you know that there are four suits (hearts and diamonds, which are red, and spades and clubs, which are black). You also know the cards are numbered 2–10 and the face cards are jack, queen, king, and ace. For example, let's say the card you pull is a 10 of clubs. If I am trying to guess which card you have pulled from the deck, I could start by asking which card it is at random. Is the card a 6 of hearts? Is the card a 4 of diamonds? You can see that I may lose track of which questions I asked. Another way is to go through each suit in order. For example, I could start in one suit, such as diamonds, begin with the lowest card, and ask if the card you pulled is a 2 of diamonds. Because the answer is no, I would ask about the next card, a 3 of diamonds. The answer is no to all questions about diamonds, so I could then ask about the hearts, and then, in turn, about the spades and clubs. You can see this process seems logical, but it is flawed in that I am forced to ask about 40 questions before I get a correct answer.

A better method of getting the correct answer faster is to think of question (tests) that will eliminate as many possibilities as possible. For example if I first ask if the card is red your answer is no, but now I have eliminated all the diamonds and all the hearts. In other words, with one question I have eliminated 26 cards and have drastically reduced the possibilities. The next question is to

ask if the card is a spade. Again your answer is no, but again I have eliminated another 13 possible cards.

You can see with this method, I have eliminated 75% of the cards with two questions. This lets me focus on the remaining 13 cards. My next question would be, Is the card higher than an 8 of clubs? This time your answer is yes. So I have eliminated the 2–8 of clubs and I know the card is 9 or higher. You can see that each time my process is to determine the midpoint and ask a question that determines if the card is above or below the midpoint.

You can also see that because I have eliminated so many of the cards, at this point I could ask if the card is the 9, if the card is the 10, etc., and I would not waste too much more time. I could also continue to make questions that eliminate half of the remaining cards.

You might wonder what the card-trick example has to do with troubleshooting. The point you should see is that if you simply start your troubleshooting of a large machine by just guessing if the hydraulic pump is bad, if the fuse is blown, or if the start switch is bad, you could end up making over a hundred tests and still not be any closer to solving the problem. By making a well-defined plan, each test you make should eliminate half of the possible problems, and each succeeding test should eliminate half of the remaining problems.

Using this method, you will narrow the possibilities down to the point where you can focus on one or two components that are causing the problem. Many technicians do not think it is worth the time to make a plan and test in this manner, but you will find that this process is very methodical, and it goes very fast, so that you can quickly begin working on the part of the machine that is causing the malfunction.

The remainder of this chapter walks you through the troubleshooting process for four very typical problems.

20.6
Assessing Four Typical Problems

20.6.1 No Power to the Machine

One of the most common problems you will encounter is when the machine has lost all power and will not operate. When you try to run the machine, you will notice that it will not try to run and it will not make any sounds. One quick check you should make is to determine if the power outage is localized to this machine or if it extends to a larger portion of the factory. If you see that several machines in this area have stopped, you can change your test from this machine to the source of power for all the machines affected. If this is the only machine that is stopped, you can work on the power source for this machine.

Figure 20–1 An electrical diagram for the machine-control circuit.

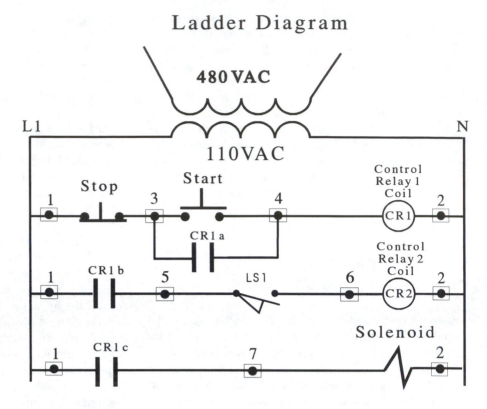

Ladder Diagram

The test that you want to make should determine if you have power to the disconnect of the machine. If power is present at the line side of the disconnect, the problem is between the disconnect and somewhere in the machine. If power is not present at the line side of the disconnect, you must locate the source of voltage for this disconnect and go back to it.

If you have power at the line side of the disconnect but not at the load side, then you have a blown fuse or a faulty switch in the disconnect. If you have voltage at the load side of the disconnect, you have eliminated it as a problem and you can move to the machine for the next tests. When you are testing for power in the machine, you must locate the terminals where the power wires are connected. Test these terminals for three-phase voltage. The voltage should be the same amount as in the disconnect. If you do not have voltage at these terminals, you have lost power between the disconnect and the machine. Turn off power and test the wires for continuity. You should have accounted for all possibilities of power loss with these tests.

20.6.2
The Machine Has Power but It Will Not Start

Another problem that you will encounter is that the machine will have power, but its relays will not energize when the start push button is depressed. Figure 20–1

shows a three-line diagram for this machine. You can see that it has a control relay that energizes a hydraulic pump motor and a solenoid that energizes the air system for the machine. In this case, the control relay coil in the first line will not energize when the start button is depressed.

In this diagram a transformer is used to step 480 VAC down to 110 V. Check the secondary voltage of the transformer, because it supplies the control voltage. If you have voltage at the primary side of the transformer but not at the secondary, the transformer is bad and should be replaced. If you have secondary voltage, you will need to test the start-stop circuit that controls the control relay or motor starter. When you are testing the control circuit, you can check for voltage across the terminals of the coil of the relay (CR1). If you have voltage at the coil of the relay when the start push button is depressed and the relay does not pull in, the relay is faulty and should be replaced. If you do not measure voltage at the coil, then you will need to test the start button, stop button and wires. Figure 20–2 shows this circuit and the test you should make.

When you are making the test of the start and stop buttons and relay, you should look for a loss of voltage. This test is used because it is the fastest way to test the components and the wire in the circuit. When you start this test, set your voltmeter to the 110 VAC range, place one lead of the meter on terminal 2, and leave it there for the entire test. Terminal 2 will be a reference point

Figure 20–2 The electrical diagram for the control relay and the start and stop buttons.

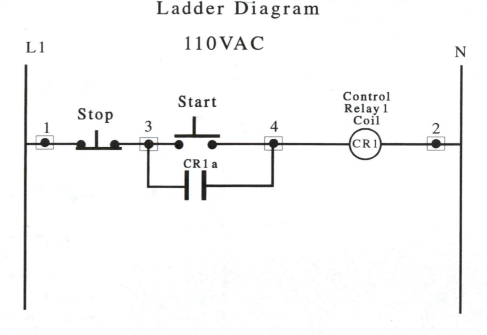

Ladder Diagram

110VAC

Stop | Start | Control Relay 1 Coil | CR1

CR1a

for all voltage tests. Place the other lead of the meter on terminal 1 and then move to terminals 3 and 4. Record the voltage at each point as you press the start push button. If you have voltage at terminal 1, it shows that the circuit has power between line 1 and terminal 2. If you do not have voltage at terminal 3, it indicates that the stop push button is bad or the wire between the stop button and its terminal is bad. At this point turn off all power to the machine and test the start button and the two wires connected to it for continuity.

If you have voltage at terminal 3, it indicates the stop button is operating correctly and passing power to terminal 3. If you have voltage at terminal 3 but not at terminal 4, this test indicates the start button is the problem. Again, turn off power to the machine and test the stop button and the two wires connected to it for continuity.

If you have voltage at terminal 4, it indicates the start button is working correctly. If you have voltage at terminal 4 but not across the leads of the relay coil, then one of the two wires between the terminals and the relay coil is broken. Turn the power off and test the wires for continuity.

Another problem that you may find is that the relay coil will energize when you depress the start push button but will not remain on when you release the start button. If this occurs, you can suspect the set of contacts that are connected across the start push button. These contacts are used to seal in the circuit, and either they are faulty or one of the wires connecting them to terminals 1 and 3 is open. If the circuit will not seal in, the problem must be in the seal in circuit contacts.

20.6.3
The Control Relay Pulls in, but the Motor Starter Does Not Energize

The next problem you may encounter is that the control relay pulls in, but the motor starter coil in the second line of the diagram does not energize. Figure 20–3 shows the diagram for this problem. To verify the problem, you have depressed the start push button and the control relay coil energizes and stays on. The motor-starter coil in the second line does not energize, so you can test for voltage directly across its coils. If you have voltage at the coil of the motor starter and the motor starter does not pull in, you can suspect the coil is bad and you can turn off power to the machine and test the coil for continuity. If the coil is bad, you can replace the coil or the entire motor starter.

If you do not measure any voltage at the coil terminals, it indicates that you have lost voltage in the second line of the diagram. You can be certain that the first line of the diagram is operating correctly, because the CR1 coil is pulled in and remains energized. Because the problem is in the second line of the diagram, you can focus on just the parts of the circuit that affect it. Again, you must leave one of the meter leads on terminal 2 so that you have an absolute reference for all of your test.

Begin the test at terminal 5. If you have voltage at this terminal, it indicates the contacts for the control relay are closed and passing power normally. If you do not have power at terminal 5, you can suspect the contacts of the relay or the two wires that are connected to them

Figure 20–3 The electrical diagram for control circuit and motor-starter coil.

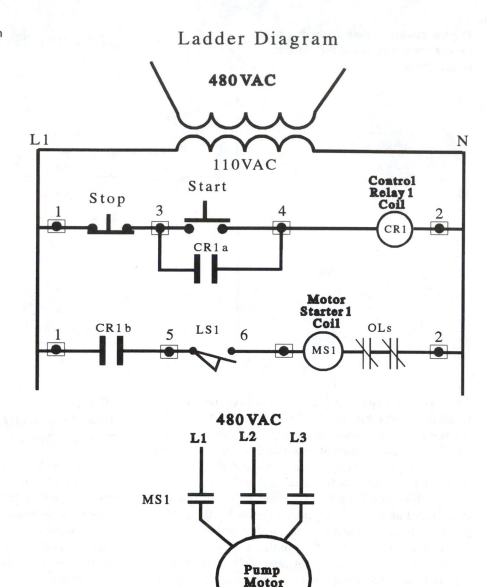

Ladder Diagram

are open. You can turn off all power to the system and test for continuity. You must "plug" the relay manually to cause the contacts to close for the continuity test, because you will not have voltage available to energize the coil. If the contacts are faulty, you can try to use a different set of contacts if they are available, or you can replace the contacts or the entire relay.

If you have voltage at terminal 5, it indicates the contacts are good and you can move to test terminal 6. If you have voltage at terminal 6, it indicates the limit switch is closed and passing power. If you do not have voltage at terminal 6, you should turn off power to the machine and test the limit switch and the two wires connected to it for continuity. Replace any faulty parts or wires that you find. If you have voltage at terminal 6 but not directly

across the terminals at the coil of the motor starter, you should suspect the overload contacts are open. If this is the case, press the reset on the motor starter and see if it resets. If the overloads are the problem, you will need to test the motor for excess current, because this is the problem that will cause the overloads to trip.

20.6.4
The Motor Starter Contacts Close, but the Motor Does Not Run

The next type of problem you may encounter is that the coil of the motor starter energizes, but the motor connected to its contacts does not run. You can see the motor is connected to the motor starter contacts in Figure 20–3. In this case, because the motor-starter coil is energized and the contacts are closed, you can forget about the control circuit and focus your tests on the motor and the load circuit.

You can begin this test by testing for 480 VAC across the line-side terminals of the motor starter. If you do not have three-phase voltage on these terminals, you can test for voltage lost between these terminals and the power terminals in the cabinet. If you have voltage at the line side but not at the load side and the contacts are closed, you can suspect that one or all of the contacts are faulty. Turn the power off and visually inspect the contacts as well as testing them for continuity. Replace the motor starter if necessary.

If you have voltage at the load side of the motor starter and the motor does not run, you need to continue to test for voltage directly to the terminals of the motor. If you have lost voltage between the motor starter line side terminals and the motor, then you need to test each of the wires that have lost voltage for continuity.

If you have voltage at the motor terminals and the motor does not run, you can turn off power to the motor and test its windings for continuity. You need to replace the motor if you locate an open.

20.7
Testing the Mechanical Part of the System

It is possible for the electrical side of system to be operating correctly and the mechanical side to be broken. You will need to check for belts, chains, or hydraulic problems if you have motors running, but the machine does not have any motion. You may need to trace the source of power through the machine until you find the point where power is lost. In some cases the power transmission is not completely lost; instead it becomes impaired and begins to vibrate or cause other problems.

20.8
Keeping Records

The last part of troubleshooting is keeping records. This is an important process for tracking the life of motors and pumps in all the machines in the factory. These records indicate problems where you have faulty components or systems that are not sized correctly, which cause parts to wear out prematurely. Records also help you determine mean time between failure. This information gives you the length of time a motor or pump should last, and you can plan scheduled maintenance and change parts before they cause downtime.

Questions for This Chapter

1. What is one of the first things you should do when you are called to fix a machine that is broken?

2. Identify two types of subsystems a machine may have.

3. Explain why you should try to operate a machine that is identified as being broken.

4. Explain why you should try to set up tests on a machine that will rule out a number of possibilities.

5. Identify two problems a machine may have if it has electrical power supplied to it but it will not start.

True or False

1. One simple problem a machine may have is a loss of voltage due to a blown fuse.

2. If an overload has tripped on a motor starter, the motor connected to it is probably drawing too much current.

3. It is not important to have a plan for safety when you are troubleshooting a machine because the most important point to remember is that you must get the machine repaired and back in production as soon as possible.

4. It is not important to try to operate a broken machine, because if it has a problem, it probably will not run anyway.

5. It is important to question the operator of a machine if possible, because the operator could tell you the problems the machine has had and how the machine is supposed to operate.

Multiple Choice

1. When you find that a machine has lost all electrical power, you need to know the flow of electricity is from _____.
   a. the switch disconnect to the power bus bar to the electrical panel on the machine.
   b. the power bus bar to the switch disconnect to the electrical panel on the machine.
   c. the power bus bar to the electrical panel on the machine to the switch disconnect.

2. If the machine has power at L1 and N but will not start, the most likely problem when you look at the diagram in Fig. 20–1 is _____.

   a. the fuses may be blown in the disconnect switch so that 480 V is not provided to the transformer.

   b. the transformer may have an open on its secondary side.

   c. the control relay may have an open in its coil.

3. When you check the circuit in Fig. 20–2 for the loss of voltage, it is important that you leave one of the meter leads _____ to provide a reference to all readings in the circuit.

   a. on line 1, terminal L1

   b. on the neutral terminal N

   c. anywhere in the circuit at terminal 1, 2, 3, or 4

4. If the overload on the motor starter that controls the hydraulic pump in Figure 20–3 is tripped, the most likely problem is _____.

   a. the hydraulic pump is drawing too much current.

   b. there are not 480 V supplied to the motor starter.

   c. the coil of the motor starter has a short.

5. If the fuses that supply 480 V to the machine are all blown, you would find that _____.

   a. you would not have 480 V at the terminals to the motor starter.

   b. you would not have 110 V at the secondary terminals of the transformer.

   c. both A and B are true.

Problems

1. In Section 20.5, an example of asking a series of questions to determine which card is pulled from a deck of cards is best solved by first asking if the card is red or black. Explain what this example has to do with making a plan for troubleshooting any machine that is broken.

2. Explain the plan you would have to troubleshoot a machine that has no electrical power at its electrical panel.

3. Devise a plan for troubleshooting the machine shown in the diagram in Figure 20–3, where there are 480 V at the terminals at the top of the motor starter, but the hydraulic pump will not run.

4. Explain what you would suspect in the diagram shown in Figure 20–2 if the control relay becomes energized every time the start push button is pressed, but it becomes deenergized when you lift your hand off the start push button.

5. Explain why you should always try to talk to the operator of a machine that has a fault and try to operate the machine.

◀ **Index** ▶